MICROBIOLOGY RESEARCH ADVANCES

MICROBIAL CATALYSTS

VOLUME 2

MICROBIOLOGY RESEARCH ADVANCES

Additional books and e-books in this series can be found on Nova's website under the Series tab.

MICROBIOLOGY RESEARCH ADVANCES

MICROBIAL CATALYSTS

VOLUME 2

SHADIA M. ABDEL-AZIZ
NEELAM GARG
ABHINAV AERON
CHAITANYA KUMAR JHA
S. CHANDRA NAYAK
AND
VIVEK KUMAR BAJPAI
EDITORS

nova
science publishers
New York

NOTICE TO THE READER

The Publisher has taken reasonable care in the preparation of this book, but makes no expressed or implied warranty of any kind and assumes no responsibility for any errors or omissions. No liability is assumed for incidental or consequential damages in connection with or arising out of information contained in this book. The Publisher shall not be liable for any special, consequential, or exemplary damages resulting, in whole or in part, from the readers' use of, or reliance upon, this material. Any parts of this book based on government reports are so indicated and copyright is claimed for those parts to the extent applicable to compilations of such works.

Independent verification should be sought for any data, advice or recommendations contained in this book. In addition, no responsibility is assumed by the Publisher for any injury and/or damage to persons or property arising from any methods, products, instructions, ideas or otherwise contained in this publication.

This publication is designed to provide accurate and authoritative information with regard to the subject matter covered herein. It is sold with the clear understanding that the Publisher is not engaged in rendering legal or any other professional services. If legal or any other expert assistance is required, the services of a competent person should be sought. FROM A DECLARATION OF PARTICIPANTS JOINTLY ADOPTED BY A COMMITTEE OF THE AMERICAN BAR ASSOCIATION AND A COMMITTEE OF PUBLISHERS.

Additional color graphics may be available in the e-book version of this book.

Library of Congress Cataloging-in-Publication Data

ISBN: 978-1-53616-088-8

Published by Nova Science Publishers, Inc. † New York

CONTENTS

Contents

PREFACE

Enzymes were discovered in the second half of the nineteenth century, since then have been extensively used in several industrial processes. Enzymes from microorganisms are extremely efficient and highly specific biocatalysts. Biocatalytic potential of microorganisms have been employed for centuries to produce bread, wine, vinegar and other common products. Microbial catalysts (*biocatalysts*) are enzymes contribute to many reactions and widely used in food and industrial products. An enzyme is a protein that catalyzes a specific reaction. The action of enzymes depends on their ability to bind the substrate at a domain of enzyme molecule called "active site." Enzymes are more specific macromolecular biological catalysts that accelerate or catalyze chemical reactions. Some enzymes are not, however, reactive on their own, and often need cofactors. Chemically, enzymes like any catalyst, are not consumed in chemical reactions, nor do alter the equilibrium of a reaction. Enzyme activity can be affected by other molecules such as inhibitors and activators. Inhibitors are molecules that decrease enzyme activity, while activators are molecules that increase the enzyme activity. Many drugs and poisons are enzyme inhibitors. Microbial enzymes are known to play a crucial role as metabolic catalysts, leading to their use in various applications and industrial fields. With the advancement in biotechnology over last three decades, especially in the area of genetics and protein engineering, enzymes have found their way into many new industrial processes.

Global market for industrial enzymes was estimated about$4.2 billion in 2014 and expected to develop at a compound annual growth rate (CAGR) of approximately 7%over the period from 2015 to 2020 to reach nearly $6.2billion.Over 500 industrial products are being made using enzymes. Microorganisms have served and continue to serve as one of the largest and useful sources of many enzymes. In the biocatalytic processes, natural catalysts, such as protein enzymes, perform chemical transformations on organic compounds. Microbial catalysts are environmentally friendly, greatly cost-effective, and consume lower energy. There is an increasing demand to replace some traditional chemical processes with biotechnological processes involving microorganisms or their enzymes such as pectinases, xylanases, cellulases, mannanase, α-galactosidase, laccases and ligninases, which not only provide an economically viable alternative but are also more environmentally friendly. Uses of microbial enzymes in foods, pharmaceuticals, textile, paper, leather, and other industries are numerous and increased rapidly. Supplementary enzymes are added to the dough to ensure high bread quality in the form of a uniform crumb structure and better volume. Special enzymes can also increase the shelf life of bread by preserving its freshness longer. Enzymes have contributed greatly to the development and improvement of modern household and industrial detergents, the largest application area for enzymes today. Moreover, one of the prime roles of enzymes is to improve the quality of leather.

Over the past century, there has been a tremendous increase in awareness of the effects of pollution. Enzymes can minimize the impact of wastes on the environment. Replacement of some traditional chemical processes with biotechnological processes involving microorganisms and enzymes is an advantage and challenge. All types of living organisms, where metabolic reactions occur, produce enzymes. A wide range of sources is used for production of commercial enzymes. Out of the total enzymes being used industrially, over half are extracted from fungi and yeast, one third are obtained from bacterial systems, and the remaining from animal (8%) and plant (4%) sources. Microorganisms are generally preferred than plant and animal as sources of industrial enzymes because their production cost is low, the enzyme content is more predictable and controllable, and more so because of the easy availability of raw materials with constant composition for their cultivation.

The current book consists of 14 different chapters. All the chapters discuss microbial catalysts with respect to the enzymes, their function, and its benefits for human kind in general and to biotechnology as a subject of which ENZYMOLOGY is an integral part. Through this book we wish to bridge the gap between various literature that exists on enzymes, which is either too high on the caliber of written technical language or too low to be of scientific interest. Therefore, experts across the globe were invited to contribute in this comprehensive collection making it unique to be understandable, lucid, and easy to increase our understanding of ENZYMOLOGY.

Collection of articles included in this book can be a useful material for Research Students, Post Graduate Students, Lecturers, Professors, and Academicians related to Microbiology, Biotechnology, Biochemistry, Environmental Sciences, Fermentation Technology, Health Professionals, Agricultural Sciences, Applied Microbiologists and many more alike.

Shadia M Abdel Aziz   S. Chandra Nayak
Neelam Garg   Vivek Kumar Bajpai
Abhinav Aeron
Chaitnaya Kumar Jha   28 March 2019

In: Microbial Catalysts. Volume 2
Editors: Shadia M. Abdel-Aziz et al.

ISBN: 978-1-53616-088-8
© 2019 Nova Science Publishers, Inc.

Chapter 1

BIOPROSPECTING FOR MICROBIAL TYROSINASES

Greta Faccio*, Michael Richter and Linda Thöny-Meyer

Empa, Swiss Federal Laboratories for Materials Science and Technology -
Laboratory for Biomaterials, St. Gallen, Switzerland

ABSTRACT

In microbes, tyrosinase is involved in the first reaction of the biosynthetic pathway leading to the formation of the dark pigment melanin by hydroxylating and oxidizing L-tyrosine. However, tyrosinase activity is not limited to small phenolic compounds but also includes proteins carrying exposed tyrosyl side chains that are crosslinked with other proteins or substrates. Traditionally, tyrosinase-producing strains have been isolated based on the intense dark colour of the organism due to melanin production. Nowadays, novel tyrosinases can be identified by searching the sequenced genomes, and produced using the tools of molecular biology. Intracellular or secreted, with varying degrees of stability, and working under different conditions, many microbial tyrosinases have been characterised. This chapter provides an overview on microbial tyrosinases and how their natural activity towards phenolic compounds can find application in different fields.

1. INTRODUCTION

Within living organisms, enzymes are the proteins accelerating most of the basic chemical reactions necessary for life. Enzymes have also developed during evolution as a response to external selection pressure in order to improve the fitness of the organism.

The ability of some microbes to produce melanin has been achieved during evolution, as it is not strictly necessary for microbial growth. In humans, the adaptive production of melanin is very apparent as it is responsible for the darkening of the skin after exposure

---

* Corresponding Author's E-mail: greta.faccio@gmail.com.

to sunlight. Similarly in some microbes, the production of melanin results in the formation of dark colonies such as the ones commonly seen on tomatoes, apples and shower curtains [1]. Melanin production seems to be a clear response of adaptation as microbes growing in areas under intense solar irradiation and radioactive areas, as for example the surrounding of Chernobyl, produce more melanin than microbes living in less exposed areas [2, 3].

The presence of melanin increases the resistance to enzymatic lysis and to extreme heat and cold temperatures. Melanin is also involved in the pathogenicity of many bacteria and fungi. Melanin is a high-molecular weight polymer with exceptional features. It is insoluble in water and organic solvents, resistant to degradation by concentrated acids and to bleaching by oxidizing agents such as hydrogen peroxide. In addition, melanin is a radioprotector and a bioabsorber as it can absorb, and thus neutralise, toxic metals, e.g., copper, zinc, iron, aluminium [1].

The first step in the biosynthesis of melanin is catalysed by the multicopper enzyme tyrosinase (EC 1.14.18.1). Tyrosinases are ubiquitous, i.e., synthesized by organisms of all kingdoms of life, and participate in the secondary metabolism of both prokaryotic and eukaryotic organisms. Tyrosinases oxidise mono- and di-phenolic compounds to the corresponding *ortho*-quinones and reduce one molecule of oxygen to water (Figure 1).

Dopaquinone, the product of the tyrosinase-catalysed reaction is converted non-enzymatically to dopachrome and further polymerized to the dark-brownish eumelanin. Instead the reaction of the oxidised quinone with a thiol-containing compound such as cysteine or glutathione leads to the production of the pinkish pheomelanin that is common, for example, in freckles and red-hair people [4].

Based on their substrate specificity for phenolic compounds, tyrosinases are sometimes associated with laccases (EC 11.10.3.2) and catechol oxidases (EC 1.10.3.1) and collectively named polyphenol oxidases. Tyrosinases and catechol oxidases share high similarity in their primary and tertiary structure and are easily distinguishable from laccases on this basis. Differently from tyrosinases, laccases oxidise a wide range of aromatic and non-aromatic molecules and their reaction proceeds through the formation of radicals. Moreover, unlike tyrosinases, catechol oxidases do not oxidise mono-phenolic compounds.

2. Discovery of Novel Tyrosinases

The presence of tyrosinase activity can be visually detected with ease. Strains producing tyrosinases can easily be identified by growing the microbe in the presence of L-tyrosine and observing the formation of a dark colour in the presence of oxygen. As an example, the Gram-positive bacterium *Bacillus megaterium* is a producer of tyrosinase that was identified by growing the bacterium on solid medium containing 0.1% L-tyrosine and detecting a black-brown colour forming around the colonies [5]. This visual screening can be easily applied to isolate novel producers from environmental samples and to discriminate among strains producing different amounts of tyrosinase, e.g., to screen for better producers after mutagenesis [6]. Results obtained in solid medium need confirmation by cultivation in liquid medium and the addition of copper or possible inducers such as L-tyrosine or methionine. Even harsh conditions such as heat or osmotic stress, can sometimes trigger an increased tyrosinase production [7]. This screening approach has, however, some limitations as only microbial strains that can be cultivated in laboratory conditions can be analysed.

Novel tyrosinases can also be identified by genomic and metagenomic approaches. This strategy allows the identification of candidate proteins also from species that cannot be cultivated in laboratory conditions and it constitutes a very powerful tool. Tyrosinases have a characteristic, conserved amino acid sequence around their active site (see below) and this can be used to retrieve genes coding for potential tyrosinases. For example, the use of degenerated oligonucleotides for the characteristic sequence motifs of tyrosinases can be used to amplify genes coding for tyrosinases from complex environmental samples. Moreover, bioinformatic tools can be used to search the available sequenced genomes for genes coding for proteins carrying the tyrosinase-specific sequence motif (entry IPR002227 at the Interpro database http://www.ebi.ac.uk/interpro and entry Pfam PF00264 at the Sanger Institute http://pfam.sanger.ac.uk). Providing the necessary computational capacities, whole genomes can been searched for novel tyrosinases in minutes. This way, genes coding for tyrosinases have been identified in the genomes of not only bacteria but also fungi and archaea. It is important to remember, that tyrosinases and the other members of the so-called type-3 copper protein family share high structural similarity, and thus the use of the mentioned tyrosinase-specific sequence motif can retrieve also catechol oxidases. These can be distinguished only by experimentally defining the substrate specificity of the enzyme. To give an idea of the number of tyrosinases identified and studied to date, more than 20 enzymes have been characterised only from bacteria and 8245 annotated tyrosinases are stored at the NCBI Database (http://www.ncbi.nlm.nih.gov) as of April 2013.

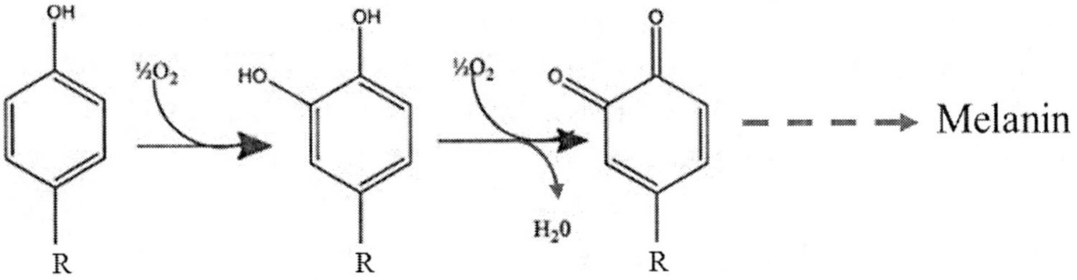

Figure 1. Oxidation of phenolic compounds by tyrosinases to *o*-quinones (arrows) that further react not-enzymatically (dashed arrows) to produce melanin.

3. STRUCTURAL AND MOLECULAR PROPERTIES OF MICROBIAL TYROSINASES

Known tyrosinases are either monomeric or multimeric and have very different sizes. Microbial tyrosinases comprise a central catalytic domain with conserved sequence motifs and often a C-terminal domain of different length and amino acid composition. Tyrosinases can also carry an N-terminal signal sequence and be secreted by the organism (Figure 2). Tyrosinases with a C-terminal domain have been identified in both bacteria and fungi. A truncated 35-40 kDa active form can often be distinguished from the 60 kDa precursor that, when isolated, is less active and retains the C-terminal domain. This is the case for fungal tyrosinases such as the ones from *Agaricus bisporus, Trichoderma reesei* and has been shown *in vitro* for the *Verrucomicrobium spinosum* tyrosinase (Figure 2) [8, 9, 10]. Activation of the precursor protein is generally achieved by proteolytic digestion in order to remove the ~20 kDa portion of the protein that covers the active site and prevents access of the substrate. Cleavage takes place in a protease-sensitive linker region located between the active core of the enzyme and the regulatory C-terminal domain. Additionally, activation can be obtained *in vitro* by loosening the structure of the precursor, for example by the addition of denaturants or detergents. Shorter tyrosinase sequences lacking the C-terminal domain and most of the linker region are found mainly in streptomycetes, where their production in a functional form requires the presence of a second helper protein. The helper or "caddie" protein is made of less than 200 amino acids and it has been found necessary for the correct folding and incorporation of the copper cofactors. This is the case for the tyrosinase from *Streptomyces castaneoglobisporus* and *Streptomyces antibioticus* (for review see 8).

Microbial tyrosinase can be intracellular or secreted in both bacteria and fungi. As an example, the tyrosinase from *S. antibioticus* and as well as the one from *T. reesei* are secreted, whereas the enzymes from *Bacillus megaterium* and *A. bisporus* are intracellular. Secreted enzymes are generally thought to have higher stability features and be more robust if compared to the intracellular enzymes that have evolved to work under the controlled cellular conditions and thus might be less useful for application in industrial processes.

Tyrosinases belong to the type-3 copper proteins together with catechol oxidases and the oxygen-binding proteins hemocyanins that are found in molluscs and insects. Similarly to these other proteins, the active site alternates between three oxidative states by interacting either with molecular oxygen or the phenolic substrate. When the enzyme is in the *oxy* state, oxygen is bridged as peroxide ion between the two copper ions and the enzyme can react with the monophenolic substrate and convert to the *met* form. Tyrosinase in the *met* form turns into the *deoxy* form by oxidising diphenolic compounds. The *oxy* state is restored when the *deoxy* form binds oxygen. Alternation between these forms can be monitored by UV-Vis spectroscopy as the *met*-form weakly absorbs at 330–350 nm, and the *oxy*-form strongly at 345–350 nm and weakly at 600 nm [11].

The active site of tyrosinase harbours two copper ions, each of them held in place by a triad of histidine residues (Figure 3). These histidine residues are highly conserved among tyrosinases from different organisms and they recur in the conserved pattern H–Xn–H–X8-H for the first copper binding region (CuA) and H–X3–H–Xn-H, where Xn is a stretch of n not defined amino acids, for the second copper binding region (CuB). For example, this motif is H–X15-27–H–X8-H and H–X3–H–X20-23-H for the selected tyrosinases reported in Figure 2. Some tyrosinases are post-translationally modified, i.e., the second histidine of the first copper binding region is autocatalytically bound to a nearby conserved cysteine residue *via* a thioether bond. This is found in the tyrosinases from *A. bisporus* and *Neurospora crassa,* in the structurally similar haemocyanin from *Octopus dofloeini* and in the catechol oxidase from the sweet potato *Ipomoea batatas.* This covalent bond is autocatalytically formed during protein folding and it is thought to confer structural rigidity to the active site [12]. However, it is absent in the available structure of bacterial tyrosinases. Analysis of the three-dimensional structures that have been solved for tyrosinases revealed additional conserved residues such as an N-terminal conserved arginine and the C-terminal tyrosine-motif Y/F-X-Y that, located at the ends of the active core domain, were found to interact in the three-dimensional structure bringing together the carboxy- and amino terminus of the molecule and conferring stability [13].

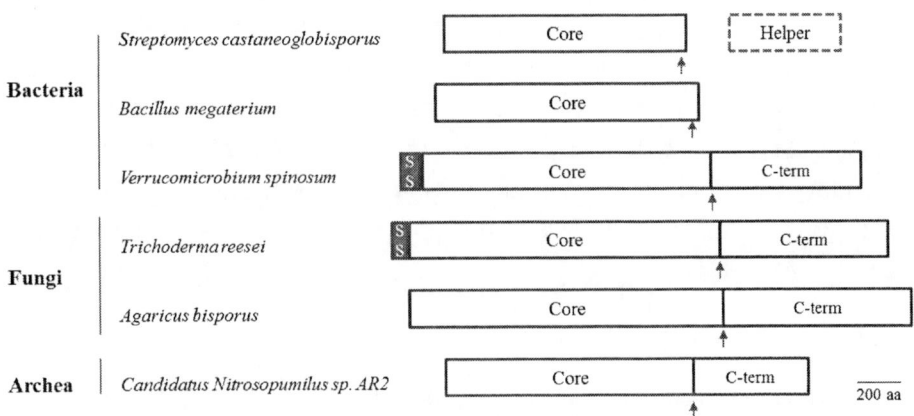

Figure 2. Diversity of domain organisation of representative microbial tyrosinases. Signal sequences (SS) are indicated by a grey box and the core regions are defined as the region between the signal sequence and the conserved tyrosine motif (pointed by an arrow). Protein sequences can be found at NCBI (http://www.ncbi.nlm.nih.gov/) under the accession numbers AAP33665.1 for the tyrosinase from *Streptomyces castaneoglobisporus* and AAP33666.1 for its helper protein (dashed box) , ACC86108.1 for the tyrosinase from *Bacillus megaterium,* ZP_02925214.1 for the tyrosinase from *Verrucomicrobium spinosum* DSM 4136, CAL90884.1 for the tyrosinase from *Trichoderma reesei*, Q00024.1 for the tyrosinase from *Agaricus bisporus* and AFS81917.1 for the uncharacterised tyrosinase from *Candidatus Nitrosopumilus* sp. AR2.

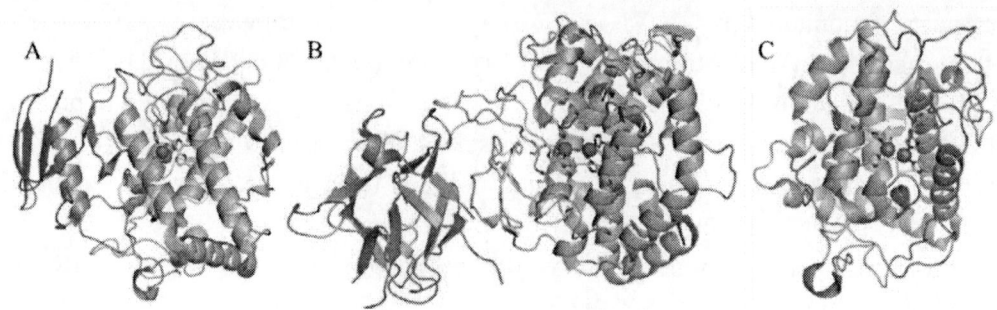

Figure 3. Three-dimensional structures of the tyrosinases from (A) *Streptomyces castaneoglobisporus* (PDB ID: 2HK), (B) *Agaricus bisporus* (PDB ID: 2Y9W), and (C) *Bacillus megaterium* (PDB ID: 3NM8). Tyrosinase form *S. castaneoglobisporus* co-crystallised with the helper protein ORF378 and the tyrosinase from *A. bisporus* with a lectin-like protein orf239342. The catalytic tyrosinase domain is in orange and accessory proteins in blue. The histidines interacting with the copper ions (red spheres) are shown as sticks. The figure was prepared with the software Pymol (The PyMOL Molecular Graphics System, Version 1.2r3pre, Schrödinger, LLC).

No three-dimensional structure of a tyrosinase precursor with the C-terminal domain is yet available. The three-dimensional structures of tyrosinases show the active enzyme alone, as in the case of the enzyme from *Bacillus megaterium* [14] or in association with the helper protein ORF378 as for the tyrosinase from *S. castaneoglobisporus* [15] (Matoba et al. 2006). The active core of the tyrosinase from *A. bisporus* co-crystallised with a 150 amino acid long protein (orf239342) showing a typical lectin-like fold [16]. In all cases, the active core of tyrosinases is mainly composed of alpha-helical elements and short beta-strands. Long beta-sheets are found in helper proteins (Figure 3), in the lectin-like protein interacting with the tyrosinase from *A. bisporus*, and are predicted to be found in the C-terminal domains. Similarly, mollusc hemocyanins are composed of a copper-binding alpha-helical domain and a beta-rich domain that covers the entrance to the active site such that the protein acquires polyphenol oxidase activity once this latter domain is removed [17].

4. BIOCHEMICAL CHARACTERISTICS AND PRODUCTION OF MICROBIAL TYROSINASES

High stability to temperature and pH, high catalytic efficiency, high specificity, and good production levels are some of the desired features of industrial enzymes. The large number of microbial tyrosinases that has been studied offers a wide range of enzymes with different optimal working conditions and production levels (for a selection see Table 1).

Table 1. Biochemical properties of some characterised microbial tyrosinases

Organism (NCBI identifier[a])	MW (kDa)	Substrates and features	Three-dimensional structure (PDB ID[b])	Reference
Agaricus bisporus (C7FF04.1)	43 (active form)	small phenolic and protein, isolated from fruiting bodies, commercially available	2y9w, 2y9x	[16, 18; 19]
Trichoderma reesei (CAL90884.1)	43 (active form)	small phenolic and protein, over-expressed in the native host (1 g/l in bioreactor)	no	[9, 20]
Bacillus megaterium (ACC86108.1)	31	active in organic solvents, overexpressed in *E. coli*	3npy	[5, 14]
Verrucomicrobium spinosum (ZP_02925214.1)	57 (precursor) 37 (active form)	small phenolic and protein, overexpressed in *E. coli* (3 g/l in bioreactor)	no	[21, 22, 23]

a http://www.ncbi.nlm.nih.gov/, b http://www.rcsb.org.

Tyrosinases working at acidic, neutral and basic pH conditions have been reported [7, 24]. Tyrosinase from *N. crassa*, for example, has maximum activity at pH 5, tyrosinase from *B. megaterium*, *V. spinosum* and *R. etli* are most active at pH 7 whereas the tyrosinases isolated from *Thermomicrobium roseum* and *Bacillus thuringiensis* have pH optima of pH 9-9.5. Concerning temperature conditions, the tyrosinase from *S. castaneoglobisporus* has an optimum temperature of 40°C and the enzymes from *R. etli* and *B. megaterium* of 50°C. The fungal tyrosinase from the white-rot fungus *Pycnoporus sanguineus* is active in a pH range of 6–7, between 30 and 70°C, and it is stable up to 60°C [6]. Tyrosinases from *A. bisporus* and *T. reesei* have optimum pH ranges of 6-7 and 9 and are stable up to 40 and 30°C [25], respectively. Remarkably, the tyrosinases isolated from *T. roseum* and *B. thuringiensis* show maximum activity at high temperature (optimum of 70-75°C).

Tyrosinases oxidise monophenolic compounds such as L-tyrosine, di-phenolic compounds such as L-DOPA, which is usually the model substrate in activity measurements, and poly-phenolic compounds such as pyrogallol and catechins. Moreover, microbial tyrosinases are active on small peptides and proteins without a rigid secondary structure, such as caseins from milk. When tyrosinase oxidises the phenolic groups of a tyrosine residue, this can non-enzymatically react with other nucleophilic groups such as amines and introduce novel covalent bonds between the proteins: this is the basis for the crosslinking activity of tyrosinases [26]. Proteins that are easily crosslinked by tyrosinase are milk caseins, e.g., alpha and beta casein, as they lack defined elements of secondary structure. The crosslinking activity of tyrosinase is usually analysed by visualising the higher-molecular weight covalent aggregates formed by SDS PAGE or size-exclusion chromatography (Figure 4).

The ability of tyrosinases to oxidise proteins is not surprising as in higher organisms this enzyme is involved in the maturation of the phenolic proteins that constitute natural glues such as the ones produced by mussels and marine worms. These proteins have an intrinsically unstructured fold and are rich in tyrosine residues, which tyrosinase converts to DOPA (L-dihydroxyphenylalanine), and the protein can eventually contain up to 25 mol% DOPA [27]. It is important to notice that catechol is not only an intermediate of the crosslinking reaction but it can also directly chelate metals and contribute to the attachment to surfaces [28]. In enzymes and globular proteins in general, tyrosine as a hydrophobic amino acid tends to be buried inside the molecule and is inaccessible to tyrosinase. Therefore, the use of tyrosinase to crosslink any protein of choice is limited by the exposure of tyrosine residues in unstructured, accessible regions of the protein.

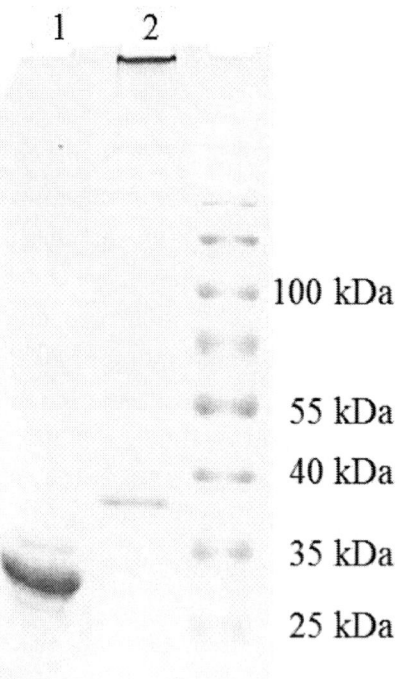

Figure 4. Crosslinking of bovine β-casein (lane 1 native form, lane 2 after crosslinking) with the tyrosinase from *Verrucomicrobium spinosum*, visible as a protein band at 37 kDa. The crosslinked product is indicated by an arrow 4.

The production of tyrosinases, as mentioned above, can require a proteolytic step for activation. This does not constitute an issue when a naturally producing strain is used or the enzyme is produced in the native host. When the precursor tyrosinase is overexpressed in a heterologous host, the proteolytic processing might not take place and the enzyme be produced in an inactive form, requiring activation during the downstream process. The direct production of the enzyme in a recombinant form lacking the C-terminal domain constitutes an alternative solution. However, the C-terminal is currently considered to be responsible for keeping the enzyme in an inactive form that is harmless for the cell and that is proteolitically activated during or after secretion. The production of tyrosinase in an active truncated form might affect the fitness of the culture and compromise the production level. Highest productivity levels have been reached in the production of the precursor form of the bacterial tyrosinase from *V. spinosum* (3 g/l in bioreactor) intracellularly in *E. coli* and of the tyrosinase from *T. reesei* (1 g/l in bioreactor) that was secreted in the active form by the native host upon overexpression [29] Ren et al. 2013;9].

5. APPLICATIONS OF MICROBIAL TYROSINASES

Tyrosinases have been tested in processes where the oxidation of small phenolics or phenolic-containing polymers is required. The interest for this family of enzymes is reflected in the number of enzymes that have been subjected to patent applications, e.g., the tyrosinases from *V. spinosum*, *R. etli*, *S. antibioticus* and *Pseudomonas* sp. DSM13540 [7]. Potential applications span over different fields such as cosmetics, dye production, and medicine and this section will focus on some applications including food engineering and wastewater treatment (Figure 5).

This section will deal with few of the most promising applications and further information about the minor fields can be found in the specific publications and review articles [7, 8, 30].

Although most studies have been developed using the commercial fungal enzyme from common mushroom, all microbial tyrosinases constitute good alternatives and may provide with their different stability features, catalytic efficiency and wider/narrower substrate specificities, the possibility to perform the process of interest under alternative conditions.

A direct application of tyrosinase is in the development of biosensors for the detection of phenolic compounds in complex mixtures, not only wastewaters but also food products such as wine, tea and fruit juices. For this purpose, tyrosinases, and phenol oxidases in general, have been immobilised to different solid materials such as silica gels, nylon, chitosan, montmorillonite and graphite. Biosensors developed with the tyrosinase from common mushroom and *Streptomyces antibioticus* allowed the detection of nanomolar amounts of phenols [7, 31, 32, 33 and references therein].

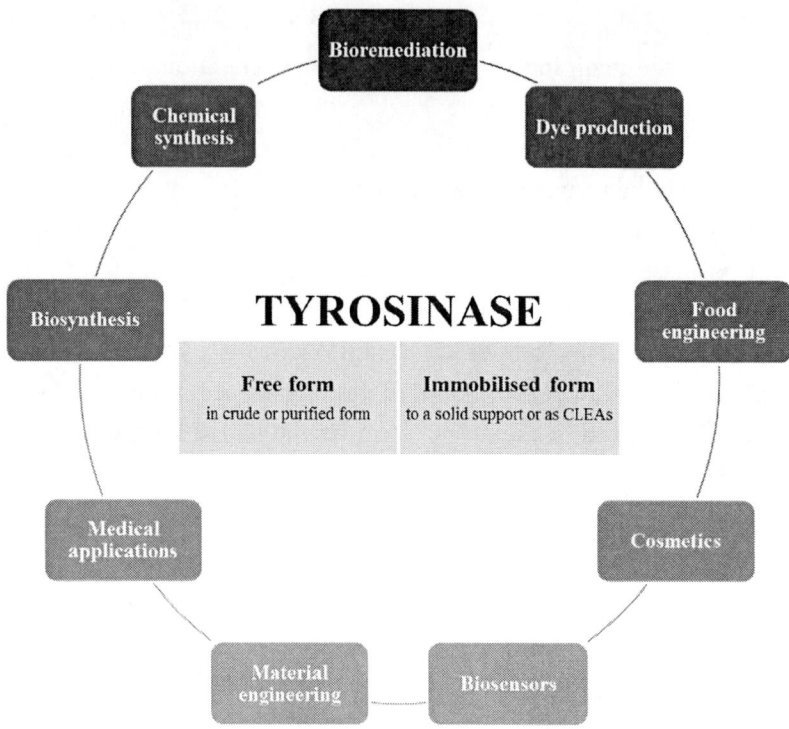

Figure 5. Overview of the different fields in which tyrosinase, as free or immobilised enzyme, can find application.

The immobilisation of industrial enzymes is a convenient way to recycle these catalysts and allows their use for several processes. This can be achieved by crosslinking the enzyme to a solid support such a bead, or by assembly into high-molecular weight covalent multimers in the presence of a crosslinking agent, also called cross-linked enzyme aggregates (CLEAs). Tyrosinase itself has been immobilised on different substrates, including multi-walled carbon nanotubes, chitosan, chitosan-clay composites, hydrogel beads, thermoresponsive materials, and zeolite. Such immobilization often leads to an improved thermal stability while preserving the catalytic activity. Due to the ease to detect its activity, tyrosinase has also been used as a model enzyme to assess the best conditions to be used for the immobilisation of other enzymes of interest. CLEAs are often produced by using chemical crosslinkers such as glutaraldehyde. CLEAs of tyrosinase have been applied for the removal of phenolic pollutants from wastewater, for example. On the other hand, tyrosinase has also been the crosslinking agent in the production of CLEAs. Since well-structured proteins such as enzymes are generally not the best substrates for crosslinking with tyrosinase, tyrosinase-catalysed CLEAs production usually takes advantage of the addition of mono- or di-phenolic compounds that are oxidised by tyrosinase and function either as mediators or as 'bridging agents', bringing together the molecules of substrate enzyme. In this way, tyrosinase form *V. spinosum*, for example, could produce CLEAs of *Candida antarctica* lipase in the presence of phenol. The possibility of engineering proteins by molecular biology can be exploited to introduce additional residues, such as tyrosine or tyrosine-rich peptides, to provide novel sites of attack for cross-linking with tyrosinase. This is how attachment of a reactive dye, Cy5-hydrazide, to proteins carrying the hemagglutinin tag Tyr-Pro-Tyr-Asp-Val-Pro-Asp-Tyr-Ala by mushroom tyrosinase was possible [34].

In a similar fashion, the action of tyrosinase has also been tested for the grafting of biomaterials and textile fibres to introduce antimicrobial and antioxidant properties. For example, sericin peptides and gallates have been immobilised to chitosan [35, 36].

Enzymes can also provide an alternative to traditional chemical synthesis reaction and the hydroxylation-oxidation reaction catalysed by tyrosinase has been successfully linked to a Diels–Alder reaction and a subsequent allylation of the intermediary formed bicyclic diketones in a domino reaction [37].

The use of tyrosinase in the production of dyes and melanin has extensively been studied and proposed. Melanin might find application, for example, in the cosmetic industry and in the cell culture sectors as photoprotectant. In cosmetics, the application of tyrosinase has been limited to the production of a tanning effect. Most of the research has focused in the identification of natural inhibitors for human tyrosinase, in order to attain bleached skin [38]. However, the production of synthetic melanin in an enzyme-free process based on the reaction of tyrosine with hydrogen peroxide might constitute a more cost-efficient alternative.

The crosslinking activity of tyrosinase has been successfully used to improve the rheological properties of protein-based foods. In multi-phase dynamic systems such as gels, the introduction of novel covalent bonds within and between proteins can significantly affect rigidity and mobility, thus affecting the overall elastic and viscous properties of the system. Tyrosinases have been tested successfully as a cross-linking agent in milk, meat and cereal products [39]. As an example, the texture of milk gels could be improved by the addition of tyrosinase. In meat products, the water-holding capacity of low-meat chicken breast homogenates was improved by the addition of tyrosinase. In wheat bread, tyrosinase softened the bread crumb and increased the volume of breads. Similarly, it increased the firmness of the dough and the specific volume of the gluten-free oat bread. In food, the addition of tyrosinase can also lead to improve the colour of the product. For example, the oxidation of the phenolic constituents to melanin confers a dark colour to tea. A promising field for tyrosinase application is the development of hydrogels and glues gelling in aqueous solutions, thus providing the possibility of in-situ gelling in humid conditions. Polymeric molecules are generally functionalised to carry phenolic moieties that once oxidised by tyrosinase, initiate the crosslinking reaction and the transition from sol to gel. These glues could be applied, for example, to wound treatment, tissue engineering and tissue reconstruction. A limitation to this could be that proteins naturally cured by the action of tyrosinase to form glue, for example, mussel glue, are difficult to be produced recombinantly in good amounts [40]. However, tyrosinase was shown able to modify chitosan, a natural biopolymer that could gel when incubated with tyrosinase and phenolic compounds such as dopamine [41, 42]. Moreover, mushroom tyrosinase could oxidise tyrosine residues from collagen and, when catechins were present, this resulted in protein precipitation and the formation of a gel with higher denaturation temperature and tensile strength [43]. The use of tyrosinase was successful for the production of a hydrogel from chondroitin sulphate in the presence of tyramine and showed potential for application in the development of drug release systems and tissue engineering [44].

CONCLUSION

The microbial world is a rich source of enzymes. Among these, various tyrosinases able to perform under different conditions have been reported. Moreover, good production levels have also been recently achieved. Due to the ability to oxidise small phenolics and large unstructured phenolic substrates such as some proteins, tyrosinases have been suggested for application in many fields, of which biosensor development, food engineering and grafting of materials, immobilisation of enzymes might be the most promising ones for microbial tyrosinases.

REFERENCES

[1]   Butler, M. J. & Day, A. W. (1998). Fungal melanins: A review. *Can J Microbiol*, *44*, 1115-1136.

[2]   Dadachova, E. & Casadevall, A. (2008). Ionizing radiation: How fungi cope, adapt, and exploit with the help of melanin. *Curr Opin Microbiol*, *11*, 525-531.

[3]   Sterflinger, K., Tesei, D. & Zakharova, K. (2012). Fungi in hot and cold deserts with particular reference to microcolonial fungi. *Fungal Ecology*, *5*, 453-462.

[4]   Plonka, P. M. & Grabacka, M. (2006). Melanin synthesis in microorganisms - biotechnological and medical aspects. *Acta Biochim Pol*, *53*, 429-443.

[5]   Shuster, V. & Fishman, A. (2009). Isolation, cloning and characterization of a tyrosinase with improved activity in organic solvents from *Bacillus megaterium*. *J Mol Microbiol Biotechnol*, *17*, 188-200.

[6]   Halaouli, S., Asther, M., Kruus, K., Guo, L., Hamdi, M., Sigoillot J. C. & Lomascolo, A. (2005). Characterization of a new tyrosinase from pycnoporus

species with high potential for food technological applications. *J Appl Microbiol*, *98*, 332-343.

[7] Faccio, G., Kruus, K., Saloheimo, M. & Thöny-Meyer, L. (2012). Bacterial tyrosinases and their applications. *Process Biochem*, *47*, 1749-1760.

[8] Fairhead, M. & Thöny-Meyer, L. (2012). Bacterial tyrosinases: Old enzymes with new relevance to biotechnology. *New Biotechnology*, *29*, 183-191.

[9] Selinheimo, E., Saloheimo, M., Ahola, E., Westerholm-Parvinen, A., Kalkkinen, N., Buchert, J. & Kruus, K. (2006). Production and characterization of a secreted, C-terminally processed tyrosinase from the filamentous fungus *Trichoderma reesei*. *FEBS J*, *273*, 4322-4335.

[10] Wichers, H. J., Recourt, K., Hendriks, M., Ebbelaar, C. E., Biancone, G., Hoeberichts, F. A. & Soler-Rivas, C. (2003). Cloning, expression and characterisation of two tyrosinase cDNAs from *Agaricus bisporus*. *Appl Microbiol Biot*, *61*, 336-341.

[11] Himmelwright, R. S., Eickman, N. C., LuBien, C. D., Solomon, E. I. & Lerch, K. (1980). Chemical and spectroscopic studies of the binuclear copper active site of neurospora tyrosinase: Comparison to hemocyanins. *J Am Chem Soc*, *102*, 7339-7344.

[12] Fujieda, N., Ikeda, T., Murata, M., Yanagisawa, S., Aono, S., Ohkubo, K., Nagao, S., Ogura, T., Hirota, S., Fukuzumi, S., Nakamura, Y., Hata, Y. & Itoh, S. (2011). Post-translational His-Cys cross-linkage formation in tyrosinase induced by copper(II)-peroxo species. *J Am Chem Soc*, *133*, 1180-3.

[13] Marusek, C. M., Trobaugh, N. M., Flurkey, W. H. & Inlow, J. K. (2006). Comparative analysis of polyphenol oxidase from plant and fungal species. *J Inorg Biochem*, *100*, 108-123.

[14] Sendovski, M., Kanteev, M., Shuster, B. Y. V., Adir, N. & Fishman, A. (2010). Crystallization and preliminary X-ray crystallographic analysis of a bacterial tyrosinase from *Bacillus megaterium*. *Acta Crystallographic F*, *66*, 1101-1103.

[15] Matoba, Y., Kumagai, T., Yamamoto, A., Yoshitsu, H. & Sugiyama, M. (2006). Crystallographic evidence that the dinuclear copper center of tyrosinase is flexible during catalysis. *J Biol Chem*, *281*, 8981-8990.

[16] Ismaya, W. T., Rozeboom, H. J., Weijn, A., Mes, J. J., Fusetti, F., Wichers, H. J. & Dijkstra, B. W. (2011). Crystal structure of *Agaricus bisporus* mushroom tyrosinase: identity of the tetramer subunits and interaction with tropolone. *Biochemistry*, *50*, 5477-5486.

[17] Decker, H., Schweikardt, T., Nillius, D., Salzbrunn, U., Jaenicke, E. & Tuczek, F. (2007). Similar enzyme activation and catalysis in hemocyanins and tyrosinases. *Gene*, *398*, 183-191.

[18] Espin, J. C. & Wichers, H. J. (1999). Activation of a latent mushroom (*Agaricus bisporus*) tyrosinase isoform by sodium dodecyl sulfate (SDS). Kinetic properties of the SDS-activated isoform. *J Agr Food Chem, 47*, 3518-3525.

[19] Kohashi, P. Y., Kumagai, T., Matoba, Y., Yamamoto, A., Maruyama, M. & Sugiyama, M. (2004). An efficient method for the overexpression and purification of active tyrosinase from *Streptomyces castaneoglobisporus*. *Protein Expres Purif, 34*, 202-207.

[20] Mattinen, M., Lantto, R., Selinheimo, E., Kruus, K. & Buchert, J. (2008). Oxidation of peptides and proteins by *Trichoderma reesei* and *Agaricus bisporus* tyrosinases. *J Biotechnol, 133*, 395-402.

[21] Fairhead, M. & Thöny-Meyer, L. (2010). Role of the C-terminal extension in a bacterial tyrosinase. *FEBS J, 277*, 2083-2095.

[22] Fairhead, M. & Thöny-Meyer, L. (2010). Cross-linking and immobilisation of different proteins with recombinant *Verrucomicrobium Spinosum* tyrosinase. *J Biotechnol, 150*, 546-551.

[23] Jus, S., Stachel, I., Fairhead, M., Meyer, M., Thöny-Meyer, L. & Guebitz, G. M. (2012). Enzymatic cross-linking of gelatine with laccase and tyrosinase. *Biocatal Biotrans, 30*, 86-95.

[24] Ito, M. & Oda, K. (2000). An organic solvent resistant tyrosinase from *Streptomyces* sp. REN-21: purification and characterization. *Biosc, Biotech Bioche, 64*, 261-267.

[25] Selinheimo, E., NiEidhin, D., Steffensen, C., Nielsen, J., Lomascolo, A., Halaouli, S., Record, E., O'Beirne, D., Buchert, J. & Kruus, K. (2007). Comparison of the characteristics of fungal and plant tyrosinases. *J Biotechnol, 130*, 471-480.

[26] Heck, T., Faccio, G., Richter, M. & Thöny-Meyer, L. (2013). *Enzyme-catalyzed protein crosslinking Appl Microbiol Biotechnol, 97*, 461-75.

[27] Wang, C. S. & Stewart, R. J. (2012). Localization of the bioadhesive precursors of the sandcastle worm, *Phragmatopoma californica* (fewkes). *J Exp Biol*, 21i5, 351-361.

[28] Wilker, J. (2011). Biomaterials: Redox and adhesion on the rocks. *Nat Chem Biol, 9*, 579-580.

[29] Ren, Q., Henes, B., Fairhead, M. & Thöny-Meyer, L. (2013). High level production of tyrosinase in recombinant *Escherichia coli*. *BMC Biotechnol, 13*, 18.

[30] Durán, N., Rosa, M. A., D'Annibale, A. & Gianfreda, L. (2002). Applications of laccases and tyrosinases (phenoloxidases). immobilized on different supports: A review. *Enzyme Microb Tech, 31*, 907-931.

[31] Abhijith, K. S., Kumar, P. V., Kumar, M. A. & Thakur, M. S. (2007). Immobilised tyrosinase-based biosensor for the detection of tea polyphenols. *Anal Bioanal Chem, 389*, 2227-2234.

[32] Jewell, W. & Ebeler, S. E. (2001). Tyrosinase biosensor for the measurement of wine polyphenolics. *Am J Enol Viticult, 52*, 219-222.

[33] Tembe, S., Karve, M., Inamdar, S., Haram, S., Melo, J. & D'Souza, S. F. (2006). Development of electrochemical biosensor based on tyrosinase immobilized in composite biopolymeric film. *Anal Biochem, 349*, 72-77.

[34] Long, M. J. C. & Hedstrom, L. (2012). Mushroom tyrosinase oxidizes tyrosine-rich sequences to allow selective protein functionalization. *Chem Bio Chem, 13*, 1818-1825.

[35] Anghileri, A., Lantto, R., Kruus, K., Arosio, C. & Freddi, G. (2007). Tyrosinase-catalyzed grafting of sericin peptides onto chitosan and production of protein-polysaccharide bioconjugates. *J Biotechnol, 127*, 508-519.

[36] Gübitz, G. M. & Paulo, A. C. (2003). New substrates for reliable enzymes: Enzymatic modification of polymers. *Curr Opin Biotech, 14*, 577-582.

[37] Müller, G. H., Lang, A., Seithel, D. R. & Waldmann, H. (1998). An enzyme-initiated hydroxylation-oxidation carbo diels-alder domino reaction. *Chem Eur J, 4*, 2513-2522.

[38] Seo, S. Y., Sharma, V. K. & Sharma, N. (2003). Mushroom tyrosinase: Recent prospects. *J Agr Food Chem, 51*, 2837-2853.

[39] Buchert, J., Ercili Cura, D., Ma, H., Gasparetti, C., Monogioudi, E., Faccio, G., Mattinen, M., Boer, H., Partanen, R., Selinheimo, E., Lantto, R. & Kruus, K. (2010). Crosslinking food proteins for improved functionality. *Annual Review of Food Science and Technology, 1*, 113-138.

[40] Cha, H. J., Hwang, D. S., Lim, S., White, J. D., Matos-Perez, C. R. & Wilker, J. (2008). Bulk adhesive strength of recombinant hybrid mussel adhesive protein. *Biofouling, 4*, 1-9.

[41] Chen, T., Embree, H. D., Brown, E. M., Taylor, M. M. & Payne, G. F. (2003). Enzyme-catalyzed gel formation of gelatin and chitosan: potential for *in situ* applications. *Biomaterials, 24*, 2831-2841.

[42] Yamada, K., Aoki, T., Ikeda, N., Hirata, M., Hata, Y., Higashida, K. & Nakamura, Y. (2008). Application of chitosan solutions gelled by melB tyrosinase to water-resistant adhesives. *J Appl Polym Sci, 107*, 2723-2731.

[43] Jus, S., Stachel, I., Schloegl, W., Pretzler, M., Friess, W., Meyer, M., Birner-Gruenberger, R. & Guebitz, G. M. (2011). Cross-linking of collagen with laccases and tyrosinases. *Mater Sci Eng C -Mater Biol Appl, 31*, 1068-1077.

[44] Jin, R., Lou, B. & Lin, C. (2013). Tyrosinase-mediated *in situ* forming hydrogels from biodegradable chondroitin sulfate? tyramine conjugates. *Polym Int, 62*, 353-361.

In: Microbial Catalysts. Volume 2
Editors: Shadia M. Abdel-Aziz et al.

ISBN: 978-1-53616-088-8
© 2019 Nova Science Publishers, Inc.

Chapter 2

MICROBIAL ENZYMES: APPLICATIONS AND RELEVANCE IN INDUSTRIES, MEDICINES AND BEYOND

Shivaiah Nagaraju[1], Vinod Gubbiveeranna[1],
Venkataramana Mudili[2], Mohan Chakrabhavi Dhananjaya[3]
Shobith Rangappa[4], Lakshmeesha Thimmappa Ramachandrappa[5],
and Siddaiah Chandra Nayaka[5,*]

[1]Department of Studies and Research in Biochemistry,
Tumkur University, Tumkur, India
[2]DRDO Centre, Coimbatore, Tamil Nadu, India
[3]Department of Molecular Biology, University of Mysore, Mysore, India
[4]Adichunchanagiri Institute for Molecular Medicine, Nagamangala, Karnataka, India
[5]Department of Studies in Biotechnology, University of Mysore, Mysore, India

ABSTRACT

Enzyme catalyzes different biochemical reactions/interconversions which are the basis of life sustainability. Enzymes are biomolecules which accelerate the metabolic processes with high specificity and efficiency which increases the rate of the reaction by 100 million to 10 billion times faster than normal rate of reaction. New scientific inventions such as recombinant technology and protein engineering have evolved enzymes as important molecules that have been widely used in different industrial and therapeutical purposes. Microbial enzymes have acquired much attention with rapid development in enzyme technology and they are preferred due to their economic feasibility, high yields, consistency, ease of product modification and optimization, continuous supply without seasonal fluctuation, wide range of inexpensive media,

* Corresponding Author's E-mail: moonnayak@gmail.com.

stability and greater specific catalytic activity. These play a major role in diagnosis, treatment, biochemical investigation and monitoring of various diseases. Microbial enzymes because of their catalytic property have great importance in development of industrial bioprocesses. Current applications are focused on many different markets including pulp and paper, leather, detergents and textiles, pharmaceuticals, chemical, food and beverages, biofuels, animal feed and personal care among others. Today there is need of new, improved, versatile, novel enzyme for sustainable and economically competitive production processes. Diversity in microbes and modern molecular techniques such as metagenomics and genomics are being used to discover new microbial catalysts whose catalytic propertyes can be improved/modified by different strategies based on rational, semi-rational and randon directed evolutionary methods.

Keywords: microbial enzymes and proteins, food, nutrition

1. INTRODUCTION

The predominant group of life on earth is represented by microorganisms and the proteins of microbial source constitute the main catalytic entities which underlies in major biochemical reactions/transformations. New technological developments have led to the comprehensive extraction, separation and identification of microbial proteins [1].

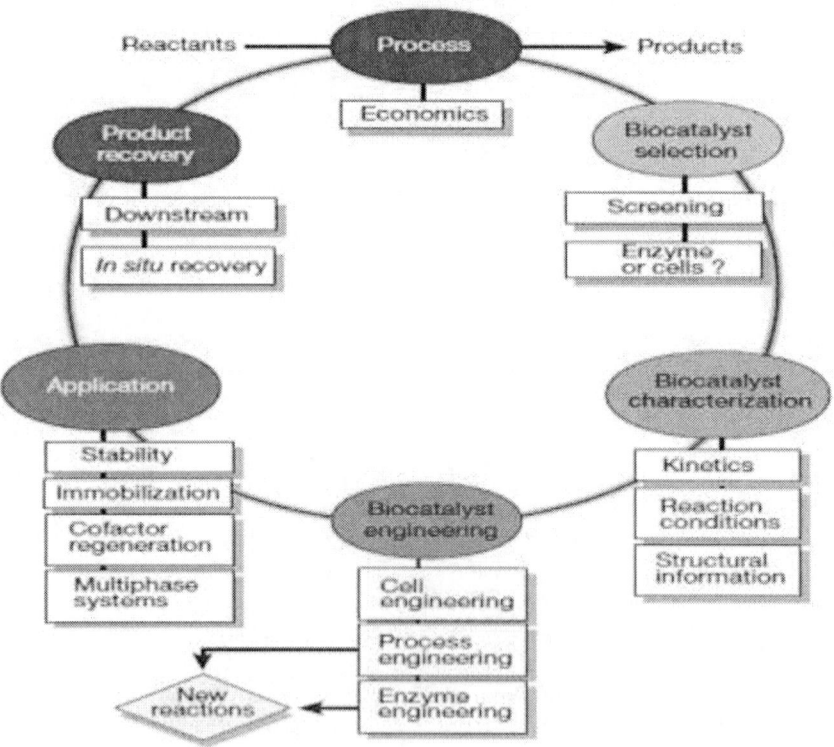

Figure 1. Microbial catalysis cycle.

Microbial enzymes play a crucial role as metabolic catalysts, making them important molecules in various industrial applications (Figure 1). Microbial enzymes gained attention due to their high substrate specificity and stability under varied conditions. With rapid development in biotechnological field these have gained large attention as microbial enzymes are considered as potential biocatalysts for a wide range of reactions. Microbial enzymes have versatility, substrate specificity, regioselectivity, chemoselectivity, enentioselectivity and catalyzes at ambient temperatures and pressures. Nowadays these have been widely used to prepare range of commercial products for pharmaceutical industry (eg: 6-aminopenicillanic acid), food and nutrition (eg: glucose and fructose syrups, L-lysine and niacinamide), as well as specialty and commodity chemicals (eg: acrylamide and acrylic acid). Microcial biocatalysts can use: (a) soluable or immobilized enzymes (b) whole cell catalysts that are not metabolically active but which still maintain one or more desirable enzyme activities or (c) direct fermentation to produce an intermediate or final product. These examples indicate the wide range of applications by microbial enzymes. The limitations of the naturally occurring microbial enzymes can be trounce by designing specific enzymes for each type of processes [2, 3].

Enzymes of natural origin have been widely used in manufacture of products such as linen, leather and indigo. These enzymes mainly derived from microorganisms or enzymes present in added preparations such as calves rumen or papaya fruit. Development of fermentation processes was aimed the production of enzymes by use of particularly selected strains, by which it is possible to produce purified, well characterized enzymes on a large scale. This development allowed the introduction of enzymes into industrial products and processes such as detergent, textiles and starch industries. The recombinant DNA technology has further improved the production processes and helped to produce enzymes commercially in large quantities. Furthermore, the developments in biotechnology, such as protein engineering and directed evolution, further revolutionized the commercialization of industrially important enzymes. This advance in biotechnology is providing different kinds of enzymes displaying new activities, adaptability to new conditions leading to their increase use in industrial purposes [4].

Majority of the currently used industrial enzymes are hydrolytic in action, these being used for degradation of various natural substances. Proteases remain the dominant enzyme type, because of their extensive use in detergent and dairy industries. Carbohydrases primarily amylases and cellulases represent the second largest group of enzyme used in starch, textile, detergent and baking industries. Microbial enzymes display regional stereospecificity, property which can be expoited for asymmetric synthesis and recemic resolution. Chiral selectivity of enzymes has been employed in synthesis of enantiomerically pure pharmaceuticals, agrochemicals, chemical feedstock and food additives.

2. ENZYMATIC BIOCATALYSIS

Microbial enzymes have been widely used in brewing and food industries for long time. These are finding new applications in industrial chemistry [5]. Biocatalysis involves the applications of microbial cells, cell extracts, purified enzymes, immobilized cells or immobilized enzymes as catalysts in many chemicals reactions/interactions [6, 7].

3. TECHNICAL APPLICATIONS

Enzymes are very important for industrial and pharmaceutical processes [8]. The total market for industrial enzymes have reached $3.3 billion in 2010 and it is estimated to reach a value of 4.4 billion by 2015 [9]. Enzymes are widely in bulk quantities in detergent, textile, pulp, paper and biofuel industries. Usage for leather and bioethanol is responsible for the highest sales figures. These enzymes had revenues of nearly $1.2 billion in 2011 which is expected to reach $1.5 billion in 2015 and $1.7 billion in 2016. The highest sales are expected to be in biofuels marker [10]. Food and beverages enzymes expected to account for $1.3 billion by 2015.

The use of enzymes as detergent additives represents a major application of industrial enzymes. Proteases, lipases, amylases, oxidases, peroxidases and cellulases are added to detergents where they catalyze the breakdown of chemical bonds on addition of water. To be suitable, they must be active under thermophilic (60O C and alkolophilic pH 9-11) conditions, as well as in the presence of the various components of washing powders.

Proteases constitute over 60% of the global market for enzymes, these are used to produce pharmaceuticals, foods, detergents, leather, silk and agrochemical products. Especially, in laundry detergents, they account for approximately 25% of the total worldwide sales of enzymes. Novo industry (now Novozymes) introduced the first detergent containing a bacterial protease (Biotex) in 1956. It contained an alcalase produced by Bacillus licheniformis. In 1994, Novo Nordisk introduced Lipolase, the first commercial recombinant lipase for use in a detergent, by cloning the Humicola lanuginose lipase into the A. oryzae genome. In 1995, Genencor International introduced two bacterial lipases, one from Pseudomonas mendocina (Lumafast), and another from Pseudomonas alcaligenes (Lipomax). An enzyme added recently to detergents is Mannaway, a Bacillus mannanase which removes food stains containing guar gum [11].

In the textile industry, enzymes are used to develop cleaner processes and reduce the use of raw materials and production of waste. The application of cellulases for denim finishing and laccases for decolorization of textile effluents and textile bleaching are the most recent commercial advances [12]. An alternative enzymatic process in the manufacturing of cotton has been recently developed based on a pectate lyase [13]. The process is performed at much lower temperatures and uses less water than the classical method.

Lipases, xylanases and laccases are another group of enzymes being used largely in removing pitch (hydrophobic components of wood, mainly triglycerides and waxes) in pulp industry [14]. A lipase from Candida rugosa is being used by Nippon Paper Industries to remove up to 90% of these compounds [15]. The use of enzymes as alternatives to chemicals in leather processing has proved successful in improving leather quality and in reducing environmental pollution. Alkaline lipases from Bacillus strains, which grow under highly alkaline conditions, in combination with other alkaline or neutral proteases, are currently being used in this industry.

Laccases oxidize phenolic and non-phenolic lignin-related compounds as well as environmental pollutants [16]. They are used to detoxify industrial effluents from the paper and pulp, textile, and petrochemical industries, as a medical diagnostic tool, for bioremediation of herbicides, pesticides, and explosives in soil, as a cleaning agent for water purification systems, as a catalyst in drug manufacture and as cosmetic ingredients.

Lipases play a major role in the fermentative steps during manufacturing of sausage, measure changes in long-chain fatty acid liberated during ripening and to modify the food flavour by synthesis of esters of short-chain fatty acids and alcohols (flavour and fragrance). Previously, lipases of different microbial sources were used for refining rice flavour, modifying soybean milk, and for enhancing the aroma and speed up the fermentation of apple wine [17]. By adding lipases the fat is removed while processing meat and fish, and this process is called biolipolysis.

Cellulases have been widely used in textile industries as these have gained additional consideration in the enzyme market owing to their ability in the degradation of lignocellulosic feedstocks. But the cost of cellulases is the key issue in achieving low price conversion of lignocellulosic biomass into biofuels and other products [18-20]. Filamentous fungi can produce native cellulases at levels greater than 100 g/L [21].

Cellulose hydrolyzing enzymes includes (1) endoglucanases, which break down cellulose chains in a random manner; (2) cellobiohydrolases, which liberate glucose dimers from both ends of cellulose chains; and (3) beta-glucosidases, which produce glucose from oligomer chains. Hypocrea jecorina (Trichoderma reesei) is the main industrial source of cellulases and hemicellulases used to depolymerize plant biomass to simple sugars [22, 23]. The overall action of T. reesei on cellulosic biomass is limited by its low content of beta-glucosidase. The result is accumulation of cellobiose which limits further breakdown. The expression of the beta-glucosidase gene from Pericona sp in T. reesei resulted in an increased level of beta-glucosidase, thus increasing overall cellulase activity and action on biomass residues [24]. Cellulases are formed adaptively, and several positive (XYR1, ACE2, HAP2/3/5) and negative (ACE1, CRE1) components involved in this regulation are now known [25]. In addition, its complete genome sequence has been published [26], thus making the organism susceptible to targeted improvement by metabolic engineering. It has recently been reported that the extreme thermophilic bacterium Caldicelluloseruptor bescil produces a cellulase/hemicellulase system twice as active as that from *T. reesei* [27].

Another important group of enzymes are glucosidases used to synthesize glycosides as slow release aroma compounds. Volatile flavours evaporate during storage leading to decrease in the right concentration at the moment of consumption, but if they are present in a bound, non-volatile form, they can be liberated upon heating, thus optimizing organoleptic characteristics of the consumed product. Glucosides are adequate derivatives as slow release flavours, due to their very low vapour pressures and the possibility of obtaining them as natural compounds.

3.1. Enzymes in the Feed Industry

Feed enzymes form the major global market in enzyme industry [28]. Feed enzymes increases digestibility of nutrients leading to greater efficiency in feed utilization. They can also degrade unacceptable components in feed, which are otherwise harmful or of little or no value [29]. Commercially available feed enzymes are phytases, proteases, α-galactosidases, glucanases, xylanases, α-amylases, and polygalacturonases, mainly used for swine and poultry [30]. Recent developments developed of heat stable enzymes, improved specific activity, some new non-starch polysaccharide-degrading enzymes, and rapid, economical and reliable assays for measuring enzyme activity have always been the focus and have been intensified recently. But, the use of the enzymes as feed additives is restricted in many counties by local regulatory authorizes [31] and applications vary from county to country.

3.2. Enzymes in Food Processing

Enzymes used in food processing can be divided into food additives and processing aids. Most food enzymes are considered as processing aids, with only a few used as additives, such as lysozyme and invertase. The processing aids are used during the manufacturing process of foodstuffs, and do not have a technological function in the final food.

Lipases are commonly used in the production of variety of products ranging from fruit juices, baked foods and vegetable fermentations to dairy enrichment. Fats, oils and related compounds are the main targets of lipases in food technology. Controlled lipase concentration, pH, temperature and emulsion content has to be maintained for the maximum production of flavor and fragrances. The lipase mediation of carbohydrate esters of fatty acids offers a potential market for use as emulsifiers in foods, pharmaceuticals and cosmetics. There are three recombinant fungal lipases currently used in the food industry, one from Rhizomucor miehi, one from Thermomyces lanuginosus and another from Fusarium oxysporum; all being produced in *A. oryzae* [32, 33].

Proteases is majorly used in dairy industry for the manufacture of cheese. Calf rennin had been preferred in cheese making due to its high specificity, but microbial proteases produced by GRAS microorganisms like Mucor miehei, Bacillus subtilis, Mucor pusillus Lindt and Endothia parasitica are gradually replacing it. The primary function of these enzymes in cheese making is to hydrolyze the specific peptide bond (Phe105-Met106) that generates para-k-casein and macropeptides [34]. Production of calf rennin (chymosin) in recombinant A. niger var awamori amounted to about 1 g/L after nitrosoguanidine mutagenesis and selection for 2-deoxyglucose resistance [35].

Other enzymes in the food industry include invertase from Kluyveromyces fragilis, Saccharomyces carlsbergensis and S. cerevisiae for candy and jam manufacture, β-galactosidase (lactase) from Kluyveromyces lactis, K. fragilis or Candida pseudotropicalis for hydrolysis of lactose from milk or whey, and galactosidase from S. carlsbergensis for crystallization of beet sugar. All these materials are expected to be safe, under the guidance of good manufacturing practice (GMP).

3.3. Enzymes in Chemical and Pharmaceutical Processes

Application of microbial enzymes in the chemical industry depends mainly on cost competitiveness with the existing and well-established chemical methods [36]. Lower energy demand, increased product titer, increased catalyst efficiency, less catalyst waste and by products, as well as lower volumes of wastewater streams, are the main advantages that biotechnological processes have as compared to well-established chemical processes. Phenol, pyruvate, pyridoxal phosphate and ammonium chloride are converted to L-tyrosine using a thermostable and chemostable tyrosine phenol lyase obtained from Symbiobacterium toebii. The titer produced was 130 g/L after 30 h with continuous feeding of substrate. Enzymes are useful for preparing beta-lactam antibiotics such as semi-synthetic penicillins and cephalosporins [37, 38]. Beta-lactams constitute 60%–65% of the total antibiotic market.

Preparation/synthesis of complex chiral pharmaceutical intermediates efficiently and economically is one the most important application in biocatalysis. Esterases, lipases, proteases and ketoreductases are widely applied in the preparation of chiral alcohols, carboxylic acids, amines or epoxides, among others [39]. The inherent inefficiency of kinetic resolution (maximum 50% yield) can be overcome by novel asymmetric reactions catalyzed by improved microbial enzymes which can provide a 100% yield [40]. Asymmetric reduction of tetrahydrothiophene-3-one with a wild-type reductase gave the desired alcohol ((R)-tetrahydrothiophene-3-ol), a key component in sulopenem, a potent antibacterial developed by Pfizer, but only in 80%–90% ee (enantiomeric excess). A combination of random mutagenesis, gene shuffling and ProSAR analysis was used to improve the enantio-selectivity of a ketoreductase towards tetrahydrothiophene-3-one. The best variant increased enantio-selectivity from 63% ee to 99% ee [41].

Atorvastatin, the active ingredient of Lipitor, a cholesterol-lowering drug that had global sales of US$12 billion in 2010, can be produced enzymatically. The process is based on three enzymatic activities: a ketone reductase, a glucose dehydrogenase and a halohydryn dehalogenase [42].

Kinetic resolution of racemic amines is a common method used in the synthesis of chiral amines. Acylation of a primary amine moiety by a lipase is used by BASF for the resolution of chiral primary amines in a multi-thousand ton scale [43]. Recently, asymmetric synthesis from the corresponding chiral ketones, using transaminases, is gaining attention. Some (R)-selective transaminases have been recently discovered using in silico strategies for a sequence-based prediction of substrate specificity and enantio-preference [44]. Optically pure (S)-amines were obtained using a recombinant ω-transaminase with 99% ee and 97% yield [45]. These enantiopure amines may find used as inhibitors of monoamine oxidase in the treatment of neurological disorders such as Parkinson's and Alzheimer's diseases.

Effective enzymatic process using enzyme evolution was developed by the biotechnology company Codexis, in cooperation with Pfizer, to produce 2-methyl pentanol, an important intermediate for manufacture of pharmaceuticals and liquid crystals [46]. Recently, protein engineering expanded the substrate range of transaminases to ketones. In a work developed by Merck and Codexis, the chemical manufacture of sitagliptin, the active ingredient in Januvia which is a leading drug for type 2 diabetes, was replaced by a new biocatalytic process. Several rounds of directed evolution were applied to create an engineered amine transaminase with a 40,000-fold increase in activity [47]. Such a process not only reduced total waste (by 19%), but also increased overall yield (by 13%) and productivity (by 53%). Codexis scientists also developed enzymatic processes for the production of montelukast (Singulair) and silopenem. They also developed an improved LovD enzyme (an acyltransferase) for improved conversion of the cholesterol-lowering agent, lovastatin, to simvastatin [48].

Optically active carboxylic acids usually synthesized through different enzymatic routes catalyzed by lipases, nitrilases or hydroxynitrile lyases. The synthesis of 2-arylpropanoic acids (e.g., ketoprofen, ibuprofen and naproxen) is mainly achieved through the kinetic resolution of racemic substrates by lipases from Candida antarctica or Pseudomonas sp. A process using a novel substrate, (R, S)-N-profenylazoles, instead of their correspondent esters, proved to be more efficient. (R)-o-chloromandelic acid is a key intermediate for manufacture of Clopidogrel, a platelet aggregation inhibitor, with global sales of $10 billion per year. The asymmetric reduction of methyl ochlorobenzoylformate with a versatile recombinant carbonyl reductase from S. cerevisiae expressed in E. coli yielded (R)-o-chloromandelate.

Microbial enzymes applications in several industrial bioconversions has been broadened by the use of organic solvents replacing water [49, 50], an important development in enzyme engineering. Many chemicals and polymers are insoluble in water and its presence leads to undesirable by-products and degradation of common organic reagents. Although switching from water to an organic solvent as the reaction medium might suggest that the enzyme would be denatured, many crystalline or lyophilized enzymes are actually stable and retain their activities in such anhydrous environments. Yeast lipases have been used to catalyze butanolysis in anhydrous solvents to obtain enantiopure 2-chloro- and 2-bromo-propionic acids that are used for the synthesis of herbicides and pharmaceuticals [51]. Lipase is also used in stereoselective step, carried out in acetonitrile, for the acetylation of a symmetrical diol during the synthesis of an antifungal agent [52].

There are many advantages of employing enzymes in organic, as opposed to aqueous, media [53], including higher substrate solubility, reversal of hydrolytic reactions, and modified enzyme specificity, which result in new enzyme activities. On the other hand, enzymes usually show lower catalytic activities in organic than in aqueous solution.

3.4. Enzymes in Therapeutic Applications

Therapeutic enzymes have a wide range of uses such as oncolytics, thrombolytics or anticoagulants, as replacements for metabolic deficiencies and some serve as anti-inflammatory agents (Figure 2) . The list of enzymes which have the potential to become important therapeutic agents and its microbial sources are shown in Table 1 and Table 2 respectively. A number of factors severely decrease the potential utility of microbial enzymes in the medical field due to large molecular size of biological catalyst which prevents their distribution within somatic cells, and another reason is the response of immune system of the host cell after injecting the foreign enzyme protein.

As compared to the industrial use of enzymes, therapeutically useful enzymes are required in relatively less amounts, but the degree of purity and specificity should be generally high. The kinetics of these enzymes is low Km and high Vmax so that it is maximally efficient even at low concentrations of enzymes and substrates. The sources of such enzymes should be selected with great care to prevent any possibility of undesirable contamination by incompatible material and also to enable ready purification. Therapeutic enzymes are usually marketed as lyophilised pure preparations with biocompatible buffering salts and mannitol diluent. The cost of these enzymes is high but comparable to those of therapeutic agents or treatments. As an example, urokinase is derived from human urine and used to dissolve blood clots. Major application of therapeutic enzymes is in the treatment of cancer and various other diseases.

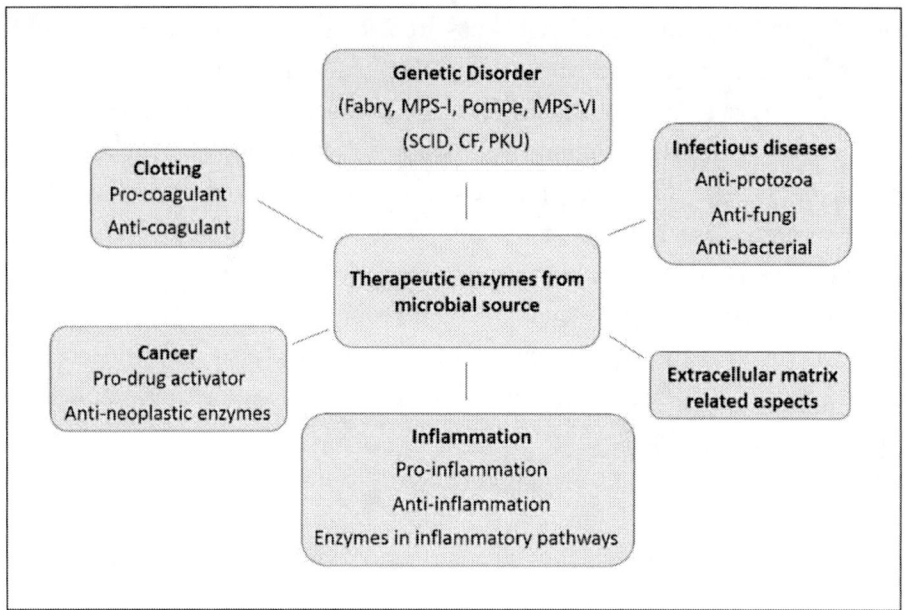

Figure 2. Application of therapeutic enzymes in different disorders and diseases.

Table. 1. Some important enzymes and their therapeutic importance

Enzymes	Reaction	Use	Sources
Asparaginase	L-Asparagine H_2O → L-aspartate + NH_3	Leukaemia	*E. coli*
Collagenase	Collagen hydrolysis	Skin ulcers	*C. perfringens*
Glutaminase	L-Glutamine H_2O →L-glutamate + NH_3	Leukaemia	*E. coli SFL-1*
Lysozyme	Bacterial cell wall hydrolysis	Antibiotic	*Homo sapiens*
Ribonuclease	RNA hydrolysis	Antiviral	*Yeast Bacteriophages*
Streptokinase	Plasminogen → plasmin	Blood clots	*Streptococci sp.*
Trypsin	Protein hydrolysis	Inflammation	*Homosapiens*and other vertebrates
Uricase	Urate + O_2→ allantoin	Gout	*A. flavus*
Urokinase	plasminogen → plasmin	Blood clots	*Bacillus subtilis*
B-Lactamase	B-Lactam ring hydrolysis	Antibiotic resistance	*Citrobacterfreundii, Serratiamarcescens Klebsiella pneumonia*
Penicillin acylase	Binding the rings of benzylppenicillin (penicillin G) and phenoxymethylpenicillin (penicillin V)	Penicillin production/broad spectrum antibiotic production	*Penicillium sp.*

Various enzymes of higher quality and purity are now in clinical trials. Proteolytic enzymes of plant and bacterial origin have been studied for the removal of dead skin of burns. Debrase gel dressing, containing a mixture of several enzymes extracted from pineapple, received clearance in 2002 from the US FDA for a Phase II clinical trial for the treatment of partial-thickness and full-thickness burns. A proteolytic enzyme (VibrilaseTM) obtained from Vibrio proteolyticus is found to be effective against denatured proteins such as those found in burned skin. The regeneration of injured spinal cord have been demonstrated using chondroitinases, where this enzyme acts by removing the glial scar and thereby accumulating chondroitin sulfate that stops axon growth [54]. Hyaluronidase has also been found to be a similar hydrolytic activity on chondroitin sulphate and may help in the regeneration of damaged nerve tissue [55].

Table. 2. List of some of the enzymes found in different species

Source	Enzyme	Microorganism
Fungal	Amylase	*Aspergillus oryzae*
	Glucosidases	*Aspergillus flavus*
	Proteases	*Aspergillus niger*
	Pectinases	*Aspergillus niger*
	Glucose oxidase	*Penicilliumnotatum*
	Catalase	*Aspergillus niger*
Bacterial	Amylase	*Bacillus subtilis*
	Protease	
	penicillinase	
Yeast	Invertase	*Saccharomyces cerevisiae*
	Lactase	*Saccharomyces fragilis*

The cell wall of various pathogenic organisms, including fungi, protozoa, and helminths is made up of chitin and is a good target for antimicrobials [56]. The lytic enzyme derived from bacteriophage is used to target the cell walls of Streptococcus pneumonia, Bacillus anthracis, and Clostridium perfringens [57]. The application of lytic bacteriophages can be used for the treatment of several infections and could be useful against new drug-resistant bacterial strains.

Cancer research has good instances of the use of microbial enzymes in the therapeutics. Recent studies have proved that arginine-degrading enzyme (PEGylated arginine deaminase) can inhibit human melanoma and hepatocellular carcinomas [58]. Another PEGylated enzyme, Oncaspar1 (pegaspargase), has shown good results for the treatment of children newly diagnosed with acute lymphoblastic leukemia and are already in use in the clinic.

Table 3. A broad spectrum of idea about using the application
of enzymes in different areas

Types of industries	Enzymes	Use
Alcohol/beverage	Amylase, glucanases, proteases, beta-glucanases, arabinoxylans, amyloglucosidase, pullulanases and acetolactate decarboxylase	Degradation of starch and polycarbonated into simple sugar. Also for degrading complex proteins into sugars thus to increase the fermentation efficiency. Production of low calorie beer
Fruit drinks	Cellulases, pectinases	Clarify fruit juice
Baby food	Trypsin	Predigest baby foods
Food processing	Amylase, protease and lactases	Degradation of starch and complex proteins, softening of meat
Dairy	Rennin, lipases and lactases	Hydrolyzing protein, cheese production (Roquefort cheese) and glucose production from lactose
Detergent	Protease, amylase, cellulases and mannanase	To remove protein after staining, remove insoluble starch in dish washing, removing oils and fats and to increase the effectiveness of detergents
Textile	Amylase, pectinase, cellulases and mannanase	To remove starch size, glue between the fiber core and the waxes, fabric finishing in denims, degrading residual hydrogen peroxide after the bleaching of cotton, wool treatment and the degumming of raw silk also known as biopolishing
Paper and pulp	Amylases, xylanases, cellulases, hemicellulose, ligninases and esterase	Degrade starch to lower viscosity, aiding sizing, deinking and coating paper. Xylanases reduce bleach required for decolorizing; cellulases and hemicellulose smooth fibers, enhance water drainage and promote ink removal; lipases reduce pitch and lignin-degrading enzymes remove lignin to softer paper, for esterification
Animal feedstock	Phytase	Increase total phosphorous content for growth, increase in phytic acid need
Rubber	Catalase	Generate oxygen from peroxide to convert latex into foam rubber
Oil and petroleum	Vellulases, ligninases and mannanase	Formation of ethanol, forming gel breaker in oil drilling
Biopolymer/plastic	Laccases, peroxidases, lipases and transglutaminases	Forming cross-links in biopolymers to produce materials in situ by means of polymerization processes
Pharmaceutical	Nitrile hydratase, D-amino acid oxidase, glutaric acid acylase, penicillin acylase, penicillin G acylase, ammonia lyase and humulin	Producing water soluble intermediates, semisynthetic antibiotics, intermediate for aspartame and biosynthetic human insulin
Molecular biology	Restriction enzymes, DNA ligase and polymerases	Used to manipulate DNA in genetic engineering, essential for restriction of digestion and polymerase chain reaction, also important in forensic science

CONCLUSION

Enzymes have been the part of human civilization and extensively used in ancient brewing and other uses. With the new scientific discoveries and inventions the scientists have studies the efficiency and specificity of enzymes leading to their wide range of applications in industries and day to day activities. Today different types of enzymes are being manufactured and used in industries like food, dairy, detergent, and chemical as well as for their important lifesaving therapeutically used medicines.

Various new technologies have been developed and are still developing to manufacture both bulk and added value products utilizing enzymes as biocatalysts, in order to meet needs such as food (e.g., bread, cheese, beer, and vinegar), fine chemicals (e.g., amino acids, vitamins), agricultural (growth hormones), and pharmaceuticals (insulin). Enzymes are also used to provide services, as in washing and environmental processes (especially clean-up processes) or for analytical and diagnostic purposes. The goal of these approaches is to design innovative products and processes that are not only competitive but also meet criteria of sustainability and economic viability.

Extensive research is being conducted to produce enzymes from microbes. Since then many microorganisms and their enzymes with unique function have also been discovered by means of extensive screening, and now they are commonly used in different industrial and medical fields. Development of these medically important enzymes has been at least as extensive as those for industrial applications thus reflecting the magnitude of the potential rewards of this sector in the near future.

REFERENCES

[1]   Asimov, Isaac. *Asimov's biographical encyclopedia of science and technology.* David & Charles, 2nd edition, 1978.

[2]   Gurung, Neelam, Sumanta Ray, Sutapa Bose, and Vivek Rai. "A broader view: microbial enzymes and their relevance in industries, medicine, and beyond." *BioMed research international* 2013 (2013).

[3]   Anbu, Periasamy, Subash CB Gopinath, ArzuColeriCihan, and Bidur Prasad Chaulagain. "Microbial enzymes and their applications in industries and medicine." *BioMed research international* 2013 (2013).

[4]   Underkofler, L. A., R. R. Barton, and S. S. Rennert. "Production of microbial enzymes and their applications." *Applied microbiology* 6, no. 3 (1958): 212.

[5]   Wohlgemuth, Roland. "Asymmetric biocatalysis with microbial enzymes and cells." *Current opinion in microbiology* 13, no. 3 (2010): 283-292.

[6]   Gong, Jin-Song, Zhen-Ming Lu, Heng Li, Jin-Song Shi, Zhe-Min Zhou, and Zheng-Hong Xu. "Nitrilases in nitrile biocatalysis: recent progress and forthcoming research." *Microbial cell factories* 11, no. 1 (2012): 142.

[7]   Schmid, A., J. S. Dordick, B. Hauer, Al Kiener, M. Wubbolts, and B. Witholt. "Industrial biocatalysis today and tomorrow." *Nature* 409, no. 6817 (2001): 258.

[8]   Sanchez, Sergio, and Arnold L. Demain. "Enzymes and bioconversions of industrial, pharmaceutical, and biotechnological significance." *Organic Process Research & Development* 15, no. 1 (2010): 224-230.

[9]   Dewan, S. "Enzymes in industrial applications: Global markets." *Market Research Report*. Wellesley, MA: BCC Research (2011).

[10]   Freedonia group. *"World Enzymes."* Cleveland, Ohio, United States of America (2011): 12-26.

[11]   Kirk, Ole, TorbenVedelBorchert, and Claus Crone Fuglsang. "Industrial enzyme applications." *Current opinion in biotechnology* 13, no. 4 (2002): 345-351.

[12]   Araujo, Rita, Margarida Casal, and ArturCavaco-Paulo. "Application of enzymes for textile fibres processing." *Biocatalysis and Biotransformation* 26, no. 5 (2008): 332-349.

[13]   Tzanov, Tzanko, Margarita Calafell, Georg M. Guebitz, and ArturCavaco-Paulo. "Bio-preparation of cotton fabrics." *Enzyme and Microbial Technology* 29, no. 6-7 (2001): 357-362.

[14]   Gutiérrez, Ana, C. José, and Angel T. Martínez. "Microbial and enzymatic control of pitch in the pulp and paper industry." *Applied microbiology and biotechnology* 82, no. 6 (2009): 1005-1018.

[15]   Farrell, Roberta L., KunioHata, and Mary Beth Wall. "Solving pitch problems in pulp and paper processes by the use of enzymes or fungi." In *Biotechnology in the Pulp and Paper Industry*, pp. 197-212. Springer, Berlin, Heidelberg, 1997.

[16]   Couto, Susana Rodríguez, and José Luis Toca Herrera. "Industrial and biotechnological applications of laccases: a review." *Biotechnology advances* 24, no. 5 (2006): 500-513.

[17]   Seitz, Eugene W. "Industrial application of microbial lipases: a review." *Journal of the American Oil Chemists' Society* 51, no. 2 (1974): 12-16.

[18]   Rubin, Edward M. "Genomics of cellulosic biofuels." *Nature* 454, no. 7206 (2008): 841.

[19]   Wilson, David B. "Cellulases and biofuels." *Current opinion in biotechnology* 20, no. 3 (2009): 295-299.

[20]   Zhang, Y-H. Percival. "What is vital (and not vital) to advance economically-competitive biofuels production." *Process Biochemistry* 46, no. 11 (2011): 2091-2110.

[21] Demain, Arnold L., and PreetiVaishnav. "Production of recombinant proteins by microbes and higher organisms." *Biotechnology advances* 27, no. 3 (2009): 297-306.

[22] Zhang, Y-H. Percival, Michael E. Himmel, and Jonathan R. Mielenz. "Outlook for cellulase improvement: screening and selection strategies." *Biotechnology advances* 24, no. 5 (2006): 452-481.

[23] Kumar, Raj, Sompal Singh, and Om V. Singh. "Bioconversion of lignocellulosic biomass: biochemical and molecular perspectives." *Journal of industrial microbiology & biotechnology* 35, no. 5 (2008): 377-391.

[24] Dashtban, Mehdi, and Wensheng Qin. "Overexpression of an exotic thermotolerant β-glucosidase in Trichoderma reesei and its significant increase in cellulolytic activity and saccharification of barley straw." *Microbial cell factories* 11, no. 1 (2012): 63.

[25] Kubicek, Christian P., Marianna Mikus, André Schuster, Monika Schmoll, and Bernhard Seiboth. "Metabolic engineering strategies for the improvement of cellulase production by Hypocreajecorina." *Biotechnology for biofuels* 2, no. 1 (2009): 19.

[26] Martinez, Natalia J., Maria C. Ow, John S. Reece-Hoyes, M. InmaculadaBarrasa, Victor R. Ambros, and Albertha JM Walhout. "Genome-scale spatiotemporal analysis of Caenorhabditis elegans microRNA promoter activity." *Genome research* (2008).

[27] Kanafusa-Shinkai, Sumiyo, Jun'ichi Wakayama, Kazumi Tsukamoto, Noriko Hayashi, Yasumasa Miyazaki, Hideyuki Ohmori, Kiyoshi Tajima, and Hiroshi Yokoyama. "Degradation of microcrystalline cellulose and non-pretreated plant biomass by a cell-free extracellular cellulase/hemicellulase system from the extreme thermophilic bacterium Caldicellulosiruptorbescii." *Journal of bioscience and bioengineering* 115, no. 1 (2013): 64-70.

[28] Sharma, Archana, and T. Satyanarayana. "Extremophiles as Potential Resource for Food Processing Enzymes." *Microbial Enzyme Technology in Food Applications* (2017): 420.

[29] Choct, M. "Enzymes for the feed industry: past, present and future." *World's Poultry Science Journal* 62, no. 1 (2006): 5-16.

[30] Selle, Peter H., and VelmuruguRavindran. "Microbial phytase in poultry nutrition." *Animal Feed Science and Technology 135*, no. 1-2 (2007): 1-41.

[31] Pariza, Michael W., and Mark Cook. "Determining the safety of enzymes used in animal feed." *Regulatory Toxicology and Pharmacology* 56, no. 3 (2010): 332-342.

[32] Méndez, Carmen, and José A. Salas. "Altering the glycosylation pattern of bioactive compounds." *Trends in biotechnology* 19, no. 11 (2001): 449-456.

[33] Okanishi, Masanori, Naoko Suzuki, and TakakiFuruta. "Variety of hybrid characters among recombinants obtained by interspecific protoplast fusion in

streptomycetes." *Bioscience, biotechnology, and biochemistry* 60, no. 8 (1996): 1233-1238.

[34] Rao, Mala B., Aparna M. Tanksale, Mohini S. Ghatge, and Vasanti V. Deshpande. "Molecular and biotechnological aspects of microbial proteases." *Microbiology and molecular biology reviews* 62, no. 3 (1998): 597-635.

[35] Dunn-Coleman, Nigel S., Peggy Bloebaum, Randy M. Berka, Elizabeth Bodie, Nancy Robinson, Gale Armstrong, Michael Ward et al. "Commercial levels of chymosin production by Aspergillus." *Nature Biotechnology* 9, no. 10 (1991): 976.

[36] Tufvesson, Pär, Joana Lima-Ramos, Mathias Nordblad, and John M. Woodley. "Guidelines and cost analysis for catalyst production in biocatalytic processes." *Organic Process Research & Development* 15, no. 1 (2010): 266-274.

[37] Volpato, G., R. C Rodrigues, and R. Fernandez-Lafuente. "Use of enzymes in the production of semi-synthetic penicillins and cephalosporins: drawbacks and perspectives." *Current medicinal chemistry* 17, no. 32 (2010): 3855-3873.

[38] Kim, Do Young, Eugene Rha, Su-Lim Choi, Jae-Jun Song, Seung-Pyo Hong, Moon-Hee Sung, and Seung-Goo Lee. "Development of bioreactor system for L-tyrosine synthesis using thermostable tyrosine phenol-lyase." *Journal of microbiology and biotechnology* 17, no. 1 (2007): 116-122.

[39] Zheng, Gao-Wei, and Jian-He Xu. "New opportunities for biocatalysis: driving the synthesis of chiral chemicals." *Current opinion in biotechnology* 22, no. 6 (2011): 784-792.

[40] Wohlgemuth, Roland. "Biocatalysis—key to sustainable industrial chemistry." *Current opinion in Biotechnology* 21, no. 6 (2010): 713-724.

[41] Liang, Jack, James Lalonde, BirtheBorup, Vesna Mitchell, Emily Mundorff, Na Trinh, D. A. Kochrekar, Ramachandran Nair Cherat, and G. Ganesh Pai. "Development of a biocatalytic process as an alternative to the (−)-DIP-Cl-mediated asymmetric reduction of a key intermediate of montelukast." *Organic Process Research & Development* 14, no. 1 (2009): 193-198.

[42] Ma, Steven K., John Gruber, Chris Davis, Lisa Newman, David Gray, Alica Wang, John Grate, Gjalt W. Huisman, and Roger A. Sheldon. "A green-by-design biocatalytic process for atorvastatin intermediate." *Green Chemistry* 12, no. 1 (2010): 81-86.

[43] Sheldon, Roger A. "E factors, green chemistry and catalysis: an odyssey." *Chemical Communications* 29 (2008): 3352-3365.

[44] Höhne, Matthias, Sebastian Schätzle, Helge Jochens, Karen Robins, and Uwe T. Bornscheuer. "Rational assignment of key motifs for function guides in silico enzyme identification." *Nature chemical biology* 6, no. 11 (2010): 807.

[45] Koszelewski, Dominik, Madeleine Göritzer, Dorina Clay, Birgit Seisser, and Wolfgang Kroutil. "Synthesis of optically active amines employing recombinant ω-transaminases in E. coli cells." *Chem Cat Chem* 2, no. 1 (2010): 73-77.

[46] Gooding, Owen W., Rama Voladri, Abigail Bautista, Thutam Hopkins, Gjalt Huisman, Stephan Jenne, Steven Ma et al. "Development of a practical biocatalytic process for (R)-2-methylpentanol." *Organic Process Research & Development* 14, no. 1 (2009): 119-126.

[47] Savile, Christopher K., Jacob M. Janey, Emily C. Mundorff, Jeffrey C. Moore, Sarena Tam, William R. Jarvis, Jeffrey C. Colbeck et al. "Biocatalytic asymmetric synthesis of chiral amines from ketones applied to sitagliptin manufacture." *Science* 329, no. 5989 (2010): 305-309.

[48] Ritter, Steve. "2012 Green Chemistry Awards." *Chemical & Engineering News* 90, no. 26 (2012): 11-11.

[49] Atomi, Haruyuki, Takaaki Sato, and Tamotsu Kanai. "Application of hyperthermophiles and their enzymes." *Current opinion in biotechnology* 22, no. 5 (2011): 618-626.

[50] Klibanov, Alexander M. "Improving enzymes by using them in organic solvents." *Nature* 409, no. 6817 (2001): 241.

[51] Kirchner, Gerald, Mark P. Scollar, and Alexander M. Klibanov. "Resolution of racemic mixtures via lipase catalysis in organic solvents." *Journal of the American Chemical Society* 107, no. 24 (1985): 7072-7076.

[52] Zaks, Aleksey, and David R. Dodds. "Application of biocatalysis and biotransformations to the synthesis of pharmaceuticals." *Drug Discovery Today* 2, no. 12 (1997): 513-531.

[53] Lee, Moo-Yeal, and Jonathan S. Dordick. "Enzyme activation for nonaqueous media." *Current opinion in Biotechnology* 13, no. 4 (2002): 376-384.

[54] Bradbury, Elizabeth J., Lawrence DF Moon, Reena J. Popat, Von R. King, Gavin S. Bennett, Preena N. Patel, James W. Fawcett, and Stephen B. McMahon. "Chondroitinase ABC promotes functional recovery after spinal cord injury." *Nature* 416, no. 6881 (2002): 636.

[55] Moon, Lawrence DF, Richard A. Asher, and James W. Fawcett. "Limited growth of severed CNS axons after treatment of adult rat brain with hyaluronidase." *Journal of neuroscience research* 71, no. 1 (2003): 23-37.

[56] Fusetti, Fabrizia, Holger von Moeller, Douglas Houston, Henriëtte J. Rozeboom, Bauke W. Dijkstra, Rolf G. Boot, Johannes MFG Aerts, and Daan MF van Aalten. "Structure of human chitotriosidase implications for specific inhibitor design and function of mammalian chitinase-like lectins." *Journal of Biological Chemistry* 277, no. 28 (2002): 25537-25544.

[57] Zimmer, Markus, NatăsaVukov, Siegfried Scherer, and Martin J. Loessner. "The murein hydrolase of the bacteriophage φ3626 dual lysis system is active against all tested Clostridium perfringens strains." *Applied and Environmental Microbiology* 68, no. 11 (2002): 5311-5317.

[58] Ensor, Charles Mark, Frederick W. Holtsberg, John S. Bomalaski, and Mike A. Clark. "Pegylated arginine deiminase (ADI-SS PEG20, 000 mw) inhibits human melanomas and hepatocellular carcinomas *in vitro* and *in vivo*." *Cancer research* 62, no. 19 (2002): 5443-5450.

In: Microbial Catalysts. Volume 2
Editors: Shadia M. Abdel-Aziz et al.

ISBN: 978-1-53616-088-8
© 2019 Nova Science Publishers, Inc.

Chapter 3

UREASE PRODUCING MICROBES TO AID BIO-CALCIFICATION: AN INTERDISCIPLINARY APPROACH

*Dweipayan Goswami[1], Janki Thakker[2]
and Pinakin Dhandhukia[3],**
[1]Department of Biochemistry-Biotechnology,
St. Xavier's College (Autonomous), Ahmedabad
[2]Department of Biotechnology, P. D. Patel Institute of Applied Sciences,
Charotar University of Science and Technology, Anand (Gujarat), India
[3]Department of Microbiology, Sheth P T Mahila College of Arts and Science
(SPTMC), Veer Narmad South Gujarat University, Surat, (Gujarat), India

ABSTRACT

Ureases (urea amidohydrolases; EC 3.5.1.5) are nickel-dependent enzymes that catalyze the hydrolysis of urea into 2 mole of ammonia and 1 mole of carbon dioxide. Urease activity tends to increase the pH of surrounding environment as it produces ammonia, a basic molecule. These enzymes are widespread in nature, being synthesized by plants, fungi and bacteria, but not by animals. Plant and fungal ureases are hexamer of single type of ~90 kDa subunit with about 840 amino acids whereas, bacterial ureases are multimers of two or three polypeptide chains that correspond to the single chain of the plant/fungal urease. Urease from the bacterium *Klebsiella aerogenes* was the first to have its tridimensional structure solved by crystallography in the year 1995. Since then ureases from several bacterial strains have been studied. First application of urease enzyme was to detect *Helicobacter pylori* in stomach and it was restricted to be used as a diagnostic tool to detect the presence of pathogens in gastrointestinal or urinary tract as these pathogens produce urease. Later, Urease conductometric biosensors were innovated for

---

* Corresponding Author's E-mail: pinakin.dhandhukia@gmail.com.

detection of heavy-metal ions which consisted of interdigitated gold electrodes and enzyme membranes were used for a quantitative estimation of heavy-metal ions in polluted water. The most astonishing use of urease producing microbes is been made in the process called Microbially induced calcium carbonate precipitation (MICP). Under natural conditions, the precipitation of carbonates occurs very slowly over long geological times. In order to produce large amounts of carbonates rapidly, microbes could employ with the ability to create conditions for precipitation of carbonates in shorter times. Urea hydrolysis by ureolytic bacteria in presence $CaCl_2$ aids this process by creating alkaline environment where high amounts of carbonates precipitate briefly describe this process. Bacterial strains that are been reported to achieve efficient carbonate precipitation are *Bacillus pasteurii, Pseudomona s*sp., *Variovorax* sp., *Leuconostoc mesenteroides, Micrococcus sp., Bacillus subtilis, Deleya halophila, Halomonas eurihalina* and *Myxococcus xanthus.* The use of MICP has gained importance due to its various applications that includes removal of heavy metals and radionucleotides, removal of calcium from wastewater and biodegradation of pollutants, atmospheric CO_2 sequestration, remediation of building materials, in cement industry to produce self-healing cement preparations, sealing cracks restoration of monuments; preparing eco-friendly bricks where soil used is bio-calcified, modifying the properties of soil thus, bacterial carbonates produced by the action of urease are serving many interdisciplinary fields.

Keywords: self healing concrete, bioaugmentation, urolysis, calcium precipitation

1. INTRODUCTION

1.1. Enzyme

Enzymes are the natural biocatalysts which aids the biochemical reaction to occur spontaneously. All enzymes are proteins which possess catalytic ability. Starting molecule on which enzyme acts is known as substrate and the molecule produced after the action of enzyme is known as product [1]. Enzymes are found in all the living organisms and they carry out biochemical reactions in the cell of the living organism [2]. Enzymes are highly specific in their action on substrates and often many different enzymes are required to bring about, by concerted action, the sequence of metabolic reactions performed by the living cell. Enzymes occur in every living cell, hence in all microorganisms. Each single strain of organism produces a large number of enzymes, hydrolyzing, oxidizing or reducing, and metabolic in nature [3].

The enzymes are been used since many centuries by humans long before the nature or function of enzymes was understood [4]. In brewing enzymes were used for conversion of starch from barley malt, for bating of hides in leather making, are examples of ancient use of enzymes. Action of enzymes responsible for bringing out such biochemical reaction started being familiar in this century. After such understanding, crude preparations from certain animal tissues such as pancreas and stomach mucosa, or from plant tissues such as malt and papaya fruit, were prepared which found applications in the textile, leather, brewing, and other industries [2]. Once the mechanics enzyme action became known, efforts were made to search more readily available sources of enzymes, less expensive enzyme with better efficiency and develop low cost enzyme preparations. It became evident that certain microorganisms produce enzymes similar in action to the amylases of malt and pancreas, or to the proteases of the pancreas and papaya fruit. This led to the development of processes for producing such microbial enzymes on a commercial scale [5]. Some microbial enzymes that gained tremendous importance on commercial scale are: Amylases, Proteases and Penicillinase which are produced by bacteria [6]; Amylases, Glucosidases, Proteases, Pectinases, Glucose oxidase and Catalase from fungi [7] whereas, Invertase and Lactase from yeast [8]. Bacterial urease initially used to detect *Helicobacter pylori* in stomach and it was restricted to be used as a diagnostic tool to detect the presence of pathogens in gastrointestinal or urinary tract as these pathogens produce urease [9]. Later, Urease conductometric biosensors were innovated for detection of heavy-metal ions which consisted of interdigitated gold electrodes and enzyme membranes were used for a quantitative estimation of heavy-metal ions in polluted water [10]. Most recent application of bacterial urease is made in the process known as 'Microbially induced calcium carbonate precipitation' (MICP). Thus, the remaining part of the chapter will focus on the microbial ureases and their role in MICP [11].

1.2. Urease

Urease (E.C. EC 3.5.1.5) belong to the superfamily of amidohydrolases. All the enzymes which act on amide bolds to hydrolyze them belong to the superfamily of amidohydrolases. Urease acts on the amide bold of urea to catalyze the production of ammonia and carbamate. The latter compound spontaneously decomposes to yield another molecule of ammonia and carbonic acid. Thus, urea is also known as urea amidohydrolase. Chemical reaction followed by urease is as follows:

$$NH_2 - CO - NH_2 + H_2O \rightarrow NH_3 + H_2CO_3$$

In solution, the released carbonic acid and the two molecules of ammonia remain in equilibrium with their deprotonated and protonated forms, respectively. Thus as a result of such reaction there is an increase in pH which is described as follows:

$$H_2CO_3 \rightleftharpoons H^+ + HCO_3^-$$

$$2NH_3 + 2H_2O \rightleftharpoons 2NH_4^+ + 2OH^-$$

In the history of science, coincidently urea was the first molecule to be synthesized [12] while urease from jack bean was the first enzyme crystallized [13]. Moreover, urease was the first enzyme to contain nickel [14, 15].

The application of urease was initially proposed by Mobley and Hausinger [16] where they hypnotized (1) urease to serve as a virulence factor in human and animal infections of the urinary and gastrointestinal tracts (2) urease to recycle nitrogenous wastes in the rumens of domestic livestock and (3) urease to play a role in environmental transformations of nitrogenous compounds, including urea based fertilizers.

Urease is a metalloenzyme that requires nickel (Ni) for its function. Urease from *Helicobacter pylori* is been studied the most and considered as a model urease enzyme. Urease enzyme is encoded by the gene cluster of seven genes designated *ureABIEFGH*. This gene cluster is of 6.13 kb located on the chromosome of *H. pylori* urease are located as a single 6.13-kb gene cluster on the chromosome of the bacterium [17, 18] (Figure 1). All the genes in the cluster is transcribed in the same direction [19, 20]. All of the genes except *ureI* share homology with urease genes of other species, including *Bacillus* sp. TB-90 [21], *Klebsiella aerogenes* [22, 23], *Proteus mirabilis* [24-26]; , *Ureaplasma urealyticum* [27, 28], *Yersinia enterocolitica* [29, 30], and the jack bean [31].

All of these seven genes can be classified as structural genes and accessory genes. The first two genes of the urease gene cluster i.e., *ureA* and *ureB* encodes for structural subunits of urease enzyme. Their gene products UreA and UreB are suspected to have molecular weight 26.5 kDa and 60.0 kD are respectively. Three copies of each subunit UreA and UreB assembles to form an inactive apoenzyme. Whereas, *ureI, ureE, ureF, ureG*, and *ureH* are accessory genes of urease gene cluster. Proteins produced by expression of these accessory genes interact with the apoenzyme and deliver nickel ions to the active site in an energy-dependent process. There is no firm evidence for the presence of regulatory gene in urease gene cluster, however *ureR* of *P. mirabilis* is thought to activate urease and unlike other genes of urease gene cluster it is transcribed in opposite direction to the rest of the genes, similar function is to be said for *ureD* gene of *Bacillus* sp. strain TB-90. Gene *ureI* is unique to *H. pylori*, the remaining accessory genes encode proteins that share homology with gene products of the urease gene clusters of other bacterial species.

On the basis of nucleotide sequence of the urease genes of *H. pylori* and other bacteria, it is certain that all ureases share a common ancestral gene. Despite the fact that ureases are composed of multiple copies of one (jack bean), two (all helicobacters), or three (all other bacterial species) distinct subunits, the amino acid sequences are well conserved. For example, UreA of *H. pylori* shares 48% and 42% amino acid sequence identity with the corresponding *N*-terminal sequences of the jack bean urease subunit and the combined sequences of UreA and UreB of *P. mirabilis*, respectively [24, 25, 31]. The entire sequence of the structural subunits of *H. pylori* shares 58% identity with the urease of *K. aerogenes* [19, 23]. *P. mirabilis, K. aerogenes, Bacillus* sp. *TB-90, Y. enterocolitica* possess three structural genes *ureU*, *ureB* and *ureC* in contrast to *H. pylori* which contains only two structural genes, *ureA* and *ureB*. *P. mirabilis, K. aerogenes, Bacillus sp. TB-90, Y. enterocolitica* possess three structural genes *ureU*, *ureB* and *ureC* in contrast to *H. pylori* which contains only two structural genes, *ureA* and *ureB*.

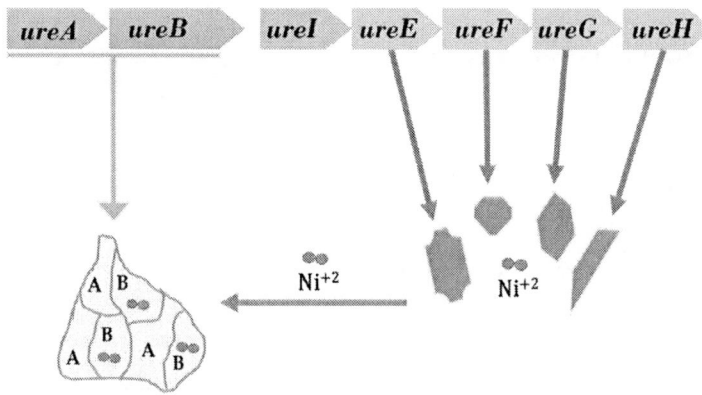

Figure 1. Model for synthesis of a catalytically active urease in *H. pylori*as described by Mobley (2001) with modification. Gene *ureA* and *ureB* produce proteins that form the structural subunit of urease enzyme which is an apoenzyme. This structural subunit gets activated only after receiving nickel ions. Gene *ureI* do not code for structural or accessory proteins. Whereas, *ureE, ureF, ureG* and *ureH* produce accessory proteins which aids in the process of delivering nickel ions to apoenzyme, where by active enzyme is formed.

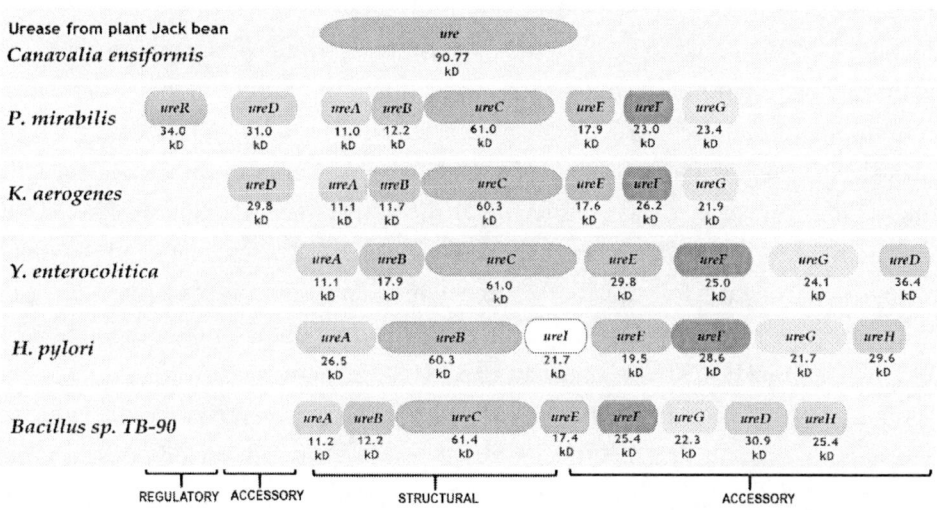

Figure 2. Description of urease gene cluster organization in various bacteria as described by Mobley et al. [33]. The genetic organization of urease gene clusters is depicted for the five bacterial species and for urease from Jack bean is described. The relative positions of each gene are assigned as predicted by the nucleotide sequence and rely on the same scale. Gene *ureR* of *P. mirabilis* is identified as a regulatory gene in the urea-inducible urease. Structural genes encode the subunit polypeptides that form the enzyme itself. Accessory genes encode polypeptides that play a role in the assembly of the nickel metallocenter. Gene *ureL* genes of *H. pylori* and *Bacillus* sp. TB-90 have no known function. *ureH* of *Bacillus* sp. strain TB-90 appears to be involved with nickel transport.

Mobley et al. [33] vividly described the organization of the urease genes of the five species (*P. mirabilis, K. aerogenes, Bacillus sp. TB-90, Y. enterocolitica* and *H. pylori*) for which the entire nucleotide sequence (Figure 2) They suggested that, there appears to be significant diversity with respect to the organization of genes among different microorganisms at first sight. Still, there are some general themes. (1) Urea inducible gene clusters begin with the *ureR* regulatory gene, which is transcribed in the direction opposite to that of the rest of the gene cluster. (2) Structural subunit genes are always aligned the same way, from smallest subunit to largest subunit. (3) Accessory genes *ureEFG* are always contiguous. (4) Gene *ureD* and its Helicobacter homolog *ureH* can either precede structural gene *urea* or follow *ureG*. From the data available, the minimal requirements for synthesis of a catalytically active urease appear to include the structural subunit genes and four accessory genes (i.e., seven genes in total for the common three-subunit urease systems). Using *P. mirabilis* or *K. aerogenes* as examples, *ureABC* are required for assembly of the catalytically inactive apoenzyme and *ureDEFG* are required *in-vivo* for assembly of the nickel metallocenter, the completion of which results in active enzyme. Additional genes such as *ureL* of *H. pylori* or *ureH* of *Bacillus* sp. strain TB-90 play specialized functions that are not required for the expression of active ureases of all species. Two ORFs, designated *ureC* and *ureD*, were originally assigned to the *H. pylori* urease gene cluster but lack of significant function and the lack of homologs in other species, dropped these two putative genes from our current model of the *H. pylori* urease gene cluster.

Subunit UreB is found to be the active site of the enzyme, where amino acid residues found throughout the primary structure are brought into proximity in the tertiary structure [34]. Amino acid numbering of UreB subunit specific for *H. pylori* [18], residues His-136, His-138, Lys-219, His-248, His-274, and Asp-362 come in direct contact with the two nickel ions, urea, or a water molecule at the active site. In addition, His-322 exists near the active site, acts as a general base in the catalysis. The mechanism by which urea is hydrolysis by urease follows the scheme which was initially described by Zerner's group for the jack bean urease [15, 16, 32]. Briefly, urea binds in *O*-coordination to one nickel ion aided by His-221. As an active base, His-322 activates a water molecule bound to the other nickel ion. Attack by the metal-coordinated hydroxide on the substrate carbon atom results in a tetrahedral intermediate that bridges the two nickel sites, a proton is transferred to the intermediate with accompanying ammonia release, and water displaces the carbamate to complete the cycle.

Later, new stains were determined to possess similar urease gene cluster. Whole genome sequence of *Pseudomonas aeruginosa* PAO1 submitted by Nouwens et al. [35] suggested the presence of urease gene cluster in the chromosomal DNA from the locus 5462763 bp to 5489011 bp. Data suggested that urease gene cluster possessed 7 genes namely *ureABCGFED*. The arrangement of genes in the urease cluster is found to be similar to *K. aerogenes*.

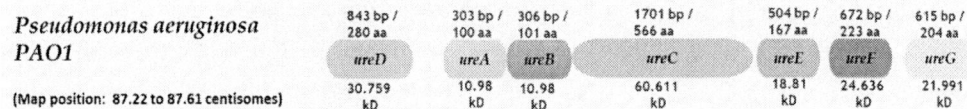

Figure 3. Organization of urease gene cluster in *Pseudomonas aeruginosa* PAO1. The arrangement and the size of the genes in the cluster is similar to *K. aerogenes* as described in Figure 1.

Not all strains of *Pseudomonas* possess ability to produce urease [36]. To identify the presence of urease coding genes in the genomic DNA of the strains belonging to *Pseudomonas* genus, we have designed primers which can be used to amplify these genes confirm their presence [37]. To identify *ureA*, the primers are UreA_FP- 5'-GCGAGAAAGACAAGCTGCTG-3' and UreA_RP- 5'-GCCCTCCATCACCTGCTC-3'where the expected product size is 200 bp. For *ureB* primers are UreB_FP- 5'-GATATCGAACTCAACGCCGG-3' and UreB_RP-5'-ATTTCCACCAGCTCCACCTC-3' where the expected product size is 218 bp. For *ureC* primers are UreC_FP- 5'-GTTCAGCATGATCAGCTCCG-3' and UreC_RP- 5'-TGTGGATCAGGTCGGTCT TC-3' where the expected product size is 524 bp. For *ureD* primers are UreD_FP-5'-CGAACTGGAAACCCGTATCC-3' and UreD_RP 5'- CCAGAGTTCGATCAGCC AGG-3' where the expected product size is 394 bp. For *ureE* primers are UreE_FP 5'-CGAAGAACTCCACCTGACCT-3' and UreE_RP 5'-ATGGTGCGAGTGGTGGTG-3' where the expected product size is 400 bp. For *ureF* primers are UreF_FP 5'-ATACAGCTATTCCCAGGGCC-3' and UreF_RP 5'- CAGGGTTTTCATCAGCACC G-3' where the expected product size is 445 bp. For *ureG* primers are UreG_FP 5'-GGACGCCTCGATCAACCT-3' and UreG_RP 5'- GATGATTTCGTCCAGCCCCT-3' where the expected product size is 346 bp. We designed these primes and detected the presence of urease gene cluster in *Pseudomonas aeruginosa* stain BG, strain also showed induction of these genes in presence of urea and the maximum urease activity detected was 17.92 ± 1.79 unit min^{-1} ml^{-1} [37]. Over the period of time several new strains are detected with an ability to produce urease enzyme and continuous effort is been made to exploit these urease producing strains in agriculture, detection of infections of the urinary and gastrointestinal tracts, recycling nitrogenous wastes in soil, etc. Thus, after having understood the basics of microbial ureases we will focus on the role of these microbes in the process of MICP.

1.3. Outline to MICP

Microbially induced CaCO$_3$ precipitation is also known as biocalcification, this process is omnipresent and plays an important cementation role in natural systems including soils, sediments, and minerals [38, 39]. Microbes that are heterotrophic in nature, obtain carbon from organic compounds by biotransformation of organic nitrogen compounds using urea hydrolysis and aids biocalcification as a secondary process. Other microbial process such as oxidation of organic compounds under denitrifying, sulfate reducing and methanogenes is also aids biocalcification [40, 41]. Furthermore, chemosynthetic and photosynthetic autotrophs viz. the organisms that can use CO_2 as their carbon source are also capable to induce precipitation of carbonates during their CO_2 removal from bicarbonate-containing solutions when there is an adequate supply of Ca^{2+} or other appropriate cations [42]. Thus any of these microbial process where soluble Ca^{2+} is been converted to insoluble $CaCO_3$ is termed as MICP or biocalcification.

1.3.1. Role of Ureolytic Bacteria in MICP

Among the different mechanisms of MICP, urea hydrolysis (or ureolysis) assisted biocalcification is most efficient out of all the processes known.

MICP induced by the hydrolysis of urea by urease producing microbes involves following 4 reactions:

$$Reaction\ 1.\ NH_2 - CO - NH_2 + H_2O \rightarrow NH_3 + NH_2COOH$$

$$Reaction\ 2.\ NH_2COOH + H_2O \rightleftharpoons NH_3 + H_2CO_3$$

$$Reaction\ 3.\ H_2CO_3 \rightleftharpoons H^+ + HCO_3^-$$

$$Reaction\ 4.\ NH_3 + H_2O \rightleftharpoons NH_4^+ + OH^-$$

$$Reaction\ 5.\ Ca^{+2} + HCO_3^- \rightleftharpoons CaCO_3 + H_2O$$

Urea is initially hydrolyzed to carbamate and ammonia in reaction 1 [16]. Spontaneously carbamate is hydrolyzed to give carbonic acid and ammonia (Reaction 2), which themselves undergo hydrolysis according to Reactions 1 and 4, with equilibrium constants of pK1 6.3 and pKa 9.3, respectively. Thus, these values suggests that there will be a net increase in pH of reaction mixture. Under alkaline conditions solubilized calcium, will gets precipitated of $CaCO_3$ (Reaction 5).

MICP is thus an complex process when it is naturally occurring in soil under non-controlled conditions which is adroitly balanced by four parameters: (1) dissolved inorganic carbon (DIC), (2) pH, (3) abundance of nucleation sites, and (4) calcium concentration. The first three parameters are directly affected by urea-hydrolyzing (ureolytic) microbial activity, as described in Reactions 1 to 4, and by bacterial cell abundance (with the bacteria providing nucleation sites) [43].

When the process of MICP is applied in soil, two major approaches implemented: (1) At a site of MICP is to be performed, a specific ureolytic bacterial strain is added together with urea, nutrients and calcium and this process of MICP is known as bioaugmentation; and (2) Another process of MICP is biostimulation, where indigenous ureolytic bacteria are first provided nutrition for their growth and once their biomass in built up they are provided with a substrate designed to stimulate $CaCO_3$ precipitation. Success rate of *In situ* bioaugmentation is low, because the strain of bacterial culture introduced in the soil often fail to survive as the indigenous flora of soil provide competition for nutrient uptake and at times the introduced bacterial strain fail to survive [44]. In contrast, Biostimulation, encourages the growth of native soil microfauna capable to produce urease through the manipulation of specific growth conditions. The only possible a possible drawback of this method is that if the initial soil concentration of ureolytic bacteria is low then it might limit the rate of ureolytic MICP at the treated site [45].

1.3.2. Bacterial Surface Assists MICP

It is hypothesized that process of microbial MICP starts on the surface of bacterial cell wall. At neutral pH, there are several negatively charged domains on the surface of bacterial cell wall which acts as a locus for the binding of charged metal ions, favoring heterogenous nucleation [41]. The process of MICP initiates as carbonate precipitates develop on the external surface of bacterial cells by successive stratification [43-45] and slowly bacterial cell gets embedded in growing carbonate crystals around them [46]. Possible biochemical reactions occurring on the cell surface of ureolytic bacteria in the process of MICP are as follows:

$$Ca^{+2} + Cell \rightarrow Cell - Ca^{+2}$$

$$Cl^- + NH_3 + HCO_3^- \rightarrow NH_4Cl + CO_3^{-2}$$

$$Cell - Ca^{+2} + CO_3^{-2} \rightarrow Cell - CaCO_3$$

Despite of research done on microbial MICP, the actual role of the bacterial precipitation of $CaCO_3$, remains a matter of debate. According to Dhami et al. [47] some authors believe this precipitation to be an unwanted and accidental by-product of the metabolism while others think that it is a specific process with ecological benefits for the precipitating organisms [48].

1.3.3. Approach to Gauge MICP

As the end product of MICP is $CaCO_3$ in its insoluble form, the most simple and the efficient way is to perform Scanning electron microscopy of the sample in which MICP is performed [47, 48]. Another approach is to perform X-ray diffraction (XRD) as the diffraction pattern $CaCO_3$ is specific and if the sample to be analyzed for XRD contains characteristic diffraction pattern of $CaCO_3$, then it confirms its presence [49]. SEM analysis of MICP studies have suggested that $CaCO_3$ is been precipitated by bacteria in different phases [50]. Calcium carbonate forms three anhydrous polymorphs viz. calcite, aragonite and vaterite, two hydrated crystalline phases viz. monohydro calcite ($CaCO_3 \cdot H_2O$) and ikaite ($CaCO_3 \cdot 6H_2O$), and various amorphous phases (ACC) with differences in short range order and degree of hydration [51, 52] (Figure 4). Despite of several polymorphs of $CaCO_3$, vaterite and calcite are the most common polymorphs produced during MICP [53], mineralization of monohydrocalcite [54] and aragonite [46] have also been reported.

Figure 4. Various polymorphs of $CaCO_3$ (http://www.ruhr-uni-bochum.de/sediment/forschung.html, http://www.ruhr-uni-bochum.de/sediment/pictures/CaCO3_web1.jpg).

The process of MICP can be easily carried out *in vitro*, when ureolytic bacteria is allowed to grow in medium containing urea and a source of calcium, generally $CaCl_2$ is used for this purpose. One of the simplest study of MICP is described by Sarda et al. [49], where they have used an ureolytic strain, *Bacillus pasteurii* NCIM 2477 to carry MICP on the surface of fired brick. Simple procedure was employed where *Bacillus pasteurii* NCIM 2477 was allowed to grow in the medium fortified with urea and $CaCl_2$. Fired brick was kept in this medium for 4 weeks. After incubation, brick was removed from the growth medium and its surface was scraped and analyzed for deposition using XRD. Result of such analysis suggested that the scraped surface of brick contained $CaCO_3$. They also achieved MICP on the surface of fired bricks using other strains *Brevibacterium ammoniagenes* ATCC 6871 and *Bacillus lentus* 2466-NCIB 8773 which were also possessing urease activity.

Similarly, we achieved MICP in shake flask conditions using *Pseudomonas aeruginosa* strain BG (KC87466) which possessed maximum of 17.92 ± 1.79 unit min^{-1} ml^{-1} urease activity [37]. To achieve MICP we performed a simple experiment where *P. aeruginosa* strain BG was allowed to grow in the half strength nutrient broth supplemented fortified with $CaCl_2$ (1% w/v) as a soluble source of calcium and 5 ml of 40% Urea in 100 ml broth. Allowing the strain to grow for one week under optimum growth conditions (27 ± 2°C under shake flask condition), the $CaCO_3$ formed and bacterial biomass was harvested by centrifugation and dried at 100°C overnight and analyzed for XRD analysis and SEM analysis.

Figure 5. The SEM image suggesting formation of calcite polymorph of $CaCO_3$ by the process of MICP using *Pseudomonas aeruginosa* strain BG.

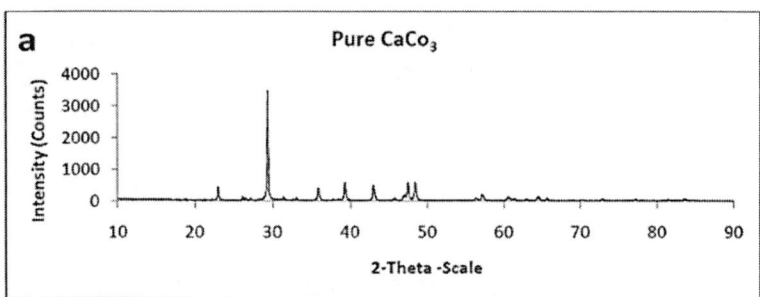

2-Theta scan for pure Calcium carbonate

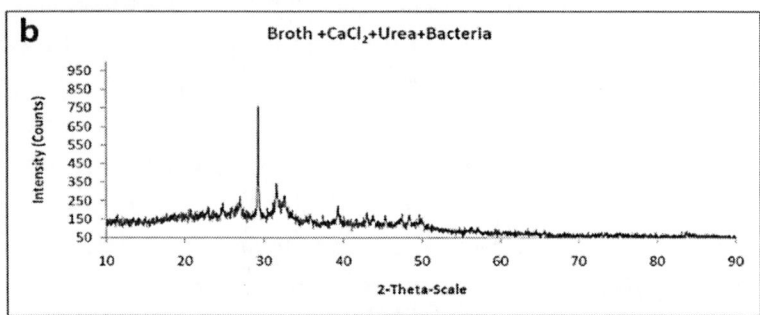

2-Theta scan for bacterial product indicating the production of calcium carbonate

Figure 6. (a) Characteristic diffraction pattern of pure $CaCO_3$ is seen with a peak at 29° under 2-Theta scale (b) is the diffraction pattern of sample in which MICP is performed, characteristic peak at 29° under 2-Theta scale of $CaCO_3$ is also observed confirming its presence.

Here, our results supports the findings of González-Muñozetal [53] who claimed that vaterite and calcite are the most common polymorphs produced during MICP. The XRD analysis of same sample was carried out to detect the presence of $CaCO_3$ formed by *P. aeruginosa* strain BG.

XRD analysis of the sample also suggests the presence of $CaCO_3$ produced by MICP (Figure 6). Thus, any strain capable of hydrolyzing urea can be tested for MICP using SEM and XRD analysis.

1.4. Application of MICP

The use of MICP in preparing microbial concrete has become the most fascinating application in civil engineering. MICP is used in enhancing durability of cementitious materials for enhanced strength, to repair of limestone monuments, for sealing cracks in concrete structures, etc. The ultimate goal for the use of MICP is to provide low cost and durable roads, high strength buildings with more bearing capacity, long lasting river banks, erosion prevention of loose sands and low cost durable housing [47, 48]. Other than using MICP in civil engineering, it has shown potential to be used in other interdisciplinary fields such as, removal of heavy metals and radionuclides,removal of calcium from wastewater, biodegradation of pollutants and atmospheric CO_2 sequestration [47].

1.4.1. MICP to Enhance Compressive Strength of Concrete

Compressive strength, permeability and corrosion the some of the properties to be examined for determining the quality of concrete. However, the primary test to determine quality of concrete is to check its compressive strength [47]. Achal and colleagues [55-58] has described the role of MICP in enhancing the compressive strength of cement, where urease producing *Bacillus* sp. CT-5 was used induce MICP in Ordinary Portland Cement under the presence of $CaCl_2$ and Urea. This treatment showed to enhance compressive strength by 36% in concrete. Further, the treated cement was analyzed for SEM imaging, were precipitated $CaCO_3$ was visualized confirms MICP to have been induced by *Bacillus* sp. CT-5. To determine the role of urease producing strain vs. urease non-producing strain in MICP, Ghosh et al. [59] studied the positive potential urease producing of *Shewanella* on compressive strength of mortar specimens and reported an increase of 17% and 25% after 7 and 28 days. But when mortar was treated with urease non-producing strain, *Escherichia coli no* noticeable increase was recorded. This concluded that urease producing microorganism plays the prime role in improvement of compressive strength of cement. Urease producing strain, *Sporosarcina pasteurii* enhanced compressive strength of cement by 17% is reported by [55-58]. Jonkers and Schlangen [60] reported the enhancement in the compressive strength of concrete up to

10 % by two urease producing strains, *Bacillus pseudofirmus* and *Bacillus cohnii*. More recent studies with the treatment of cement by *Arthrobacter crystallopoietes* showed enhancement in its compressive strength by 22% which was the higher compared to *Sporosarcina soli, Bacillus massiliensis and Lysinibacillus fusiformis* used along with [61]. As discussed earlier permeability and corrosion are other parameter to gauge quality of cement, Ramakrishnan et al. [62] showed increase in resistance of concrete toward alkali, freeze thaw attack, drying shrinkage and reduction in permeability upon application of bacterial cells. Achal et al. [58] treated mortar cubes with *Bacillus* sp.CT-5 and reported reduction in water absorption by six folds as compared to untreated specimens. Thus, research cited since late 90's till date suggests that several microbes with urease producing ability has been shown to induce MICP in cement which ultimately aids in enhancing the compressive strength of concrete.

1.4.2. MICP to Overhaul Cracks in Concrete

There are few reports that claim the use of MICP in the repair of cracks in the old concrete structures. Ramachandran et al. [63] proposed the use of MICP in concrete crack remediation. Specimens were filled with bacteria, nutrients and sand. There was an increase in compressive strength and stiffness values as compared to those without cells. Process of MICP was more prominently observed on the surface of the cracks due to better availability of air as the bacterial biomass grew more dominantly in presence of oxygen. Also, the pH of the concrete is highly basic which negatively affects the growth of bacteria, to overcome such constraints polyurethanes were used as vehicle for immobilization of calcifying enzymes and whole cells because of its mechanically strong and biochemically inert nature [47]. Further studies for the use of bacteria by their encapsulation in polyurethanes were performed by Bang et al. [64], investigation reported positive potential of microbiologically enhanced crack remediation by polyurethane immobilized bacterial cells. The study was also designed to understand the effect of immobilized bacterial cells on strength of concrete cubes by varying the concentration of immobilized cells per crack where, the highest compressive strength was obtained with cubes remediated with 5×10^9 immobilized cells per crack for 7 days while after that, increase in strength was found to be marginal. For confirmation of MICP occurring in the cracks, SEM imaging and its analysis was also followed by them. Further, in 2008 there was a report for the use of urease producing *B. sphaericus* in the process of MICP for the repair of cracks in concrete [65]. Later, Qian et al. [66] also reported that compressive strength of treated specimens could be restored to 84% upon treatment of bacterial calcite. Thus, since 2000's the research to exploit MICP in various civil engineering applications have been tried and tested which has shown fruitful results.

1.4.3. Removal of Heavy Metals and Radionucleotides

The use of MICP to remove heavy metals and radionucleotides from groundwater has vividly been described by Reddy et al. [48]. To give brief account, groundwater gets contaminated with heavy metals and radionucleotides due to their improper disposal. The previously known method to pump out and treat have been found ineffective. To this constrain, MICP act as a boon. Briefly, MICP causes precipitation of radionuclide and contaminant metals into calcite as competitive co-precipitation reaction in which suitable divalent cations are incorporated into the calcite lattice [47, 48]. Following is an example of strontium precipitation by microbes:

$$^{90}Sr^{+2} + OH^- + HCO_3^- \rightarrow SrCO_3 \downarrow + H_2O$$

These cations and radionuclides merge into the calcite structure by substitution of calcium ions in the microenvironment of the mineral precipitate. This forms insoluble strontium-carbonate minerals [67, 68]. Warren et al. [69] reported 95% capturing of the total strontium in solid phase by MICP using *S. pasteurii*. Lately, Achal et al. [57] reported positive potential in remediation of Arsenic contaminated sites by *Sporosarcina ginsengisoli CR* 5 and bioremediation of copper by *Kocuria flava* CR1.

1.4.4. Repossession of Environmental Carbon Dioxide

Human activities such as excessive use of fossil fuels, developing electricity from coal etc., is causing an increase in the concentrations of carbon dioxide (CO_2). The only natural way to recycle this CO_2 is by their uptake by plants, but the rate by which CO_2 is formed, is not met by plant uptake. The possible solution to this problem is by MICP where atmospheric CO_2 solubilized in water, which is then precipitated in carbonate minerals. To describe few examples, Dupraz et al. [70] used *S. pasteurii* to transform CO_2 to solid carbonate in artificial groundwater and termed this process as mineral trapping. Jansson and Northen [71] used cyanobacteria for point-source carbon capture and sequestration where the cyanobacteria utilize solar energy through photosynthesis to convert carbon dioxide to calcium carbonate [72].

CONCLUSION AND FUTURE PERSPECTIVES

There are several microbes which possess ability to produce ureases. Research on the strains producing ureases started back in 1960's but their use in the process of MICP is quiet recent. Despite of several strains capable to produce urease, none of them have gained title of model organism to study MICP. The process of MICP have shown several desirable applications as mentioned above but none of them are been used on large and industrial scale. Process of MICP has still to be studied in depth under *in situ* conditions for several applications such as removal of heavy metals and radionucleotides from groundwater. In the process of development biocement, MICP has shown great promise, yet the research has not come up with a defined urease producing strain, nor is specific process for using MICP in cement industry has developed. Thus, the use of urease producing strains in MICP is well understood and studied but effort has to be made in using MICP on the larger scale.

ACKNOWLEDGMENTS

Authors are thankful to Council of Scientific & Industrial Research (CSIR) for providing Senior Research Fellowship (SRF) and Department of Science & Technology (DST) for financial aid. Authors are also thankful to Charotar University of Science and Technology (CHARUSAT) and Vanita Vishram management for providing necessary facilities.

REFERENCES

[1]   Østergaard, Lars H., and Hans Sejr Olsen. "Industrial applications of fungal enzymes." In *Industrial applications*, pp. 269-290. Springer, Berlin, Heidelberg, 2011.

[2]   Duza, Mohammad Badrud, and S. A. Mastan. "Microbial enzymes and their applications–a review." *American Journal of Pharm Research* 3, no. 8 (2013).

[3]   Buchholz, Klaus, Volker Kasche, and Uwe Theo Bornscheuer. *Biocatalysts and Enzyme Technology*. John Wiley & Sons, 2012.

[4]   Gurung, Neelam, Sumanta Ray, Sutapa Bose, and Vivek Rai. "A broader view: microbial enzymes and their relevance in industries, medicine, and beyond." *BioMed Research International* 2013 (2013).

[5]   Bwanganga Tawaba, Jean-Claude, François Bera, and Philippe Thonart. "Modelling the β-amylase activity during red sorghum malting when *Bacillus subtilis* is used to control mould growth." *Journal of Cereal Science* 57 (2013): 115-119.

[6]   Underkofler, L. A., R. R. Barton, and S. S. Rennert. "Production of microbial enzymes and their applications." *Applied Microbiology* 6, no. 3 (1958): 212.

[7]   Ghorai, Shakuntala, Samudra Prosad Banik, Deepak Verma, Sudeshna Chowdhury, Soumya Mukherjee, and Suman Khowala. "Fungal biotechnology in food and feed processing." *Food Research International* 42, no. 5-6 (2009): 577-587.

[8]   Türkel, Sezai, and Nükhet Kayakent. "Isolation and Molecular Identification of New *Kluyveromyces lactis* Strains Producing High Levels of Lactase and Invertase Enzymes." *Journal of Applied Biological Sciences* 3 (2013): 51-55.

[9]   Strugatsky, David, Reginald McNulty, Keith Munson, Chiung-Kuang Chen, S. Michael Soltis, George Sachs, and Hartmut Luecke. "Structure of the proton-gated urea channel from the gastric pathogen *Helicobacter pylori*." *Nature* 493, no. 7431 (2013): 255.

[10] Tekaya, Nadèje, Olga Saiapina, Hatem Ben Ouada, Florence Lagarde, Philippe Namour, Hafedh Ben Ouada, and Nicole Jaffrezic-Renault. "Bi-Enzymatic conductometric biosensor for detection of heavy metal ions and pesticides in water samples based on enzymatic inhibition in arthrospira platensis." *Journal of Environmental Protection* 5 (2014): 441-453.

[11] Chu, Jian, Viktor Stabnikov, and Volodymyr Ivanov. "Microbially induced calcium carbonate precipitation on surface or in the bulk of soil." *Geomicrobiology Journal* 29, no. 6 (2012): 544-549.

[12] Andrews, Robert K., Robert L. Blakeley, and B. U. R. T. Zerner. "Urea and urease." *Advances in Inorganic Biochemistry* 6 (1984): 245-283.

[13] Sumner, James B. "The isolation and crystallization of the enzyme urease preliminary paper." *Journal of Biological Chemistry* 69, no. 2 (1926): 435-441.

[14] Dixon, Nicholas E., Carlo Gazzola, Robert L. Blakeley, and Burt Zerner. "Jack bean urease (EC 3.5. 1.5). Metalloenzyme. Simple biological role for nickel." *Journal of the American Chemical Society* 97, no. 14 (1975): 4131-4133.

[15] Dixon, Nicholas E., Peter W. Riddles, Carlo Gazzola, Robert L. Blakeley, and Burt Zerner. "Jack bean urease (EC 3.5. 1.5). V. On the mechanism of action of urease on urea, formamide, acetamide, N-methylurea, and related compounds." *Canadian Journal of Biochemistry* 58, no. 12 (1980): 1335-1344.

[16] Mobley, H. L., and R. P. Hausinger. "Microbial ureases: significance, regulation, and molecular characterization." *Microbiological Reviews* 53, no. 1 (1989): 85-108.

[17] Clayton, Christopher L., Mark J. Pallen, Harry Kleanthous, Brendan W. Wren, and Soad Tabaqchali. "Nucleotide sequence of two genes from *Helicobacter pylori* encoding for urease subunits." *Nucleic Acids Research* 18, no. 2 (1990): 362.

[18] Labigne, Agnts, Valeirie Cussac, and Pascale Courcoux. "Shuttle cloning and nucleotide sequences of *Helicobacter pylori* genes responsible for urease activity." *Journal of Bacteriology* 173, no. 6 (1991): 1920-1931.

[19] Cussac, Valerie, Richard L. Ferrero, and Agnes. Labigne. "Expression of *Helicobacter pylori* urease genes in *Escherichia coli* grown under nitrogen-limiting conditions." *Journal of Bacteriology* 174, no. 8 (1992): 2466-2473.

[20] Hu, Li Tai, P. A. Foxall, R. Russell, and H. L. Mobley. "Purification of recombinant *Helicobacter pylori* urease apoenzyme encoded by ureA and ureB." *Infection and immunity* 60, no. 7 (1992): 2657-2666.

[21] Maeda, Michihisa, Makoto Hidaka, Akira Nakamura, Haruhiko Masaki, and Takeshi Uozumi. "Cloning, sequencing, and expression of thermophilic *Bacillus* sp. strain TB-90 urease gene complex in *Escherichia coli.*" *Journal of Bacteriology*176, no. 2 (1994): 432-442.

[22] Lee, Mann Hyung, Scott B. Mulrooney, Michael J. Renner, Yves Markowicz, and Robert P. Hausinger. "*Klebsiella aerogenes* urease gene cluster: sequence of ureD and demonstration that four accessory genes (ureD, ureE, ureF, and ureG) are involved in nickel metallocenter biosynthesis." *Journal of Bacteriology* 174, no. 13 (1992): 4324-4330.

[23] Mulrooney, Scott B., and R. P. Hausinger. "Sequence of the *Klebsiella aerogenes* urease genes and evidence for accessory proteins facilitating nickel incorporation." *Journal of Bacteriology* 172, no. 10 (1990): 5837-5843.

[24] Jones, B. D., and H. L. Mobley. "*Proteus mirabilis* urease: genetic organization, regulation, and expression of structural genes." *Journal of Bacteriology* 170, no. 8 (1988): 3342-3349.

[25] Jones, Bradley D., and H. L. Mobley. "*Proteus mirabilis* urease: nucleotide sequence determination and comparison with jack bean urease." *Journal of Bacteriology* 171, no. 12 (1989): 6414-6422.

[26] Sriwanthana, Busarawan, Michael D. Island, and Harry LT Mobley. "Sequence of the *Proteus mirabilis* urease accessory gene ureG." *Gene* 129, no. 1 (1993): 103-106.

[27] Blanchard, A. "*Ureaplasma urealyticum* urease genes; use of a UGA tryptophan codon." *Molecular Microbiology* 4, no. 4 (1990): 669-676.

[28] Neyrolles, Olivier, Stephane Ferris, Nilane Behbahani, Luc Montagnier, and Alain Blanchard. "Organization of *Ureaplasma urealyticum* urease gene cluster and expression in a suppressor strain of *Escherichia coli.*" *Journal of Bacteriology* 178, no. 3 (1996): 647-655.

[29] Skurnik, Mikael, Stephen Batsford, A. Mertz, E. Schiltz, and P. Toivanen. "The putative arthritogenic cationic 19-kilodalton antigen of *Yersinia enterocolitica* is a urease beta-subunit." *Infection and Immunity* 61, no. 6 (1993): 2498-2504.

[30] de Koning-Ward, Tania F., Alister C. Ward, and Roy M. Robins-Browne. "Characterisation of the urease-encoding gene complex of *Yersinia enterocolitica.*" *Gene* 145, no. 1 (1994): 25-32.

[31] Riddles, Peter W., Vicki Whan, Robert L. Blakeley, and Burt Zerner. "Cloning and sequencing of a jack bean urease-encoding cDNA." *Gene* 108, no. 2 (1991): 265-267.

[32] Mobley, Harry LT. "Urease." In *Helicobacter pylori: Physiology and genetics.* ASM Press, 2001.

[33] Mobley, H. L., Michael D. Island, and Robert P. Hausinger. "Molecular biology of microbial ureases." *Microbiological Reviews* 59, no. 3 (1995): 451-480.

[34] Jabri, Evelyn, Mary Beth Carr, Robert P. Hausinger, and P. Andrew Karplus. "The crystal structure of urease from *Klebsiella aerogenes.*" *Science* 268, no. 5213 (1995): 998-1004.

[35] Nouwens, Amanda S., Stuart J. Cordwell, Martin R. Larsen, Mark P. Molloy, Michael Gillings, Mark DP Willcox, and Bradley J. Walsh. "Complementing genomics with proteomics: the membrane subproteome of *Pseudomonas aeruginosa* PAO1." *Electrophoresis: An International Journal* 21, no. 17 (2000): 3797-3809.

[36] Gaby, W. L., and C. Hadley. "Practical laboratory test for the identification of *Pseudomonas aeruginosa.*" *Journal of Bacteriology* 74, no. 3 (1957): 356.

[37] Goswami, Dweipayan, Krupa Patel, Swapnsinh Parmar, Hemendrasinh Vaghela, Namrata Muley, Pinakin Dhandhukia, and Janki N. Thakker. "Elucidating multifaceted urease producing marine *Pseudomonas aeruginosa* BG as a cogent PGPR and bio-control agent." *Plant Growth Regulation* 75, no. 1 (2015): 253-263.

[38] Monger, H. Curtis, LeRoy A. Daugherty, William C. Lindemann, and Craig M. Liddell. "Microbial precipitation of pedogenic calcite." *Geology* 19, no. 10 (1991): 997-1000.

[39] DeFarge, Christian, Jean Trichet, Anne-Marie Jaunet, Michel Robert, Jane Tribble, and Francis J. Sansone. "Texture of microbial sediments revealed by cryo-scanning electron microscopy." *Journal of Sedimentary Research* 66, no. 5 (1996): 935-947.

[40] Hamdan, Nasser, Edward Kavazanjian Jr, Bruce E. Rittmann, and Ismail Karatas. "Carbonate mineral precipitation for soil improvement through microbial denitrification." *Geomicrobiology Journal* 34, no. 2 (2017): 139-146.

[41] Aloisi, Giovanni. "A pronounced fall in the $CaCO_3$ saturation state and the total alkalinity of the surface ocean during the Mid Mesozoic." *Chemical Geology* 487 (2018): 39-53.

[42] Rowe, R. Kerry, Jamie F. VanGulck, and Sandra C. Millward. "Biologically induced clogging of a granular medium permeated with synthetic leachate." *Journal of Environmental Engineering and Science* 1, no. 2 (2002): 135-156.

[43] Gat, D., M. Tsesarsky, D. Shamir, and Z. Ronen. "Accelerated microbial-induced $CaCO_3$ precipitation in a defined coculture of ureolytic and non-ureolytic bacteria." *Biogeosciences* 11, no. 10 (2014): 2561-2569.

[44] van Veen, Johannes A., Leonard S. van Overbeek, and Jan D. van Elsas. "Fate and activity of microorganisms introduced into soil." *Microbiol. Mol. Biol. Rev.* 61, no. 2 (1997): 121-135.

[45] Tobler, Dominique J., Mark O. Cuthbert, Richard B. Greswell, Michael S. Riley, Joanna C. Renshaw, Stephanie Handley-Sidhu, and Vernon R. Phoenix. "Comparison of rates of ureolysis between *Sporosarcina pasteurii* and an indigenous groundwater community under conditions required to precipitate large

volumes of calcite." *Geochimica et Cosmochimica Acta* 75, no. 11 (2011): 3290-3301.

[46] Sánchez-Navas, Antonio, Agustín Martín-Algarra, María A. Rivadeneyra, Santiago Melchor, and José Daniel Martín-Ramos. "Crystal-growth behavior in Ca– Mg carbonate bacterial spherulites." *Crystal Growth and Design* 9, no. 6 (2009): 2690-2699.

[47] Dhami, Navdeep Kaur, Sudhakara M. Reddy, and Abhijit Mukherjee. "Biofilm and microbial applications in biomineralized concrete." In *Advanced Topics in Biomineralization*. IntechOpen, 2012.

[48] Reddy, M. Sudhakara. "Biomineralization of calcium carbonates and their engineered applications: a review." *Frontiers in microbiology* 4 (2013): 314.

[49] Sarda, Deepak, Huzaifa S. Choonia, D. D. Sarode, and S. S. Lele. "Biocalcification by *Bacillus pasteurii* urease: a novel application." *Journal of Industrial Microbiology & Biotechnology* 36, no. 8 (2009): 1111-1115.

[50] Rodriguez-Navarro C, Doehne E, Sebastian E (1999) Origins of honeycomb weathering: the role of salts and wind. *Geological Society Americal Bulletin* 111(8):1250-1255.

[51] Gebauer, Denis, Philips N. Gunawidjaja, JY Peter Ko, Zoltan Bacsik, Baroz Aziz, Lijia Liu, Yongfeng Hu et al. "Proto-calcite and proto-vaterite in amorphous calcium carbonates." *Angewandte Chemie International Edition* 49, no. 47 (2010): 8889-8891.

[52] Gower, Laurie B. "Biomimetic model systems for investigating the amorphous precursor pathway and its role in biomineralization." *Chemical Reviews* 108, no. 11 (2008): 4551-4627.

[53] González-Muñoz, Maria Teresa, Carlos Rodriguez-Navarro, Francisca Martínez-Ruiz, Jose Maria Arias, Mohamed L. Merroun, and Manuel Rodriguez-Gallego. "Bacterial biomineralization: new insights from Myxococcus-induced mineral precipitation." *Geological Society, London, Special Publications* 336, no. 1 (2010): 31-50.

[54] Krumbein, Wolfgang E. "On the precipitation of aragonite on the surface of marine bacteria." *Naturwissenschaften* 61, no. 4 (1974): 167-167.

[55] Achal, V., Abhijit Mukherjee, P. C. Basu, and M. Sudhakara Reddy. "Strain improvement of *Sporosarcina pasteurii* for enhanced urease and calcite production." *Journal of industrial microbiology & biotechnology* 36, no. 7 (2009): 981-988.

[56] Achal, Varenyam, Abhijit Mukherjee, and M. Sudhakara Reddy. "Microbial concrete: way to enhance the durability of building structures." *Journal of Materials in Civil Engineering* 23, no. 6 (2010): 730-734.

[57] Achal, Varenyam, Xiangliang Pan, Qinglong Fu, and Daoyong Zhang. "Biomineralization based remediation of As (III) contaminated soil by *Sporosarcina ginsengisoli.*" *Journal of Hazardous Materials* 201 (2012): 178-184.

[58] Achal, Varenyam, Xiangliang Pan, and Daoyong Zhang. "Remediation of copper-contaminated soil by *Kocuria flava* CR1, based on microbially induced calcite precipitation." *Ecological Engineering* 37, no. 10 (2011): 1601-1605.

[59] Ghosh, P., S. Mandal, B. D. Chattopadhyay, and S. Pal. "Use of microorganism to improve the strength of cement mortar." *Cement and Concrete Research* 35, no. 10 (2005): 1980-1983.

[60] Jonkers, Henk Marius, and Eric Schlangen. "Self-healing of cracked concrete: a bacterial approach." *Proceedings of FRACOS6: fracture mechanics of concrete and concrete structures. Catania, Italy* (2007): 1821-1826.

[61] Ghim. "Calcite-forming bacteria for compressive strength improvement in mortar." *Journal of microbiology and biotechnology* 20, no. 4 (2010): 782-788.

[62] Ramakrishnan, V., S. S. Bang, and K. S. Deo. "A novel technique for repairing cracks in high performance concrete using bacteria." In *Proc. of the Int. Conf. on HPHSC*, pp. 597-618. 1998.

[63] Ramachandran, Santhosh K., V. Ramakrishnan, and Sookie S. Bang. "Remediation of concrete using micro-organisms." *ACI Materials Journal-American Concrete Institute* 98, no. 1 (2001): 3-9.

[64] Bang, Sookie S., Johnna K. Galinat, and V. Ramakrishnan. "Calcite precipitation induced by polyurethane-immobilized *Bacillus pasteurii.*" *Enzyme and Microbial Technology* 28, no. 4-5 (2001): 404-409.

[65] De Belie, Nele, and Willem De Muynck. "Crack repair in concrete using biodeposition." In *Proceedings of the International Conference on Concrete Repair, Rehabilitation and Retrofitting (ICCRRR), Cape Town, South Africa*, pp. 291-292. 2008.

[66] Qian, Chunxiang, Ruixing Wang, Liang Cheng, and Jianyun Wang. "Theory of Microbial Carbonate Precipitation and Its Application in Restoration of Cement-based Materials Defects." *Chinese Journal of Chemistry* 28, no. 5 (2010): 847-857.

[67] Colwell, Frederick S., R. W. Smith, F. Gratn Ferris, Jani C. Ingram, A-L. Reysenbach, Yoshiko Fujita, T. L. Tyler et al. *Microbially-mediated subsurface calcite precipitation for removal of hazardous divalent cations.* Idaho National Engineering and Environmental Lab., Idaho Falls, ID; University of Idaho, Idaho Falls, ID; University of Toronto, Toronto, Ontario, Canada; Portland State University, Portland, OR; Idaho State University, Pocatello, ID (US), 2003.

[68] Mitchell, Andrew C., and F. Grant Ferris. "The influence of *Bacillus pasteurii* on the nucleation and growth of calcium carbonate." *Geomicrobiology Journal* 23, no. 3-4 (2006): 213-226.

[69] Warren, Lesley A., Patricia A. Maurice, Nagina Parmar, and F. Grant Ferris. "Microbially mediated calcium carbonate precipitation: implications for interpreting calcite precipitation and for solid-phase capture of inorganic contaminants." *Geomicrobiology Journal* 18, no. 1 (2001): 93-115.

[70] Dupraz, Sébastien, Bénédicte Ménez, Philippe Gouze, Richard Leprovost, Pascale Bénézeth, Oleg S. Pokrovsky, and Francois Guyot. "Experimental approach of CO_2 biomineralization in deep saline aquifers." *Chemical Geology* 265, no. 1-2 (2009): 54-62.

[71] Jansson, Christer, and Trent Northen. "Calcifying cyanobacteria—the potential of biomineralization for carbon capture and storage." *Current Opinion in Biotechnology* 21, no. 3 (2010): 365-371.

[72] Kamennaya, Nina, Caroline Ajo-Franklin, Trent Northen, and Christer Jansson. "Cyanobacteria as biocatalysts for carbonate mineralization." *Minerals* 2, no. 4 (2012): 338-364.

In: Microbial Catalysts. Volume 2
Editors: Shadia M. Abdel-Aziz et al.

ISBN: 978-1-53616-088-8
© 2019 Nova Science Publishers, Inc.

Chapter 4

RECENT DEVELOPMENTS OF MICROBIAL FUEL CELL AS SUSTAINABLE BIO-ENERGY SOURCES

Taslim Ur Rashid, Khandoker S. Salem, Md. Minhajul Islam, Asaduz Zaman, M. Nuruzzaman Khan, Ismat Z. Luna, Papia Haque and Mohammed Mizanur Rahman[*]
Department of Applied Chemistry and Chemical Engineering, Faculty of Engineering and Technology, University of Dhaka, Dhaka, Bangladesh

ABSTRACT

Microbial fuel cell (MFC) or modified MFC is a promising technology for organic waste treatment and sustainable bioelectricity production. In this system, microorganisms are used as biocatalysts to transform chemical energy or light energy into electricity. It is considered as a prominent source of bioenergy in the present crisis of conventional fuel sources. MFC has great potential as a source of renewable energy, wastewater treatment process, heavy metals removal and biosensor for oxygen and pollutants. Different microbial catalysts and engineering approaches have been examined to exploit the maximum effectiveness of the MFC. In this work, the recent development, applications and prospects of MFC from different aspects have been discussed briefly.

1. INTRODUCTION

The concerns of global fossil fuel depletion, uprising cost and environmental pollution from fossil fuel combustion are driving the search for carbon-neutral, non-edible feedstocks, renewable energy alternatives, the security and diversification of

---

[*] Corresponding Author's E-mail: mizanur.rahman@du.ac.bd.

energy supply. In these circumstances, global production of biomass and biofuel is growing rapidly. There are many scenarios that predict a high potential for biomass in the future and many studies have been performed in recent decades to estimate the future demand and supply of bioenergy. In recent years, one of the most advancement in bioenergy utilization is the development of fuel cells which are considered as an alternative energy technology and implemented in full-scale in numerous field of industries [1]. Fuel cells usually convert the chemical energy of fuel into electricity through chemical reactions. They are distinguished according to the types of electrolyte and catalyst used. Among different types, microbial fuel cells (MFCs) have garnered rapid increase in application as they are unique in their ability to utilize electrochemically active microorganisms as biocatalysts for converting the chemical energy of feedstock directly into electricity, when compared to enzymes or inorganic compounds or molecules [2]. Microorganisms are used in MFCs to generate electricity while accomplishing the biodegradation of organic matters or wastes. Microbes in the anodic chamber of an MFC oxidize the substrates and generate electrons and protons in the process while, carbon dioxide is produced as a by-product. However, there is no net carbon emission because the carbon dioxide in the renewable biomass originally comes from the atmosphere into the added substrates and generates electrons and protons in the process.

In recent years, this concept of energy generation from biomass using the microbial metabolism has drawn special attraction of the researchers, which has led to dramatic raise in developments, and modifications of MFCs. Many recent reviews have discussed on this prospective field of energy generation. Logan et al. provided a review of the different materials and methods used to construct MFCs and techniques used to analyze system performance [3]. They also recommended the evidence that should be included in MFC studies and the most convenient ways to present the results. Some reviews highlighted recent advances in MFC research and summarized the application of MFC in energy production in last 100 years [4]. Other reviews focused on engineering designs and studies of biological aspects of MFCs to improve interactions of anode and microorganisms [5-13]. The main limiting factors of MFC operation and remedial suggestions have been discussed in some articles [14, 15]. Applications of MFC in wastewater treatment, biorecovery and their environmental aspects have been discussed in a handful numbers of reviews [16-28].

This book chapter will examine most recent design of the MFC's, their current uses, potential future applications and the limitations to implement those applications. It will also suggest some probable scopes of improvement in design of MFCs and will recommend their potential applications for future.

2. FUEL CELL

A fuel cell is a device that converts the chemical energy of fuel into electricity through a chemical reaction of oxygen or another oxidizing agent with hydrogen or hydrocarbons, where the oxygen and a hydrogen-rich fuel combine to form water [29]. It is like a battery where the energy is released electro-catalytically instead of being combusted. Both batteries and fuel cells convert potential energy into electrical energy and also into heat energy as a byproduct. Traditional battery stores chemical potential energy within it and is discarded once the energy is depleted or an external supply of electricity is used to drive the electrochemical reaction in the reverse direction. In contrast, a fuel cell can run indefinitely using external supply of chemical energy, as long as it is supplied with a source of hydrogen and a source of oxygen (usually air). In addition, it can be made highly energy efficient, especially if the heat produced by the reaction is also harnessed for space heating, driving refrigeration cycles etc.

2.1. History

There is some ambiguity regarding who discovered the principle of fuel cells. According to the Department of Energy of the United States, German chemist Christian Friedrich Schönbein, first conducted scientific research on the phenomenon of fuel cell in 1838 [1]. It is asserted that Sir William Robert Grove introduced the concept of hydrogen fuel cell before Friedrich Schönbein could come with his discovery [30]. Grove discovered that a constant current could be found to flow between the electrodes by immersing two platinum electrodes on one end in a solution of sulphuric acid and the other two ends separately sealed in containers of oxygen and hydrogen. Later on, in 1893, Friedrich Wilhelm Ostwald, considered the founder of the chemistry-physics, experimentally determined the interconnection of various components of a fuel cell such as electrodes, electrolyte, oxidizing and reducing agents, anions and cations [31]. The first fuel cell with practical applications was developed by William W. Jacques in 1896, and in 1900, where zirconium was first used as solid electrolyte [32]. The leading researchers of the late nineteenth and early twentieth century in the field of fuel cells built the first molten carbonate fuel cell in 1921 [33].

In the early sixties, attention was focused on the catalyst platinum and acid electrolyte fuel cell in two different ways. One of them allowed working at high temperatures (150–200°C) [34, 35]. It was reported that using an electrolyte composed of a mixture of lithium carbonate, sodium and/or potassium, impregnated on magnesia sintered porous disk, operating temperature could be reached up to 650°C. A fuel cell was made of conducting ceramic oxide impregnated with zirconia which was reported to work at an operating temperature of 1000°C [36]. Present manufacturers are working on various applications of fuel cell [37].

2.2. Working Principle of Fuel Cells

Fuel cells generate electricity and heat by electrochemical reaction between oxygen and hydrogen to form the water, which, is actually the reverse electrolysis reaction. The different types of fuel cells operate with the same basic principle but the fuel cell designs vary in the chemical characteristics of the electrolyte [29, 38]. The following reaction shows the electrochemical reaction and Figure 1 depicts the operating principle of a fuel cell.

$$2H_2 + O_2 \longrightarrow 2H_2O + Energy$$

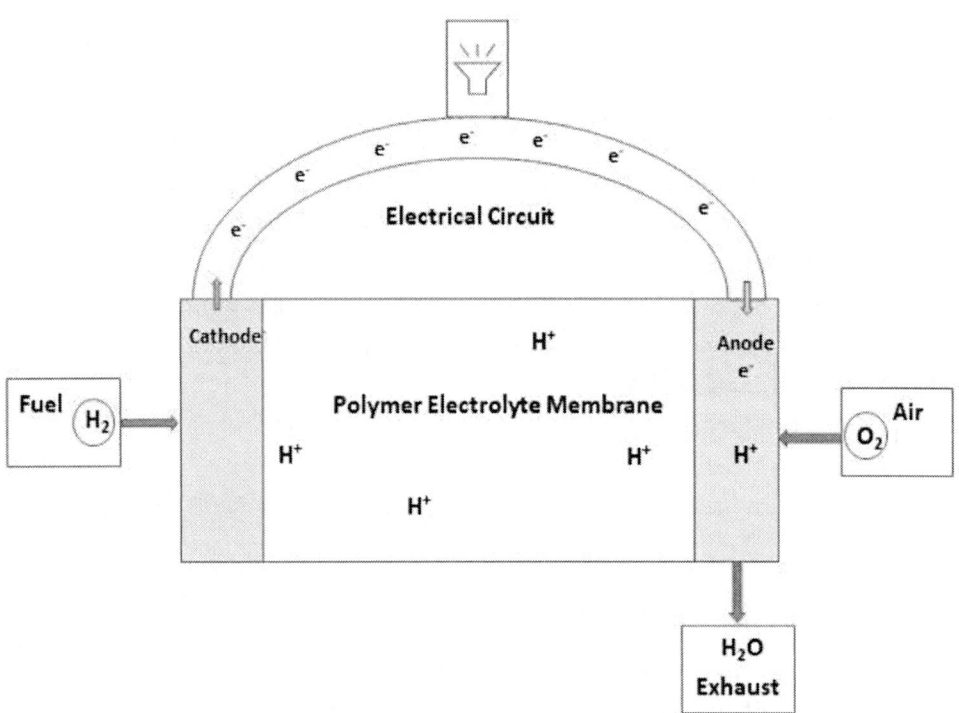

Figure 1. Operating principle of a fuel cell (modified figure from Mekhilef et al. [29]).

Anode, cathode, electrolyte and the external circuit are the four main parts of a fuel cell. Hydrogen is oxidized into protons at the cathode and oxygen is reduced to oxides at the cathode and reacts to form water. Either protons or oxide ions are transported through electron-insulating electrolyte depending on its type, while electrons travel through an external circuit to deliver electric power [39]. The reactions happening at the anode and cathode are shown below.

Anode: $2H_2 \longrightarrow 4H^+ + 4e^-$
Cathode: $O_2 + 4e^- + 4H^+ \longrightarrow 2H_2O$

2.3. Types of Fuel Cell

Table 1. Various features of different types of fuel cells

Type	Proton exchange membrane fuel cell (PEMFC)	Direct methanol fuel cell (DMFC)	Solid oxide fuel cell (SOFC)	Alkaline fuel cell (AFC)	Molten carbonate fuel cell (MCFC)	Phosphoric acid fuel cell (PAFC)
Electrolyte or Reagents	Hydrogen and oxygen or air	Methanol	Metallic oxide solid ceramic	Potassium hydroxide	Molten carbonate salt mixture	Liquid phosphoric acid
Operating Temperature	60–100°C	50–120°C	Around 1000°C	60–90°C, even at 23–70°C	Around 600°C	150–220°C
Electrical Efficiency	40–50%	30–40%	70–80%in combined cycle	60%, reaches above 80% in combined cycle	Around 60%	40–50%, reaches about 85% in combined cycle
Application	Transportation, portable and stationary applications	Portable energy purposes	In large scale power generation systems having capacity of hundreds of megawatts	To supply drinking water and electric power to the shuttle missions, transportation	Natural gas and coal based power plants, industrial and military applications	Onsite stationary applications
Advantage/ Disadvantages	Cheaper to manufacture, Longer lifetime and Require minimum maintenance	Low temperature operation, Long lifetime and Rapid refueling system	Long start up and cooling down times, Mechanical and chemical compatibility issues	Easily poisoned with carbon dioxide, Purified air or pure oxygen is required which increases operating cost	Long time is required to reach to operating temperature and generating power	High initial cost

Fuel cells are generally classified based on the nature of the electrolytes. Each type requires particular materials, fuel and operational characteristics, offering advantages to particular applications. This makes fuel cells a very versatile technology. They are classified according to the choice of fuel and electrolyte into 6 major groups as follows [29]:

- Proton exchange membrane fuel cell (PEMFC)
- Direct methanol fuel cell (DMFC)
- Solid oxide fuel cell (SOFC)
- Alkaline fuel cell (AFC)
- Molten carbonate fuel cell (MCFC)
- Phosphoric acid fuel cell (PAFC)

Various features of the above mentioned fuel cells such as, electrolyte or reagents required for running the cells, their operating temperature and efficiency, different applications and advantages–disadvantages of these fuel cells are discussed briefly in the Table 1.

2.4. Applications

The use of fuel cells is categorized into three broad areas: power for transportation, portable power generation and stationary power generation. Most of the vehicle manufacturers are currently using fuel cell vehicles for research, development or testing since it reduces CO_2 emissions which would become zero if the hydrogen is produced from renewable sources [40, 41]. Reduced fuel consumption, increased energy efficiency and less pollution are the main reasons for considering the use of fuel cells for mining trains and ships [42]. Portable fuel cells are those which are designed to be moved conveniently. Stationary fuel cells are units which provide electricity and heat as well, but cannot be moved. This fuel cell system is an independent system of the grid to generate electricity in remote or isolated areas or connected to the grid to provide additional electrical power to the plant [43-45].The overall efficiency of the system increases up to 80–95%. Other applications for stationary fuel cells are in landfills and wastewater plants [46].

3. MICROBIAL FUEL CELL AND ENERGY PRODUCTION

The fuel cell industry has faced and continues to face challenges as it comes through a period of recession and transition from R&D to commercialization. Furthermore, since many western countries seek to balance their economy towards high value manufacturing and green technologies, scientists and researchers have shown huge interest on biological fuel cells to enter a period of sustained growth. These fuel cells, generally known as Microbial Fuel Cell (MFC), have operational and functional advantages over the technologies that are currently used for generating energy from organic matter and conventional fuel cell, which are as follows – First, the direct conversion of substrate energy to electricity facilitates high conversion efficiency.

- Second, MFCs do not require extra energy input for aeration in the cathode, since it operates on general biochemical reaction, based on the function of microbial communities. This distinguishes them from all current bioenergy processes resulting in lower operation and maintenance costs.
- Third, MFC does not require any gas treatment because the off gases of MFC are mainly carbon dioxide which does not have any useful energy content. Furthermore, the redox products are usually CO_2 and H_2O, which are not considered contaminants.
- Fourth, MFCs can have a huge range of resources as their fuel as they do not need to be high quality. Wastes, mostly full of degradable organic compounds, can be used as fuel in MFCs.
- Fifth, MFCs have the potential for extensive application in the areas lacking electrical infrastructures and also to expand the diversity of fuels to meet up the energy requirements.

Although this MFC is still far from being commercialization, the opportunities for growth are very promising in the near future. The success of certain application segments in recent years has consolidated particular technologies into a standard reference design for MFCs. This has led MFCs to increasingly being developed as remarkable energy solutions having the capacity of serving several different market segments.

3.1. History of Microbial Fuel Cell

The technology of electrical current extraction from the metabolic processes of living microbes to use in external circuits has been in development for more than a century and the resulting devices are termed microbial fuel cells (MFCs) [12]. They have several potential advantages over more prominent sustainable energy technologies such as solar or wind power. Moreover, they can directly convert organic waste into electricity without pollution or inefficient intermediate steps involving mechanical generators [47].

Microbial fuel cells (MFCs), also known as, bioelectrochemical systems (BESs), were first described about 100 years ago. The first MFC, being inspired by the findings of other biologists in the late nineteenth and early twentieth century, was devised by Potter and has undergone substantial research and development [48]. The phenomenon of microbial induced electrode reduction was discovered in 1910. The first true MFC was constructed by Cohen and later modified by Davis and Yarbrough [49, 50]. The phenomena of microbial respiration with a solid electron acceptor respiration donor were further studied during 1960 which was followed by the study of use of electron transport mediators as a way to enhance power output [4, 51, 52]. Since then, the interest in electrogenic respiration in reactor systems dimmed. During the middle of the 1990s MFC regained more attention and the new interest was due to the potential use of MFC for clean, sustainable and renewable energy production in combination with the potential of a new wastewater treatment system. Even more and more potential applications have been established till now which is depicted in the Figure 2.

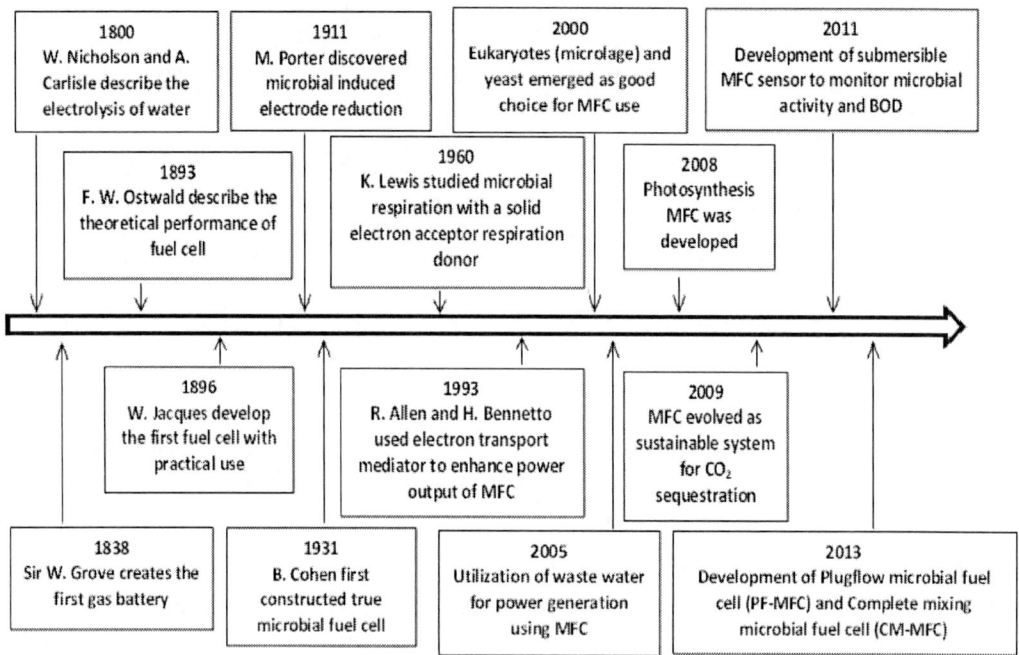

Figure 2. Historical review of microbial fuel cell.

3.2. Working Principle of Microbial Fuel Cell

An MFC is distinguished from other conventional fuel cells by the mechanism of chemical reactions. At least one of the two reactions is catalyzed by a biological component in MFCs [53]. Generally, the reaction takes place in anode, where organic carbon or another electron donor is oxidized by microorganisms and a solid state electron acceptor (i.e., the electrode) is subsequently reduced. Figure 3 shows an overview of a MFC.

Microorganisms are used as catalysts in these unique devices to transform chemical energy directly into electricity and thus generate combustionless, pollution free bioelectricity directly from the organic matter in biomass. Energy stored in organic matter is converted to electrical energy through enzymatic reactions by microorganisms which are associated with the normal living processes of bacteria.

In a typical MFC configuration as shown in Figure 3, the microorganisms are situated in anodic compartment and use the biomass for growth to produce electrons and protons [12, 54]. The electrons are directly expelled by some microorganisms or transported to an electrode using redox mediators for reducing the substrate. The protons or H$^+$ ions are diffused through the electrolyte to the cathode and is oxidized to water.

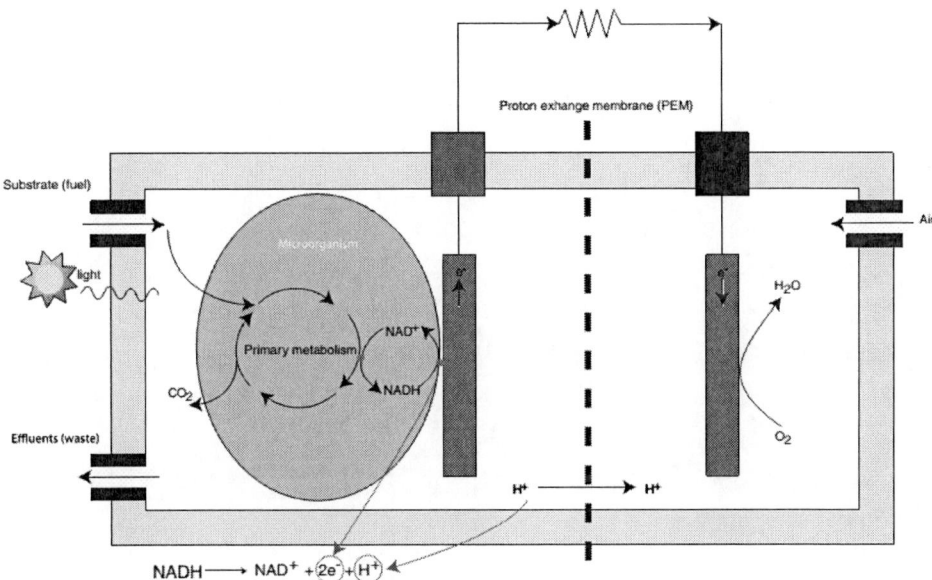

Figure 3. The working principle of a microbial fuel cell. Bacteria in anode compartment obtains electron from an electron donor (glucose or light in the case of photosynthetic organisms) and transfers to the anode electrode. Protons, produced in excess during electron production, flow through the proton exchange membrane (PEM) into the cathode chamber. The electrons flow from the anode to the cathode through an external resistance (or load) and react with the final electron acceptor (oxygen) and protons [12].

The cathode can be placed in a separate chamber (i.e., double chambered MFCs) or in the same chamber (i.e., single chambered MFCs). In a single chambered MFC, the need for the cathodic chamber is eliminated by exposing the cathode directly to the air where the only byproduct released is carbon dioxide, which can be fixed by plants for photosynthesis.

The driving force of a typical MFC using glucose as substrate can be expressed through the redox reaction taking place at each electrode as follows:

$$C_6H_{12}O_6 + 6H_2O \longrightarrow 6CO_2 + 24H^+ + 24e^-$$
$$6O_2 + 24H^+ + 24e^- \longrightarrow 12H_2O$$

There are three different concepts of MFCs which accommodates all the processes where a few processes can be placed in one or more concepts.

MFCs are required to run under predefined conditions for optimum growth and living conditions of the used microorganisms. Thus, the following factors affect the MFC's efficiency –

- Electrode material
- pH buffer and electrolyte
- Proton exchange system and
- Operating conditions in both the anodic chamber and the cathodic chamber

MFCs are usually operated at ambient temperature and atmospheric pressure. The pH conditions are kept neutral or slightly acidic. MFCs harness electrons from these systems in three main operational modes which are:

- Mediated electron transfer (MET)
- Direct electron transfer (DET) and
- Product mode

Photosynthetic MFCs use photosynthesis as the electron source and can also be operated in the same modes.

4. TYPES OF MICROBIAL FUEL CELL

4.1. Mediator Microbial Fuel Cell

Most of the microbial cells are electrochemically inactive. Thus, electron transfer from microbial cells to the electrode is facilitated by mediators. Thionine, methyl viologen, methyl blue, humic acid, neutral red etc. have been extensively used as mediator in MFCs [55]. Most of the mediators are expensive and toxic.

4.2. Mediator Free Microbial Fuel Cell

Mediator free microbial fuel cell is a more recent area of research and they use electrochemically active bacteria to transfer electrons to the electrode and thus, do not require a mediator. Shewanella putrefaciens [56], Aeromonas [57] and a few others are well known as electrochemically active bacteria. Some bacteria, having pili on their external membranes, can transfer their electron via these pili. For optimum efficiency, the factors such as the strain of bacteria used in the system of ion-exchange membrane, and system conditions (temperature, pH, etc.) are required to be controlled properly.

Mediator free MFCs can run on wastewater and also derive energy directly from certain plants like, reed sweetgrass, cordgrass, rice, tomatoes, lupines and algae [58]. The latter is known as a plant microbial fuel cell. The power thus derived from living plants (in situ-energy production) can provide additional ecological advantages.

4.3. Soil Based Microbial Fuel Cell

Soil based microbial fuel cells follow the same basic MFC principles, where the soil acts as the nutrient rich anodic media, the inoculum and proton exchange membrane (PEM). The anode is placed at a certain depth into the soil. The cathode rests on top the soil with contact of oxygen in the air above it. Soils are naturally packed with a diverse consortium of microbes, which includes electrogenic microbes needed for MFCs, and are full of complex sugars and other nutrients. Moreover, the aerobic microbes present in the soil act as an oxygen filter, much like the expensive PEM materials used in MFC systems, which decrease the redox potential of the soil with greater depth. For these natural facilities soil based MFCs are becoming popular day by day.

4.4. Phototrophic Bio-Film Microbial Fuel Cell

Phototrophic bio-film MFCs (PBMFCs) are the ones which use anode having a phototrophic bio-film holding photosynthetic microorganism like chlorophyta, cyanophyta etc. These microorganisms could carry out photosynthesis and thus, act as both organic metabolite producers and electron donors. Though PBMFCs yield one of the highest power densities, difficulties have been faced in increasing their power density and long term performance to make it cost effective MFC [59]. The PBMFCs using purely oxygenic photosynthetic material at the anode is also known as biological photovoltaic systems [60]. PBMFCs convert sunlight into electricity by the metabolic reaction of MFC. Several photosynthetic organisms like blue green algae, green algae and few higher algae have been used directly or along with MFC to generate bioelectricity. Photosynthetic cyanobacterial species like Anabaena [61], Synochocystis and some other microalgae are being used in outdoor bioreactors [62]. The photosynthetic chlorella algae are used with heterotrophic bacteria to get a synergistic effect on reaction to generate electricity [63-65].

Microalgae and several species of marine algae have also been investigated as artificial redox mediator for power generation in MFC with mediator like HNQ (hydroxy 1, 4 napthaquinone) to transport the electron from microorganism to the anode [66].

Similarly, power generation can be achieved by growing Saccharomyces for glucose oxidation and blue green algae for electron generation in micro-machined MFC and photosynthetic electrochemical cells respectively [67]. Most of these studies observed an increase in power generation during dark phases as, the oxygen production by algae, limits power production during light phases.

5. ENGINEERING DESIGN AND BIOLOGICAL ASPECTS

MFCs although, have not yet exceeded the laboratory being a very promising technology, a wide variety of materials have been investigated for electrode purposes. However, the most commonly used material, for both the anode and cathode, is modified carbon. Materials used for the construction of the system are usually nonconductive glass reactor flasks or plastic frames. The following sections will give detail descriptions of different materials and methods of working of MFCs.

5.1. Electrode Materials

Electrode materials for MFCs require a high specific surface area to create a high volumetric current density. The surface areas are generally determined by means of N_2 gas adsorption, but in case of three-dimensional electrodes, the surface available is determined by biological interaction. Different types of carbon and stainless steel have mostly been used for both anode and cathode electrodes where the carbon materials are frequently applied as granular, felt, cloth, brushes and solid blocks [68-71]. Carbon is generally inert and various attempts have been made to increase the low catalytic activity of carbon electrodes. The use and effectiveness of catalysts on an electrode depend on the catalyst loading, stability and the operating conditions of the electrode compartment [72, 73]. The use of undefined streams along with a microbial catalyst leads to clogging of the chemical coating on the electrode. This fouling on electrode can be removed by treating the activated carbon felt with ethylenediamine and nitric acid. These increase the power density by 25% and 58%, respectively, which is due to modified surface characteristics of the original material and decrease the start-up time of the bio-anode [74]. Tungsten carbide modified anodes in combination with a pyrolyzed iron (II) phthalocyanine-modified cathode also improve both anodic and cathodic reactions which is remarkable since tungsten is cheaper than platinum [75, 76]. A graphite anode doped with Mn^{4+} and a graphite cathode doped with Fe^{3+} atoms increase the power density by 1000-fold in comparison to woven graphite control electrodes [77]. The modification of the cathode with either a soluble iron chelated compound or a fixed iron complex was investigated [78]. Iron-EDTA coated on the cathode can be a sustainable alternative. However, the highest power and current densities are found using non-sustainable hexacyanoferrate iron complex. Unmodified carbon and stainless steel felt are also suitable materials although operational parameters such as salinity and bio-film have strong impact on its performance [73]. Granular graphite can also efficiently catalyze the oxygen reduction reaction without any biological catalyst [79]. A range of carbon materials including unmodified carbon cloth, paper and sponge, unmodified graphite, and reticulated vitreous carbon (RVC) are used for application as a cathode in marine sediment fuel cells. Researchers also developed modified materials such as, carbon paper coated with Fe and Fe-Co tetra methoxy phenyl porphyrin (TMPP), platinized carbon and platinized titanium electrodes [80]. Cathode catalyzed with Fe-CoTMPP resulted higher power output when compared with unmodified carbon sponge. Thus, the choice of material is very crucial since performance vary from material to material [81].

In recent years research has been focused on finding materials for efficient hydrogen gas generation at the cathode. Metal based electrodes such as stainless steel, nickel, and platinum coated with NiO_x catalyst showed good performance. However, that long term stability of the catalyst needs improvement [82].

Recent advances in nanofabrication provide a unique opportunity to use nanoparticles and carbon nanotubes as efficient materials for electrodes in MFCs [73]. Nickel powders, palladium nanoparticles and nanostructured manganese oxide have been used for hydrogen production in an MFC [82-85]. Using graphite disks with gold and palladium nanoparticles, the current density can be improved by 20-fold when compared with a plain graphite anode [86].

5.2. Membrane

Membrane is an important part of MFC as it is in a chemical fuel cell; however membrane-less MFC also showed promising results and can be a reasonable alternative [87, 88]. The main purposes of a membrane are –

i. To separate reactants in order to prevent internal short circuiting and
ii. To provide a mean of internal charge balancing between the anode and cathode reactions

Membranes, used in MFCs, include dialysis membranes, cation exchange membranes (CEMs), proton-exchange membranes (PEMs) and anion-exchange membranes (AEMs); and among these AEMs is found to be the best choice for decreasing internal resistance and increasing internal charge transport [89, 90]. Moreover, non-ion-exchange separators like cloth, filtration membranes, bipolar plates and baked are also used [91-94]. The use of any kind of membrane has the drawback that it promotes a pH gradient between the two compartments, which interferes with transport of OH$^-$ and H$^+$ ions and leads to a voltage loss.

5.3. Configurations and Design

Most MFCs are designed like conventional fuel cells, i.e., a membrane separating two flat electrode compartments. Design parameters are adjusted from the traditional design which includes electrode spacing, flow patterns, reactor volumes and electrode surface areas. The original H-type fuel cells are generally used for studies of physiological parameters of the biocatalyst and not for optimizing output [95].

Moreover, tubular up flow designs have been developed to deal with high internal resistances to charge transfer [96, 97]. Some research has been focused on miniaturizing a MFC. It is based on the concept that smaller MFCs have lower internal resistance and when electrically connected can deliver larger volumetric power outputs [89, 98]. Miniaturization also makes systems easier to set up and has been studied in detail [99-103]. Most of these works has so far been focused on the anode compartment. Conductive electrode material coated on a membrane, i.e., membrane electrode assembly (MEA) is mostly used in small milliliter scale fuel cells. This has led to a one chamber design that contains the anode compartment while the cathode electrode is open to the air [104, 105].

Large scale MFCs, mostly stacked systems, have also been studied to evaluate performance efficiencies [106]. Owing to the biological nature, the phenomenon of stack reversal can occur and leads to a decrease in usable output [107, 108]. To date, MFCs with sizes up to 20 L have been characterized [6, 109-112].

5.4. Microbial Catalysts

Microorganisms are used as catalyst in microbial electro-catalysis reactions occurring at electrodes. They are engaged in extracellular electron transfer (EET), a process of electrons transport in and out of the cell, and can catalyze both oxidation and reduction reactions [113, 114]. They are able to lower the over potentials (lower energy loss) at both anodes and cathodes giving an increased performance of the system which confirms the catalytic properties of microorganisms [115, 116]. Nevertheless, since part of the substrate or electron donor is consumed for growth, they cannot be considered as true catalysts. Till date, most research has been done on biological anodes, as cathodes are mainly abiotic and of minor interest. In the last few years, it has been realized that the bio-cathode still is one of the weakest points of this technology which has led to an exorbitant increase in the scope of both chemical and biological possible electrode materials and cathode reactions.

5.4.1. Anode Reactions

In bio-anodes, microorganisms oxidize organic or inorganic electron donors and simultaneously liberate electrons and protons. The electrons travel through the internal electron transport chain of the microorganisms and are deposited on the anode. The energy levels of the electrons deposited on the electrode depend on the terminal electron transfer molecules.

Two strategies are used in bio-anodes for energy conservation by microorganisms: Shewanella spp. uses substrate level phosphorylation whereas; Geobacter spp. uses oxidative phosphorylation for energy conservation. Electrons pass through a cascade of cytochromes, quinones and other electron transfer molecules from the electron donor through the inner cell membrane and periplasmic space and reach a terminal electron transfer molecule in the outer cell membrane [117, 118]. This is the case for Gram-negative microorganisms. However, the exact electron transfer mechanisms have not yet been elucidated for Gram-positive bacteria.

5.4.1.1. Electron Donors

A large variety of substrates that can be handled by bio-anodes have been reported and different types of wastewaters have been found suitable to drive electron donating reactions [16]. Moreover, sulfide containing inorganic waste streams, in addition to organic electron donors, can be applied as a substrate for the anode [119]. More recently, even sunlight has been reported as an energy source for MFCs which requires photosynthetically active plants or microorganisms having additional advantage that CO_2 is fixed from the atmosphere [59].

Attention should be paid to the type of substrate used and the loading rate since they not only influence the composition of the microbial community, but also influence the performance, which includes power density and coulombic efficiency [70, 120].

5.4.1.2. Biocatalyst for Anode

The term biocatalyst relates to the microorganisms which play a catalytic role in the liberation and transportation of electrons conserved in the substrate to the electrode. But, they cannot be considered as true catalysts because part of the substrate is consumed for growth and maintenance purposes.

Substantial research has been carried out to establish the key players in bioanode processes, i.e., electrochemically active microorganisms capable of respiring with insoluble materials which is known as electricigens or anode respiring bacteria (ARB). In addition to Geobacter and Shewanella spp., frequently used in bioanodes, a broad spectrum of microorganisms comprising both Gram-positive and Gram-negative bacteria are found to live in anode bio-films and respiring with electrodes [103, 121-123]. The thickness of bio-films on anodes typically ranges from 10–50 μm and even thicker bio-films have been reported [124]. But the bacteria densities within bio-films are difficult to quantify due to the amorphous nature of the various electrodes materials.

5.4.1.3. Electron Transfer Mechanisms

One of the most interesting aspects of the biocatalysts is the mechanism of electron transport to the anode by the microorganisms. Several electron transfer mechanisms either directly or indirectly can take place. Furthermore, all of these mechanisms occur simultaneously within the microbial community to maximize the use of the substrate present for microbial benefit. Further analysis revealed that the Gram-negative bacteria have much stronger electron transfer capacities than the Gram-positive bacteria and therefore, all the proposed transfer mechanisms are based on the studies with Gram-negative isolates, frequently been found in microbial community of MFCs [125-127].

5.4.1.3.1. Direct Electron Transfer Mechanisms

Direct electron transfer (ET) is the transfer of electrons to the electrode without the need for a mobile component which can diffuse to and from the cell for electron transport [113]. Physical contact between the bacterial cell membrane and the electrode surface makes this happen. Direct ET is based on pure culture of metal reducing Gram-negative bacteria such as Geobacter sulfurreducens and Shewanella oneidensis which involve at least a series of periplasmatic and outer membrane complexes [95, 128]. Cytochromes, among all complexes play a pivotal role, for example OmcS and OmcZ, regarding Geobacter spp., are the most important electron transferring cytochromes in the final electron transfer step [129-131]. Additionally, appendages of the cell wall, pili, also seemed to be involved in the transport of electrons [132] and have also been proposed as possible electron transport mechanisms between different microorganisms [133] and as a means for oxic metabolism in anoxic sediments [134]. Direct electron transport eliminates diffusion limitations inherent in indirect transport to certain extent and simplifies solid liquid separations as the biocatalyst is immobilized in the reactor [113].

5.4.1.3.2. Indirect Electron Transfer Mechanisms

Indirect electron transfer (ET) uses electron transports which physically transfer electrons from the cell to the electrode [113]. Commonly applied mediators include humic substances such as anthraquinone 2, 6-disulfonate (AQDS) which can be expensive, toxic and prone to wash-out of the system [114], [135]. Some microorganisms, in addition to artificial redox mediators, are able to produce their own mediators such as secondary metabolites like phenazines and flavins [103, 126, 136]. Besides, primary metabolites like sulfur species and hydrogen gas can also convey electrons toward electrodes [137, 138]. Overall, mediators can increase the active range of the electrode beyond the bio-film into the bulk liquid by enhancing the electrical interconnectivity between electrochemically active microorganisms and electrode and care should be taken when applying them artificially.

5.4.2. Cathode Reactions

In cathodes, the electrons and protons come from anodic oxidation reaction and are used to carry out reduction reactions. These reactions can be either purely chemical or biologically catalyzed. Typical chemical cathode reactions are oxygen [139] and proton reduction [140] to water and hydrogen gas, respectively. Hexacyanoferrate is a chemical electron acceptor commonly used in cathodes to investigate bio-anode processes without the need of paying much attention to the cathode. However, its application is limited by replenishment and possible toxic effects.

In addition to chemical cathodes, biocathodes have also been developed. Since they fit much better with the sustainable nature of MFCs than chemical cathodes, an increasing interest in the development of biologically driven cathode reactions has been demonstrated in recent years. The biological aspects of electrochemical systems, biocathodes and their electron transfer mechanisms are discussed further.

5.4.2.1. Biocatalysts for Cathode

Microorganisms play a pivotal role in catalyzing reduction reactions in biocathodes. Although the Gram-positive bacteria are able to play role in the electron transfer, most of the electrochemically active bacteria in biocathodes have been found to be Gram-negative. This indicates that, there are potentially wide capability of bacteria to catalyze electrode reactions.[141, 142].

5.4.2.2. Electron Transfer Mechanisms

However, numerous reports are available on microbial aided electron transfer towards bioanodes, only little information is available on the reverse process. In particular, the microbial electron uptake from cathode by microorganisms requires in-depth investigation.

Recent biocathode studies have shown that by changing the environmental and operating conditions electrochemically active bioanodes may be turned into biocathodes and hence some known anodic ET mechanisms were re-investigated and their potential role was evaluated in cathodic ET mechanisms [142],[143, 144]. It was concluded that both direct and indirect mechanisms take place in cathode and are very similar to the processes at the bioanode and differ in the sense that the redox active components operate at higher redox potentials.

Additionally, it has not yet been established that biocatalyzed electron transfer yields energy for electrochemically active microorganisms by a respiratory process [17, 142]. For instance, it is shown that biological oxygen reduction at the cathode can occur without active involvement of the bacterium [145]. Nevertheless, an electron transfer mechanism with the generation of a proton in the microorganisms present in denitrifying biocathodes has been speculated [146].

5.4.2.2.1. Direct Electron Transfer Mechanisms

Cytochromes play an important role in direct cathodic ET similarly as they do in direct anodic ET mechanisms which has already been discussed. However, these compounds were found to cover a broad redox active range and were investigated successfully for their involvement in electron transfer [17, 142]. Furthermore, it has become apparent that hydrogenase containing microorganisms are also capable of accepting electrons directly from polarized electrodes [142].

5.4.2.2.2. Indirect Electron Transfer Mechanisms

During indirect cathodic electron transfer (ET), artificial or naturally produced mediators, as reported in indirect anodic ET studies, can be applied which results in the similar advantages and possible negative effects to the cathode side [114]. However, it is found that manganese oxides wash out slowly due to their solid nature. This advantage has led manganese oxides to be applied for transportation of electrons for oxygen reduction.

5.4.2.2.3. Electron Acceptors

A large variety of possible reactions have been established in MFC operating modes for biocathodes. The possible specific reduction reactions that can take place in biocathodes have been studied which indicates that these biocathodes show potential for use in a wide range of applications [17, 142].

5.4.3. Pure Cultures and Mixed Microbial Communities

To establish biological anodes or cathodes both mixed and pure cultures can be used. Mixed culture electrodes are found to be more robust and resilient and a higher power output has been observed comparing with pure culture systems. This may be due to either synergistic effects or the presence of a more productive exoelectrotroph in the mixed culture biofilms or even may be due to pH effects and underdeveloped bio-films of the pure cultures [147].

Pure cultures used in MFCs are ideal for studying fundamental aspects or for the catalysis of species specific reactions and can stimulate microbial communities to use electric current [114]. Furthermore, pure isolates can be used for high quality cathodic processes for specific generation of products [113].

5.4.4. Photosynthetic Biocatalysts

The concept of using photosynthetic microorganisms or plants in combination with an MFC existed for several years. The phenomenon of direct involvement of photosynthetic bacteria, i.e., direct electron transport with electrodes has already been suggested. For example, the Cyanobacterium Synechocystis sp. (PCC 6803) has been reported to possess potentially conductive nano-wires which could be involved in direct electron transfer with electrodes [128]. Bicarbonate reduction with light in a photo biocathode strongly indicated direct electron transport. Since these autotrophic microorganisms did not generate oxygen, electrocatalysts were absent and flushing of the cathode to wash out soluble mediators did not affect the current production [148]. Further research with pure photosynthetic bacterial culture is required to substantiate the direct involvement with electrodes.

5.4.5. Biological Limitations

Applying microorganisms in a technology is associated with some biological limitations. Proper physical parameters like pH (around neutral) and temperature (15−30°C) should be maintained for good microbial metabolism. However, some bacteria are able to function under slightly different conditions such as, the microbial reductive dechlorination of perchlorate was found to be the highest at pH 8.5 [149]. In addition, shear forces also influence the biofilm thickness and density as well as the power output and is an important physical parameter [150]. The flux of substrates and products within the biofilm are affected by biofilm thickness, structure, composition, and density. The products within the biofilm result in large overpotentials and negatively impact the performance of the system. Thicker anodic biofilms are observed to produce higher power, whereas, reverse effect has been observed in cathodic biofilms [94].

There are also some intrinsic limitations associated with the microbial metabolism like, their growth rate, uptake rate of nutrients and electron transfer capacities, even if they are not limited by their environment. As a result, cathodic communities encounter harsh conditions as they have to invest energy for CO_2 fixation in autotrophic microorganisms and have to take up electrons from the cathode at a high potential.

6. OPERATIONAL PARAMETERS AFFECTING THE MICROBIAL FUEL CELL

6.1. Thermodynamics and the Electromotive Force

Electricity is generated in MFC when the overall reaction is thermodynamically favorable. The maximal work that can be derived from the reaction can be expressed in terms of Gibbs free energy in units of Joules (J) and calculated as

$$\Delta G_r = \Delta G_r^0 + RT\ln(\Pi) \tag{1}$$

where, ΔG_r (J) is the Gibbs free energy for the specific conditions, ΔG_r^0(J) is the Gibbs free energy under standard conditions, R is the universal gas constant, T (K) is the absolute temperature and Π (unit less) is the reaction quotient calculated as the activities of the products divided by those of the reactants [151].

For MFC calculations, the reaction is evaluated in terms of the overall cell electromotive force (emf), E_{emf} (V). It is defined as the potential difference between the cathode and anode and is related to the work, W (J), produced by the cell and is given by the following Eq. 2

$$W = E_{emf}Q = -\Delta G_r \tag{2}$$

where, $Q = nF$ is the charge transferred in the reaction, expressed in Coulomb (C), n is the number of electrons per reaction mol and F is Faraday's constant. Combining these two equations, we have

$$E_{emf} = -\ \Delta G_r / nF \tag{3}$$

at standard conditions the reactions can be written as

$$E^0_{emf} = -\ \Delta G_r^0/nF \tag{4}$$

Here, E^0_{emf} is the standard cell emf. Therefore, in terms of the potentials the overall reaction can be expressed as

$$E_{emf} = E^0_{emf} - RT\ln(\Pi)/nF \tag{5}$$

The Eq. 5 is positive for a favorable reaction and produces an emf value for the reaction. This emf provides an upper limit for the cell voltage and due to various potential losses, the actual potential derived from the MFC will be lower.

6.2. Standard Electrode Potentials

The reactions occurring in the MFC can be written as separate reactions occurring at the anode and the cathode, i.e., as half-cell reactions and are expressed as a reduction potential. The standard potentials are reported relative to the normal hydrogen electrode (NHE) having a potential of zero at standard conditions. Eq. 5 is used to express the theoretical anode and cathode potential, E_{an} and E_{cat}, respectively, under specific conditions, where the activities of the different species assumed to be equal to their concentrations. The cell emf is calculated as

$$E_{emf} = E_{cat} - E_{an} \qquad (6)$$

Here although an oxidation reaction is occurring, but the minus sign is given according to the definition of the anode potential as reduction reaction. Eq. 6 represents that different cell voltages and different levels of power output can be obtained using the same anode in a system with different cathode conditions. Therefore, the power produced by an MFC depends on the choice of the cathode.

6.3. Open Circuit Voltage (OCV)

The open circuit voltage (OCV) is the cell voltage measured after some time in the absence of current and the cell emf is a thermodynamic value that does not consider internal losses. Though, theoretically, the OCV should approach the cell emf, in practice the OCV is substantially lower than the cell emf. This is due to various potential losses like energy loss occurring at the cathode. This energy loss is often referred to the difference between the potential under equilibrium conditions and the actual potential, i.e., over potential. The size and nature of energy losses is identified by application of thermodynamic calculations.

6.4. Identifying Factors that Decrease Cell Voltage

Due to number of losses, the measured MFC voltage is considerably lower than the theoretically achievable maximum MFC voltage. The maximum MFC voltage achieved in an open circuit is 0.80V, which remains below 0.62V during current generation [96, 152]. This difference, i.e., overvoltage is the sum of the overpotentials of electrodes and the ohmic loss of the system

$$E_{cell} = E_{emf} - (\sum \eta_a + \left| \sum \eta_c \right| + IR_\Omega)$$

Where, $\sum \eta_a$ and $\left| \sum \eta_c \right|$ are the overpotentials of the anode and the cathode respectively and IR_Ω is the sum of all ohmic losses. The ohmic losses are proportional to the generated current (I) and ohmic resistance of the system (R_Ω). The overpotentials of the electrodes are generally current dependent and can be classified as follows: (i) activation losses, (ii) bacterial metabolic losses and (iii) mass transport or concentration losses. In MFCs, the measured cell voltage is generally a linear function of current and can be represented as

$$E_{cell} = OCV - IR_{int} \qquad (7)$$

Where, IR_{int} is the sum of all internal losses of the MFC. They are proportional to the generated current (I) and internal resistance of the system (R_{int}). The overpotentials of electrodes occurring under open circuit conditions are given by OCV, while the current dependent overpotentials of the electrodes and ohmic losses of the system are included in the value of IR_{int}. MFC systems show a maximum power output when the internal and external resistances are equal to each other [153]. The performance of MFC can be evaluated in terms of both overpotentials and ohmic losses or in terms of OCV and internal losses.

6.4.1. Ohmic Losses

The ohmic losses, also known as ohmic polarization in an MFC, include the resistance to the flow of electrons through the electrodes and interconnections. In addition, the resistance to the flow of ions through the cation exchanges membrane CEM (if present) and the anodic and cathodic electrolytes are also responsible for ohmic losses [154]. These losses can be reduced by using a membrane having low resistivity, minimizing the electrode spacing, checking thoroughly all contacts and increasing solution conductivity to the maximum which the bacteria can tolerate.

6.4.2. Activation Losses

The activation losses, i.e., activation polarization occur during the electron transfer from or to a compound reacting at the electrode surface due to the activation energy needed for an oxidation or reduction reaction. This compound can be present as the final electron acceptor reacting at the cathode, or at the bacterial surface as a mediator in the solution. Activation losses often strongly increase low currents and when current density increases the losses show a steady increase. Activation losses can be lowered by increasing the electrode surface area, operating temperature, improving electrode catalysis and by establishing an enriched bio-film on the electrode(s).

6.4.3. Bacterial Metabolic Losses

Bacteria transport electrons from a substrate at a low potential through the electron transport chain to the final electron acceptor at a higher potential and generate metabolic energy. In an MFC, the anodic potential determines the energy gain for the bacteria. Metabolic energy gain for the bacteria increases with the increase in the difference between the redox potential of the substrate and the anode potential, but the MFC voltage changes inversely. Therefore, the potential of the anode should be kept as low (negative) as possible to get maximum MFC voltage. On the other hand, electron transport will be inhibited if anode potential becomes too low and may provide greater energy for the microorganisms by fermentation of the substrate.

6.4.4. Concentration Losses

The rate of mass transport of a species to or from the electrode limits current production is responsible for concentration losses (concentration polarization) [154]. It may occur mainly at high current densities which limit mass transfer of chemical species by diffusion to the electrode surface. A limited discharge of oxidized species from the electrode surface or a limited supply of reduced species toward the electrode causes concentration losses at the anode. This increases the ratio between oxidized and reduced species at the electrode surface and increases in the electrode potential. At the cathode side, a drop in cathode potential is observed. Diffusional gradients may also arise in the bulk liquid in poorly mixed systems and limit the substrate flux to the bio-film, which is another type of concentration loss. The onset of concentration losses can be determined by recording polarization curves.

7. POTENTIAL APPLICATION OF MICROBIAL FUEL CELL TECHNOLOGY

7.1. Current Research Practices

Microbial Fuel Cell (MFC) has been used in numerous fields and it has been becoming a prominent sector of researcher's attraction since availability of common energy sources became limited. Applications of microbial fuel cell are summarized by Franks et al. in their recent review [155]. It should be noted that many of these envisaged applications are not currently feasible and require significant improvements if they are to become viable technologies [122, 156, 157]. One of the most active areas of MFC research is the production of power from wastewaters combined with the oxidation of organic or inorganic compounds. The use of an anode as a final electron acceptor by microorganism has led to the possibility of a wide range of applications. Studies have demonstrated that any bacteria degradable compound can be converted into electricity [16]. The range of compounds include, but by no means limited to, acetate [158, 159], glucose [160], starch [161], cellulose [162], wheat straw [163], pyridine [164], phenol [165], *p*-nitrophenol [166] and complex solutions such as domestic waste [104, 167], brewery waste [168], land file leachate [169], chocolate industry waste [170], mixed fatty [171] and petroleum contaminates [172]. Within these systems, less biomass is generally produced than their equivalent aerobic processes and without the need for an energy intensive aeration process, less energy is required [173]. MFCs for the large scale treatment of wastewaters still face problems of scale up from laboratory experiments and slow rates of substrate degradation. Recent developments of such prominent applications of MFC have been discussed below.

7.2. Renewable Energy Generation or Power Generation

MFCs are capable of converting the chemical energy stored in the chemical compounds in a biomass to electrical energy with the help of microorganisms like bacteria. Carnot cycle with a limited thermal efficiency is avoided and a much higher conversion efficiency can be achieved (>70%) just like conventional chemical fuel cells as the chemical energy from the oxidization of fuel molecules is converted directly into electricity instead of heat. Chaudhury and Lovley reported that *R. ferrireducens* could generate electricity with yield as high as 80% [174]. Higher electron recovery such as, electricity up to 89% was also reported [175]. An extremely high Coulombic efficiency of 97% was reported during the oxidation of formate with the catalysis of Pt black [176]. However, MFC power generation is still considered very low and that means the rate of electron abstraction is very low [177, 178]. One feasible way to solve this problem is to store the electricity in rechargeable devices and then distribute the electricity to end-users [179]. Broad demands and high operational sustainability can be satisfied by microbial-cathode MFCs, or cathodes using only bacterial catalysts (biocathodes) which have gained considerable attention in recent years [17]. These versatile biocathodes allow to use not only oxygen but also contaminants as possible electron acceptors, allowing nutrient removal and bioremediation in conjunction with electricity generation.

A simple and low-cost modification was developed to improve the power generation performance of inexpensive semicoke electrode in microbial fuel cells (MFCs) [180]. After carbonization and activation with water vapor at 800–850°C, the MFC with the activated coke (modified semicoke) anode produced a maximum power density of 74 Wm^{-3}, 17 Wm^{-3}, and 681 mWm^{-2} (normalized to anodic liquid volume, total reactor volume, and projected membrane surface area, respectively), which was 124% higher than MFCs using a semicoke anode (33 Wm^{-3}, 8 Wm^{-3}, and 304 mWm^{-2}) and 211% increase can be achieved when they were used as biocathode materials. Another group of researchers developed cost-effective activated carbon (AC) material for the development of gas-diffusion cathode employed in membraneless single chamber microbial fuel cells (SCMFCs) treating different feeding solutions [181]. The electrocatalytic activity of AC cathodes was monitored in synthetic wastewater containing phosphate buffer saline solution and sodium acetate (PBS and NaOAc) and compared with several types of wastewaters (e.g., fresh urine (FU), hydrolyzed urine (HU), wastewater and sodium acetate (WW + NaOAc) and raw wastewater (WW)). The results showed that the urine fed SCMFCs generated 3 times higher power densities than those with raw WW and 25%

higher than those with WW + NaOAc, most likely due to the high amount of electrons generated from organic substances.

Schröder et al. reported a microbial fuel cell that continuously generates a current output more than one order of magnitude larger than the known microbial fuel cells (up to 1.5 mAcm^{-2}) (Figure 4) [138]. The novel fuel cell concept uses polymer-modified catalytically active anodes which shuttle electrons from the bacterial suspension to the anode.

Table 2. Mean current densities at differently modified pyrolytic carbon

Electrode modification	Current density [mA^{-1}cm^{-2}]
Polyaniline	0.29
platinum black	0.84
platinum black + polyaniline overlay	1.45

The electrode was placed in a stirred anaerobic culture of Escherichia coli K12 in a standard glucose medium (Cglucose=0.55 mmolL-1). The potential applied to the electrode was 0.2 V.

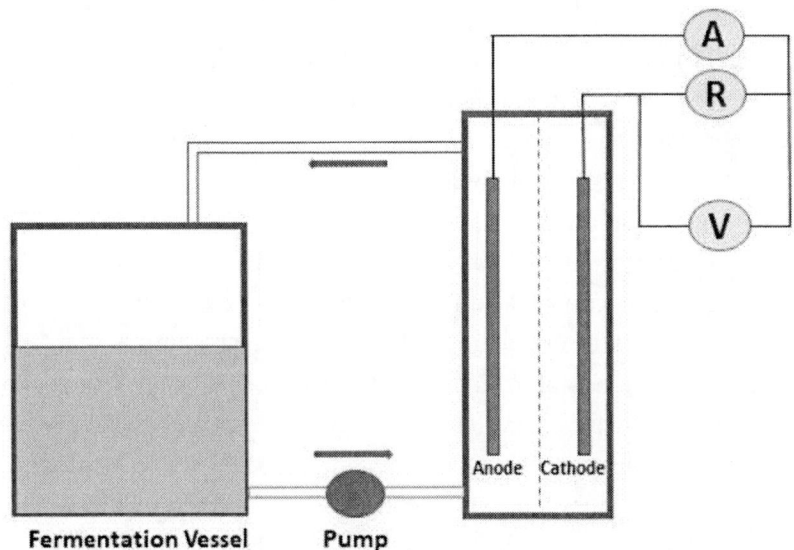

Figure 4. Schematic drawing of the novel microbial fuel cell (Redrawn from Schröder et al. [138]).

The conductive polymer has a protective function as well as directly contributes to the current flow. Table 2 shows different current densities obtained by using three different types of polymers. At purely polyaniline-modified carbon electrodes, current densities of 0.29 mAcm^{-2} can be achieved. This result means that whereas the platinum black accesses the oxidation of metabolites, the conductive polymer layer fulfils a multitude of tasks. Consisting of molecular units similar to the conventionally used redox mediators, the polymers form a redox active biocompatible layer that takes the function of the dissolved mediators, thus banishing these artificial compounds from the bacterial medium. Because of their reversible redox behavior and electronic conductivity, the polymers are also involved in the oxidation of excreted metabolites.

In near future, MFCs will have to compete with more mature renewable-energy technologies, such as wind, hydro and solar power. The operating costs needed for electricity production with MFCs will probably be too great if the substrate for the MFC is grown as a crop in a manner similar to that for ethanol production from corn. Platinum (Pt) is the most commonly used catalyst on the cathode, but its high cost prohibits its use for commercial MFC applications. Hao et al. examined various cathode catalysts prepared from metal porphyrines and phthalocyanines for their oxygen reduction activity in neutral pH media [182]. Iron phthalocyanine supported on KJB (FePc-KJB) carbon demonstrated higher activity towards oxygen reduction than Pt in neutral media. Effects of platinum loadings on the cathodic reactions in Single Chamber Microbial Fuel Cells (SCMFCs) were investigated and a cost-effective MFC operational protocols were developed [183]. The power generation of SCMFCs was examined with different Pt loadings (0.005–1 mgPt/cm^2) on cathodes. The results showed that the power generation of the SCMFCs with 0.5–1 mg Pt/cm^2 were the highest in the tests, decreased 10–15% at 0.01–0.25 mg Pt/cm^2, and decreased further 10–15% at 0.005 mg Pt/cm^2 whereas the SCMFCs with Pt-free cathode (graphite) had the lowest power generation. Another important feature of the study was that the SCMFCs (with 0.5–1 mg Pt/cm^2 at cathode) fed with combination of wastewater and sodium acetate (NaOAc) gave the highest power generation (786 mW/m^2) comparing with wastewater only.

Renewable energy production from waste biomass is likely to be a more viable route for near-term energy recovery. Great interest exists in using wood based materials for renewable energy production. Steam explosion is currently the most cost-effective treatment process for the production of soluble sugars from solid lignocellulosic materials, such as agricultural residues and hardwoods [184]. The use of a neutral hydrolysate, produced by steam explosion of corn stover in an MFC, has recently shown to be feasible, producing as much as 933 mW/m^2 in MFC [185]. A single chamber air cathode mediator-less microbial fuel cells was reported which utilized mixed monosaccharides found in the hydrolysates of lignocellulosic biomass to generate power energy [186]. The different monosaccharide utilization rates ranging from 212 mg L^{-1} h^{-1} to 389 mg L^{-1} h^{-1} indicate the presence of preferential utilization of different monosaccharides where three volatile fatty acids (VFAs), including acetic, propionic and butyric acids were detected as the main intermediates, which were generated mainly through a fermentation process. Pant et al. used a large number of substrates as feed where the major substrates were various kinds of artificial and real wastewaters and lignocellulosic biomass to get power energy [16]. Although the current and power yields were relatively low, they expected large amount of electric current from these systems through providing a sustainable way of directly converting lignocellulosic biomass or wastewaters to useful energy. The effect of different substrates (e.g., acetate, glucose, or butyrate) on the performance and microbial community of two-chamber microbial fuel cells (MFCs) was investigated by Zhang et al [187]. They suggested that the type of substrate fed to MFC is a very important parameter for reactor performance and microbial community, and significantly affected power generation in MFCs. Thus, MFC technologies appear to be technically feasible for energy recovery from this and other waste biomass materials.

Oh et al. investigated the effects of different sludge pretreatment methods (ultrasonic vs. combined heat/alkali) with varied sources of municipal sewage sludge (primary sludge (PS), secondary excess sludge (ES), anaerobic digestion sludge (ADS)) on electricity generation in microbial fuel cells (MFCs). Introduction of ultrasonically pretreated sludge (PS, ES, ADS) to MFCs generated maximum power densities of 13.59, 9.78 and 12.67 mW/m^2 and soluble COD (SCOD) removal efficiencies of 87%, 90% and 57%, respectively which shows better results than sludge pretreated by combined heat/alkali (0.04 N NaOH at 120°C for 1 h). Higher SCOD by sludge pretreatment enhanced performance of the MFCs and the electricity generation was linearly proportional to the SCOD removal, especially for ES [188]. Zhao et al. mentioned that the thermodynamics and the kinetics of the electrocatalytic oxygen reduction are severely affected by the physical and chemical environment in the microbial fuel cells [72]. The neutral pH in combination with low buffer capacities and low ionic concentrations strongly affect the cathode performance and limit the fuel cell power output.

Cheng et al. showed that the cathode surface area is always important for increasing power [189]. Doubling the cathode size can increase power by 62% with domestic wastewater, but doubling the anode size increases power by 12%. Cathode's specific surface area is the most critical factor for scaling-up MFCs to obtain high power densities. They systematically varied substrate concentration using acetate, and conductivity, and then examined the effects on power using different cathode specific surface areas and also examined trends in cathode sizes using the acetate and wastewater solutions based on the available data, combined with results from others. The 250-mL MFC (1 g/L acetate in 100 mM PBS) was used to investigate the effect cathode area on power density (using LSV). The power increased by a factor of 5, or from 15 to 78 W/m^3, when the cathode surface area was increased from 24 to 96 cm^2 by increasing the number of cathodes from 1 to 4 [189].

MFCs are especially suitable for powering small telemetry systems and wireless sensors that have only low power requirements to transmit signals such as temperature to receivers in remote [190, 191]. MFCs themselves can serve as distributed power systems for local uses, especially in underdeveloped regions of the world. Locally supplied biomass can be used to provide renewable power for local consumption. Applications of MFCs in a spaceship are also possible since they can supply electricity while degrading wastes generated onboard. Some scientists envision that in the future a miniature MFC can be implanted in a human body to power an implantable medical device with the nutrients supplied by the human [192]. The MFC technology is particularly favored for sustainable long-term power applications only after potential health and safety issues brought by the microorganisms in the MFC are thoroughly solved. MFCs are viewed by some researchers as a perfect energy supply candidate for Gastrobots by self-feeding the biomass collected by themselves [193]. Realistic energetically autonomous robots would probably be equipped with MFCs that utilize different fuels like sugar, fruit, dead insects, grass and weed. Capacitors were used in their biologically inspired robots named EcoBot I to accumulate the energy generated by the MFCs and worked in a pulsed manner whereas robot EcoBot-II solely powers itself by MFCs to perform some behavior including motion, sensing, computing and communication [194-196]. Zhiyong et al. tested binary culture of *C. cellulolyticum* and *G. sulfurreducens* in MFCs for *in situ* electricity production directly by converting cellulose into electricity without enzymatic pretreatment or an exogenous catalyst [2]. This coculture achieved maximum power densities of 143 mW/m^2 (anode area) and 59.2 mW/m^2 from 1 g/L carboxymethyl cellulose (CMC) and MN301 cellulose, respectively. In this defined system, *C. cellulolyticum* fermented cellulose mainly into acetate, ethanol, hydrogen, and carbon dioxide, and *G. sulfurreducens* transferred electrons from some of these fermentation products to the anode via anaerobic respiration. In scanning electron micrographs, the binary culture showed tight microcolony structures, while the mixed culture from activated sludge showed heterogeneous cell morphologies and a looser biofilm structure. Lower magnification images showed that the coculture biofilm had patchy coverage of the anode, while the mixed culture formed a more uniform biofilm over the surface of the graphite plate.

7.3. Biohydrogen

The architecture and operation of MFCs can be readily modified to produce hydrogen instead of electricity. Under normal operating conditions, protons released by the anodic reaction migrate to the cathode and combine with oxygen to form water. A bio-electrochemically assisted microbial reactor (BEAMR) is established where oxygen is removed from the cathode and a small additional voltage is applied to the circuit, hydrogen gas is evolved from the cathode [197-199]. Another option of biohydrogen production is simply the bacterial electrolysis of organic because the protons and electrons are derived from the organic matter and not water [140]. In biohydrogen production using MFCs, oxygen is no longer needed in the cathodic chamber. Thus, MFC efficiency improved because oxygen leak to the anodic chamber is no longer an issue. Another advantage is that hydrogen can be accumulated and stored for later usage to overcome the inherent low power feature of the MFCs. Therefore, MFCs provide a renewable hydrogen source that can contribute to the overall hydrogen demand in a hydrogen economy [200].

Research has shown that the process works with domestic wastewater, but H_2 recoveries in current reactor designs are still too low to make H_2 production with BEAMR likely to be as viable as electricity production [199]. Hydrogen generation from the protons and the electrons produced by the metabolism of microbes in an MFC is thermodynamically unfavorable. Liu et al. applied an external potential to increase the cathode potential in a MFC circuit and thus overcame the thermodynamic barrier [197]. In this mode, protons and electrons produced by the anodic reaction are combined at the cathode to form hydrogen. The energy comes from the biomass oxidation process in the anodic chamber is utilized to provide the required external potential for an MFC which is theoretically 110 mV, much lower than the 1210 mV required for direct electrolysis of water at neutral pH. MFCs can potentially produce about 8–9 mol H_2/mol glucose compared to the typical 4 mol H_2/mol glucose achieved in conventional fermentation [197]. Bacteria produce an anode working potential of ~ –0.3 V. The protons and electrons that are produced at the anode can combine at the cathode to produce H_2 with only an additional total potential of 0.11 V. In practice, however, 0.25V or more must be put into the circuit to make H_2, because of overpotential at the cathode [140, 197, 198].

An integrated hydrogen gas production process from cellulose consisting of a dark fermentation reactor and microbial fuel cells (MFCs) as power sources for a microbial electrolysis cell (MEC) was investigated by Wang et al. [201]. The method described a single integrated process, with fermentation and MECs linked together for hydrogen production and fermentation of effluent used in both MFCs and MECs so that power generated in the MFCs could be used to power an MEC. The advantage of this combined fermentation and MFC–MEC system was the production of hydrogen at higher yields than fermentation alone, without the need for an exogenous electrical power input. In order to test the system with a renewable substrate, cellulose was used as the starting substrate. One to three MFCs were used to power the MEC in order to examine the effect of different applied voltages on MEC performance. A maximum of 0.43 V using fermentation effluent as a feed, achieving a hydrogen production rate from the MEC of 0.48 m^3 H_2/m^3/d (based on the MEC volume), and a yield of 33.2 mmol H_2/g COD removed in the MEC. The overall hydrogen production for the integrated system (fermentation, MFC and MEC) was increased by 41% compared with fermentation alone to 14.3 mmol H_2/g cellulose, with a total hydrogen production rate of 0.24 m^3 H_2/m^3/d and an overall energy recovery efficiency of 23% (based on cellulose removed) without the need for any external electrical energy input.

7.4. Wastewater Treatment

Energy recovery at a wastewater treatment plant could lead not only to a sustainable system based on energy requirements but also to production of a net excess of energy. The MFCs were considered to be used for treating waste water early in 1991 [202]. Electricity generation will not justify MFC operation, but the BOD removal with this more sustainable technology is attractive [203]. Municipal wastewater contains a multitude of organic compounds that can fuel MFCs. An MFC would be used in a treatment system as a replacement for the existing energy-demanding bioreactor (such as an activated sludge system), resulting in a net energy-producing system. The amount of power generated by MFCs in the wastewater treatment process can potentially halve the electricity needed in a conventional treatment process that consumes a lot of electric power in aerating activated sludge and they yield 50–90% less solids to be disposed of [200]. MFCs can enhance the growth of bio-electrochemically active microbes during wastewater treatment thus they have good operational stabilities. Continuous flow and single-compartment MFCs and membrane-less MFCs are favored for wastewater treatment due to concerns in scale-up [204-206]. Sanitary wastes, food processing wastewater, swine wastewater, biodiesel waste and corn stover are all great biomass sources for MFCs because they are rich in organic matters [104, 207-209]. Up to 80% of the COD can be removed in some cases [104, 208] and a coulombic efficiency as high as 80% has been reported [210]. Furthermore, organic molecules such as acetate, propionate, and butyrate can be thoroughly broken down to CO_2 and H_2O. A hybrid incorporating both electrophiles and anodophiles are especially suitable for wastewater treatment because more organics can be biodegraded by a variety of organics. MFCs using certain microbes have a special ability to remove sulfides as required in wastewater treatment [119]. An MFC system could even be useful for individual homes or other small applications, although power production would probably be too low to warrant recovery of electricity. Septic tanks are typically used for single- to multiple-house applications, but they are inefficient systems for removing BOD or nutrients. An MFC-based system, however, might provide the opportunity for better removal of BOD, COD and even nutrients [8, 208]. MFC applications may be particularly useful in areas where septic tanks cannot be used because of the need for high BOD removal. Such applications are currently carried out by small aerobic systems that consume energy, often in remote areas with little power available to run them.

Energy recovery wastewater was found to depend significantly on the operational conditions (flow mode, temperature, organic loading rate, and HRT) as well as the reactor [211]. Temperature was an important parameter for treatment efficiency and power generation. The highest power density of 422 mW/m^2 (12.8 W/m^3) was achieved under continuous flow and mesophilic conditions, at an organic loading rate of 54 g COD/L-d, achieving 25.8% COD removal. The results demonstrate that the main advantages of using temperature-phased, in-series MFC configurations for domestic wastewater treatment are power savings, low solids production, and higher treatment efficiency.

Khan et al. conducted several investigations to enhance the current and voltage generation of Membrane-Less Microbial Fuel Cell (ML-MFC) in different operating conditions, such as direct effluent of food-processing industries, adding drainage sludge concentration, aeration in cathode compartment, increasing the electrode area [212]. In addition, COD removing capability of the ML-MFC was also studied. The study documented a maximum power density of 7.11874 mW/m^2 with the current density of 97.34 mA/m^2. COD removal was observed 47% to 74% in all experiments. Sevda et al. presented a novel method for simultaneous bioelectricity generation and wastewater treatment by operating an air–cathode MFC with membrane-electrode assembly over three batch cycles (total of 160 days) and results indicated that molasses mixed sewage wastewater (high strength wastewater) containing 9978 mg/L of chemical oxygen demand (COD) could be used as substrate to produce bioelectricity with this system [27]. Three different compositions of wastewater were used as substrate. The original wastewater, half-diluted wastewater and centrifuged wastewater were used as substrate in MFCs. Maximum voltage output of 762 mV and maximal power density of 382.5 mW/m^2 (5.06 W/m^3) were obtained with the original wastewater by the 14th day of operation.

Hazardous Hg^{2+} can be removed from industrial waste effluent as an electron acceptor of a microbial fuel cell (MFC) [213]. A maximum power density of 433.1 mW/m^2 was achieved from 100 mg/L Hg^{2+} at pH 2 during this removal process. The higher the concentration of Hg^{2+} in the effluent the higher the power density obtained. Still now wastewater treatment in MFC is hindered by the prohibitive cost of cathode material, especially when platinum is used to catalyze oxygen reduction. Recycled scrap metals could be used efficiently as cathode material in a specially-designed MFC [214]. In terms of cost and long term stability, Inconel 718 was the preferred choice. Operating the MFC in full-loop mode option allowed reaching 99.7% acetate removal while generating a maximum power of 36 $W\ m^{-3}$ at an acetate concentration of 2535 mg L^{-1}. Under these conditions, the energy produced by the system averaged 0.1 kWh m^{-3} of wastewater treated. A tubular air-cathode microbial fuel cell (MFC) stack with high scalability and low material cost was constructed by Zhuang et al [215]. They studied the ability of simultaneous real wastewater treatment and bioelectricity generation under continuous flow mode and found two organic loading rates (ORLs) tested (1.2 and 4.9 kg COD/m^3 d), five non-Pt MFCs connected in series and parallel circuit modes treating swine wastewater can enable an increase of the voltage and the current. A novel microbial fuel cell (MFC) and membrane bioreactor (MBR) integrated system, which combines the advantages of both the systems, was proposed for simultaneous wastewater treatment and energy recovery [216]. The system favored a better utilization of the oxygen in the aeration tank of MBR by the MFC biocathode, and enabled a high effluent quality. Continuous and stable electricity generation, with the average current of 1.9 ± 0.4 mA, was achieved with a maximum power density of 6.0 W m^{-3}. Another new hybrid, air-biocathode microbial fuel cell-membrane bioreactor (MFC-MBR) system was developed to achieve combined advantages of wastewater treatment and ultrafiltration to produce water for direct reclamation [25]. This biocathode system could remove 97% of the soluble chemical oxygen demand, 97% NH$_3$–N, and 91% of total bacteria (based on flow cytometry).

7.5. Biosensor and Bioelectronics

A potential application of the MFC technology is to use it as a sensor for pollutant analysis and *in situ* process monitoring and control [164, 217-227].The proportional correlation between the coulombic yield of MFCs and the strength of the wastewater makes MFCs possible to act as biological oxygen demand (BOD) sensors [228]. Chang et al. and Kim et al. showed a good linear relationship between the coulombic yield and the strength of the wastewater in a quite wide BOD concentration range which is the accurate method to measure the BOD value of a liquid [226, 228]. However, a high BOD concentration requires a long response time because the coulombic yield can be calculated only after the BOD has been depleted unless a dilution mechanism is in place. Several attempts have been taken to improve the dynamic responses in MFCs [229]. A submersible microbial fuel cell (SUMFC) type sensor has been developed for *in situ* monitoring of microbial activity and biochemical oxygen demand (BOD) in groundwater [217]. The microbial activity and BOD content of contaminated groundwater could be successfully detected in less than 3.1 h by using biofilm-colonized anode which was fitted well with the one measured by the standard methods, with deviations ranging from 15% to 22% and 6% to 16%, respectively. A low BOD sensor monitoring mode can be applied to real-time BOD determinations for either surface water, secondary effluents or diluted high BOD wastewater samples [230]. Excellent operational stability and good reproducibility and accuracy of MFC-type of BOD sensors made them advantageous over other types of BOD sensor. An MFC-type BOD sensor constructed with the microbes enriched with MFC can be kept operational for over 5 years without extra maintenance, far longer in service lifespan than other types of BOD sensors reported in the literature [228].

A novel silicon-based MFC is used to detect toxic matters in industrial effluent. This device is capable of detecting the variation on the current produced by the cell when toxic compounds are present in the medium [218]. A proton exchange membrane placed between two micro-fabricated silicon plates is used to act as current collectors. The device has been tested as a toxicity sensor by setting it at a fixed current while monitoring changes in the output power. A drop in the power production is observed when a toxic compound is added to the anode compartment. The compact design of the device makes it suitable for its incorporation into measurement equipment either as an individual device or as an array of sensors for high throughput processing. Stein et al. investigated the influence of the magnitude of the external resistance on the sensitivity and recovery time of the MFC-based biosensor to detect toxicity and monitor quality of drinking water [231]. They concluded that a low resistance resulted in a large change in signal and a more sensitive sensor, while a high resistance resulted in a shorter recovery time. The recovery times of the sensors under both anode potential control and current control were found to be longer than when an external resistor was used. This type of biosensor should be operated at controlled anode potential, controlled pH and saturated substrate concentrations to reach a stable baseline current under nontoxic conditions [219]. Currently available models for describing polarization curves of MFC based biosensors, cannot describe the effect of the presence of toxic components. A bioelectrochemical model combined with enzyme inhibition kinetics, that describes the polarization curve of an MFC-based biosensor, was modified to describe four types of toxicity [232]. Simulations the four modified models were performed to predict the overpotential that gave the most sensitive sensor. A stable and sensitive sensor has been established by controlling the overpotential. These simulations were based on data and parameter values from experimental results under non-toxic conditions. The parameter values from experimental results, controlling the overpotential at 250 mV leads to a sensor that is most sensitive to components that influence the whole bacterial metabolism or that influence the substrate affinity constant.

The sensitivity of biosensor to detect nickel in drinking water was investigated [222]. There was no delay in the response of the sensor and the sensitivity was $0.0027 \text{ A/m}^2/\text{mg}$ Ni/l at an anode potential of -0.4 V. The effect of four types of ion exchange membranes (cation exchange, anion exchange, monovalent cation exchange and bipolar membranes) on the sensitivity was not significant.

The toxicity shocks (sudden change in toxins concentration) of acute-toxic heavy metal (chromium), low-toxic metal (iron), common nutrient (nitrate), and organic contaminant (sodium acetate) in wastewater were examined in a novel self-sustained single chamber batch-mode cube microbial fuel cell (CMFC) biosensor [220]. The results showed that the CMFC was able to distinguish shocks of toxins from non-toxins based on voltage signal changes. A (MFC)-based biosensor was recently developed for the detection of toxic components in water [221]. In this biosensor, substrate consumption rate and metabolic activity of bacteria are directly related to the electric current. Analysis shows that a weighted least squares method is necessary to secure a good fit at the overpotentials where current is most sensitive to changes in kinetic parameters. A protocol for on-line detection of toxicity and for detection of the type of kinetic inhibition is provided by the team.

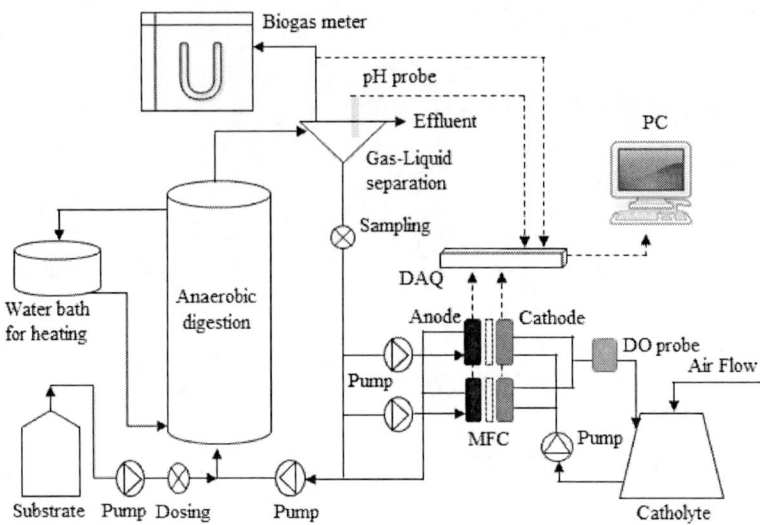

Figure 5. Scheme of anaerobic biodegradation system integrated with MFC based sensor (modified drawing from Liu et al. [238])

Systems based on the microorganism as sensors for quantifying the biological oxygen demand in sewage [227-229, 233, 234] which can readily be expanded to detect other compounds those act as electron donors for electricity production, such as hydrogen [235] or aromatic contaminants [236]. A BOD biosensor, based on the microbial fuel cell principle, was tested for online and *in situ* monitoring of biodegradable organic content of domestic wastewater and a stable current density 282 ± 23 mA/m^2 was obtained with domestic wastewater containing a BOD$_5$ of 317 ± 15 mg O$_2$/L at 22 ± 2°C, 1.53 ± 0.04 mS/cm and pH 6.9 ± 0.1 [223]. The current generation from the BOD biosensor was dependent on the measurement conditions such as temperature, conductivity and pH. Thus, a correction factor should be applied to the measurements done under different environmental conditions from the one used in the calibration. Three important organic pollutants: aldicarb, dimethyl-methylphosphonate (DMMP), and bisphenol-A (BPA) can be detected by artificial neural network (ANN) processing with microbial fuel cell (MFC)-based biosensor [225]. MFCs can also possibly be used to power sensor devices, particularly in river and deep-water environments where it is difficult to routinely access the system to replace batteries. Sediment fuel cells are being developed to monitor environmental systems such as creeks, rivers, and the ocean [177, 237]. Power for these devices can be provided by organic matter in the sediments although power densities obtained in sediment fuel cells is very low because of both the low organic matter concentrations and their high intrinsic internal resistance. Systems developed to date are limited to producing less than 30 mW/m^2 [177]. However, the low power density can be offset by energy storage systems that release data in bursts to central sensors [191]. MFC based biosensor array can also be capable of measuring individual acetate, propionate and butyrate concentrations with sensitivity down to 5 mg l^{-1} and up to 40 mg l^{-1} to monitor bioproceses such as an aerobic digestion [224] A biofilm based wall-jet MFC biosensor has been developed with a character of being portable, short hydraulic retention time (HRT) for sample flow through and convenient for continuous operation which can be used to monitor anaerobic digestion (AD) (Figure 5) [238]. It was an upflow anaerobic fixed-bed (UAFB) reactor fed with OECD wastewater, which was equipped with data acquisition hardware, data record and analysis software and sensor units for real time on-line monitoring, containing electrical signal via MFC based sensor, biogas signal via in-house developed gas flow meter and proton signal via commercial available pH meter. This research suggest that the MFC signal can reflect the dynamic variation of AD and can potentially be a valuable tool for monitoring and control of bioprocess.

7.6. Biorecovery

An MFC can be modified in interesting and useful ways, and this can lead to new types of fuel-cell-based technologies. One such application is the modification of the basic two-electrode system for bioremediation. The MFC is not used to produce electricity; instead, power can be put into the system to drive desired reactions to remove or degrade chemicals, such as converting soluble U(VI) to insoluble U(IV)[239]. Bacteria are not only able to donate electrons to an electrode but can also accept electrons from the cathode. Fe(II) in acid mine drainage (AMD) reacts with dissolved oxygen to produce iron oxide precipitates, resulting in further acidification, discoloration of stream beds, and sludge deposits in receiving waters. It is possible to use AMD types of soluble iron solutions, an important contributor to surface water pollution, in microbial fuel cell-based technologies to create spherical nano-particles of iron oxide (ferrihydrite) that are transformed to goethite(a-FeOOH) upon drying as well as to generate electricity [240]. Particle diameters were controlled to be in the range of 120-700 nm by varying the conditions in the fuel cell, such as current density. These results provided a method to produce iron oxide particles that can be used in pigments and other products. Gregory et al. were able to precipitate uranium directly onto a cathode because of bacterial reduction [239]. Nitrate can also be converted to nitrite when electrodes are used as electron donors [141]. Electrolytic cultivation has been used to extend the growth rates of suspensions of iron-oxidizing bacteria in the laboratory [241]. In a recent work, a microbial fuel cell system has been integrated with supported liquid membrane (SLM) technology to simultaneously treat organic- and heavy metal containing wastewaters for simultaneous removal of organic and zinc contamination in parallel effluent streams [242]. The MFC/SLM combination produces a synergistic effect which enhances the power performance of the MFC significantly. The change in the substrate removal efficiency and coulombic efficiency (CE) compared to the controls is small. The system demonstrates that within 72 h, 93 ± 4% of the zinc ions are removed from the feed phase, hence the Zn^{2+} removal rate is not significantly affected and is comparable to the SLM control (96 ± 1%), while MFC power output is significantly increased.

7.7. Desalination

Most of the water desalination techniques (reverse osmosis, ultrafiltration, ion exchange, electrodialysis etc) are cost effective [243]. An integrated desalination with wastewater treatment and electricity production in a microbial desalination cell (MDC) was studied by utilizing the mutual benefits among the above functions [244]. When using wastewater as the sole substrate, the power output from the MDC (8.01 W/m^3) was four times higher than a control MFC without desalination function. In addition, the MDC removed 66% of the salts and improved COD removal by 52% and coulombic efficiency by 131%. Desalination in MDCs improved wastewater characteristics by increasing the conductivity by 2.5 times and stabilizing anolyte pH, which therefore reduced system resistance and maintained microbial activity.

A recirculation microbial desalination cell (rMDC) was designed and operated to allow recirculation of an solutions between the anode and cathode chambers [245]. This recirculation avoided pH imbalances that could inhibit bacterial metabolism. The salt solution (20 g/L NaCl) was reduced in salinity by 34 ± 1% (50 mM) and 37 ± 2%(25 mM) with recirculation (rMDC), and by 39 ± 1% (50 mM) and 25 ± 3% (25 mM) without recirculation (MDC) as well as generating maximum power density of 931 ± 29 mW/m^2 with a 50 mM phosphate buffer solution (PBS) and 776 ± 30 mW/m^2 with 25 mM PBS. A continuously operated microbial desalination cells (MDCs) has been developed for the purpose of salt removal [246]. A high charge transfer efficiency of 98.6% or 81% was achieved at hydraulic retention time (HRT) 1 or 4 d. These results demonstrated the potential of the MDC as either a sole desalination process or a pre-desalination reactor for downstream desalination processes. Another new technology (CDI-MFC) that combined capacitive deionization (CDI) and microbial fuel cell (MFC) was developed to treat low-concentration salt water with NaCl concentration of 60 mg/L [247]. The water desalination rate was 35.6 mg/(L h), meanwhile the charge efficiency was 21.8%. Water desalination was accomplished in a three-chamber MFC with two membranes (an AEM placed adjacent to the anode and a CEM positioned next to the cathode). When current was produced by bacteria at the anode, ionic species in the middle chamber were transferred into the two electrode chambers, thereby desalinating the water in the middle chamber [16]. This new method of desalination was able to remove 90% of the salt content without pressurizing or use of an external power source.

8. CURRENT DIRECTIONS OF MICROBIAL FUEL CELL RESEARCH

The current worldwide energy crisis requires concerted efforts from researchers to search for all possible energy solutions. MFCs can potentially be an attractive part of bioenergy because they can utilize low-grade organic carbons in wastewater [248]. But this technology has to compete with the mature methanogenic anaerobic digestion technology that has seen wide commercial applications because they can utilize the same biomass in many cases for energy productions [249]. The power density output of an MFC does not reach that of a chemical fuel cell because the latter uses energy intensive fuels such as hydrogen and methanol, while MFCs typically use low grade organic matter in wastewater. So in term of energy generation MFC should be a prominent sector. The fuel in an MFC is often a rather dilute biomass (as in wastewater treatment) in the anodic chamber that has a limited energy (reflected by its BOD). Another limitation is the inherent naturally low catalytic rate of the microbes. Even at their fastest growth rate microbes are relatively slow transformers. Although coulombic efficiency over 90% has been achieved in some cases, it has little effect on the crucial problem of low reaction rate. Although some basic knowledge has been gained in MFC research, there is still a lot to be learned in the scale up of MFC for large-scale applications.

By using MFCs for wastewater treatment, a significant energy saving may be achieved. A major disadvantage of MFCs is their reliance on biofilms for mediator-less electron transport, while anaerobic digesters such as up-flow anaerobic sludge blanket reactors eliminate this need by efficiently reusing the microbial consortium without cell immobilization [250]. New developments in MFC research have found more uses of MFCs in the form of MECs for production of biomaterials apart from biohydrogen. The various MFC reactor types and operating conditions reviewed in this work were aimed at enhancing MFC performance while lowering costs. Super-bug biofilm consortia, engineered through mutation or genetic engineering, increase the possibilities of practical MFC deployment beyond powering small sensors. It is likely that any practical deployment of MFCs for locally distributed power generation or wastewater treatment will be membrane-less, because a membrane poses a major mass transfer resistance and a significant cost factor in reactor design and maintenance. Despite the major advances made in the past decade, MFCs and MECs still face considerable challenges for large-scale real-world applications.

MFCs are capable of converting biomass and with low substrate concentrations, both of which are problematic for methanogenic digesters [250]. To improve the power density output, new anodophilic microbes that vastly improve the electron transport rate from the biofilm covering an anode to the anode are much needed [251]. Mutagenesis and even recombinant DNA technology can conceivably be used in the future to obtain some "super bugs" for MFCs. Microbes may be used as a pure culture or a mixed culture forming a synergistic microbial consortium to offer better performance. One type of bacterium in a consortium may provide electron mediators that are used by another type of bacterium to transport electrons more efficiently to an anode [252]. It is possible in the future that an optimized microbial consortium can be obtained to operate an MFC without extraneous mediators or biofilms while achieving superior mass transfer and electron transfer rates. As aforementioned, MFCs can potentially be used for different applications. When used in wastewater treatment, a large surface area is needed for biofilm to build up on the anode. A breakthrough is needed in creating inexpensive electrodes that resist fouling. It is unrealistic to expect that the power density output from an MFC to match that of conventional chemical fuel cell such as a hydrogen-powered fuel cell.

9. ENVIRONMENTAL ASPECTS

The production of different types of products and processes such as dyes, explosives, pesticides, paper printing, color photography, pharmaceuticals, cosmetics, and leather goods are not readily degradable under natural conditions and are typically not removed from wastewater by conventional physicochemical methods, such as adsorption, coagulation–flocculation, and especially advanced oxidation as they are energy and cost intensive. MFC technology can be considered as a reliable alternative to these processes. It has been demonstrated that microorganisms can degrade many types of toxic and refractory compounds, such as phenols, indoles, azo dyes, and halogenated compounds, and generate electricity simultaneously [53]. Different investigations found positive results to degrade furfural and acid orange [211], azo dye [253], nitrobenzene [254], antibiotic (ceftriaxone) [255], 1,2-dichloroethane [256], Biogenic palladium nanoparticles [84].

MFC is also used to remove heavy metals, toxic metals, salts and different organic matters from waste effluent and contributes to mitigate the environment pollution [213, 240, 242, 244, 246]. Heavy metals that can be removed by MFC include- copper, nickel, cobalt, zinc, uranium and mercury from mining and metallurgical wastewaters [213, 239, 257-259]. Moreover MFCs can be used potentially to mitigate greenhouse gas [260].

Most of the BOD and COD present in the effluent can be efficiently removed by MFC while generating useful electricity or other energies (most of them has been discussed in section 7.4). Generally, for wastewater treatment, COD removal is the prime parameter of importance. This COD removal can be expressed as columbic efficiency. However, the added value of the produced power can be as important in terms of economic feasibility. This implies that the resistance over the microbial fuel cell cannot be too high, neither too low.

10. LIMITATIONS OF MICROBIAL FUEL CELLS

Microbial fuel cells (MFCs) are a promising technology for sustainable production of alternative energy and waste treatment. But low power output hinders its broad use. The performance of microbial fuel cells is severely affected by limitations based on irreversible reactions and processes in the anode and the cathode compartments [261]. The three major irreversibilities that affect MFC performance are- activation losses, ohmic losses, and mass transport losses. These losses are defined as the voltage required to compensate for the current lost due to electrochemical reactions, charge transport, and mass transfer processes that take place in both the anode and cathode compartments [262]. Poor kinetics of oxygen reduction at neutral pH and low temperatures hinder the improvement of MFC performance [204, 263]. Another limitation of the MFC system is the reduction of molecular oxygen by the cathode. Various metals have typically been used to catalyze the cathodic reaction but reduction of oxygen at the cathode is currently an important limiting factor in a MFC [76, 139]. To reduce these cathodic limitations researchers have increased the cathode to anode ratio [264] and used biological catalysts [112]. Moreover, Pt which is an effective cathode in MFC may concern with environmental aspects.

Applying microorganisms in a technology brings some biological limitations also. First, proper physical parameters such as pH (around neutral) and temperature (15–30°C) need to be maintained for good microbial metabolism although some bacteria are able to function under slightly different conditions. For instance, the microbial reductive dechlorination of perchlorate was the highest at pH 8.5 [149]. Shear forces also seem to be an important physical parameter as they influence the biofilm thickness and density, and also the power output [150]. Biofilm thickness, structure, composition, and density affect the flux of substrates and products within the biofilm. The latter can result in large overpotentials, which have a negative impact on the performance of the system. In the case of bioanodes, higher power production was observed from thicker anodic biofilms [159]. Strikingly, the reverse effect has been observed in cathodic biofilms [94]. Finally, even if the microorganisms are not limited by their environment, there are some intrinsic limitations associated with the microbial metabolism such as their growth rate, uptake rate of nutrients, and electron transfer capacities. Certainly, cathodic communities seem to encounter harsh conditions as they have to invest energy for CO_2 fixation in their biomass (autotrophic microorganisms), and have to be able to take up electrons from the cathode at a high potential.

11. PROSPECTS FOR TECHNOLOGY IMPROVEMENT AND INNOVATION

Researchers in this field have overcome many of the limitations for real-life applications but, there are many others that are still to be overcome. Improving the stability of the biocatalyst, overpotential at cathode surface, chemical and mechanical stability of the electrode materials are some of the challenges that should be given emphasize in future research. Current state of the art and the rate of development in power output and energy production warrants a thorough rethinking of the applied value and niches for MFC systems in practice. Therefore, it is argued that the energy concept is beyond the reach of current possibilities. Notwithstanding the great advances that have been made in the past and will be made in the coming decades in MFC and extracellular electron transport research, the current research should be guided along two distinct paths. On the one hand, there is the need for more fundamental knowledge which can be gained by an in-depth study of the mechanisms and catalysts involved and warranted the use of small (ml) scale and defined reactor setups. Bioelectrochemical systems should be developed and integrated in sustainable green technology, i.e., that not necessarily produces hard currency but renders less tangible benefits in terms of improved environmental quality [4].

The MFCs, currently in existence, are exciting systems for studying microbial communities and improving the understanding of how bacteria transfer electrons to solid substrates. Harnessing that power in an economical manner, however, remains a greater challenge. It is generally considered that the first applications of MFCs will be as power sources for monitoring devices in the environment and for water treatment [265]. Several tests of large systems have already shown sufficient power production by sediment MFCs for long-term, unattended power sources for data-collecting devices [115, 177, 237].The replacement of electricity-consuming wastewater-treatment bioreactors with electricity-producing MFC-based systems shows great promise because of the high cost of operating existing systems. The cost of the materials for construction of MFCs has been brought up in numerous articles but an overall economic analysis is premature at this stage because the designs and choice of materials continue to rapidly improve. Expensive materials (e.g., Pt and Au), although not feasible in large-scale, may give useful insight into reaction mechanisms. The suitability of less expensive materials has not been unequivocally examined [261]. So, there are scope to study intensively to minimize the cost of the MFC's materials. Using MFCs as the treatment reactor makes good economic sense because the total infrastructure needed for power generation does not have to pay for itself on the electricity produced alone only the cost of replacing the existing reactor with an MFC. Accomplishing treatment at a lower overall cost by producing a valuable product (i.e., electricity) saves money and, therefore, makes the process more economical. Renewable electricity production using MFC-based technologies seems further off in the future but, with advances, MFCs might become practical as a method for producing a mobile fuel from renewable biomass sources. Ethanol is currently made from refined sugar, which limits its usefulness and economic potential, although it is hoped that one day cellulosic materials could be used for ethanol production [266, 267].With further development, MFCs or modified MFCs could become practical methods for hydrogen production from the same materials envisioned for ethanol production (sugar and cellulosic wastes) and from other biodegradable materials. Either the electricity produced by MFCs can be used for the electrolysis of water or, more efficiently, hydrogen could be produced directly from biomass sources using the BEAMR process. Any material that produces electricity in the MFC process can be used to produce hydrogen in the BEAMR process, although hydrogen yields cannot yet be reliably predicted and must be measured. Taken together, these biotechnology-based MFC approaches hold great promise as methods for renewable energy production. New methods can be developed to extract heavy metals more efficiently from hazardous effluents. Development in biorecovery sector may add extra dimension in potential use of MFCs.

CONCLUSION

MFCs are a promising technology for the production of electricity from organic material and wastes. Currently limited applications are possible because of low MFC power output. But if we consider the energy balance and compare the energy production and consumption in an MFC, it might open the door to a new method for renewable and sustainable energy production. This is based on the assumption that electric energy is the only energy produced in an MFC; in fact, when MFCs treat high-strength wastes, hydrogen, methane etc. can be produced and should be included in the energy production, and if other energy contents such as algal biomass are coproduced, they should be included, as well. Wastewater treatment or bioremediation are some current applications of MFCs which may be much more promising than the electrical production of the MFC itself. Moreover, if the two major problems, i.e., proton accumulation within the biofilm and over potential at the cathode in power generation from MFC are addressed properly and remedial actions can be developed then it will open the door on a new world of exploration and insight into complex microbial communities and the evolution of life.

REFERENCES

[1]   Andujar, J.M. and F. Segura, Fuel cells: History and updating. A walk along two centuries. *Renewable and Sustainable Energy Reviews,* 2009. *13*(9): p. 2309-2322.

[2]   Ren, Z., T.E. Ward, and J.M. Regan, Electricity production from cellulose in a microbial fuel cell using a defined binary culture. *Environmental science & technology,* 2007. *41*(13): p. 4781-4786.

[3]   Logan, B.E., et al., Microbial fuel cells: methodology and technology. *Environmental science & technology,* 2006. *40*(17): p. 5181-5192.

[4]   Arends, J. and W. Verstraete, 100 years of microbial electricity production: three concepts for the future. *Microbial biotechnology. 5*(3): p. 333-346.

[5]   Du, Z., H. Li, and T. Gu, A state of the art review on microbial fuel cells: a promising technology for wastewater treatment and bioenergy. *Biotechnology advances,* 2007. *25*(5): p. 464-482.

[6]   Logan, B.E., Scaling up microbial fuel cells and other bioelectrochemical systems. *Applied microbiology and biotechnology. 85*(6): p. 1665-1671.

[7]   Rosenbaum, M., Z. He, and L.T. Angenent, Light energy to bioelectricity: photosynthetic microbial fuel cells. *Current opinion in biotechnology. 21*(3): p. 259-264.

[8]   Villasenor, J., et al., Operation of a horizontal subsurface flow constructed wetland-Microbial fuel cell treating wastewater under different organic loading rates. *Water research. 47*(17): p. 6731-6738.

[9]   Dutta, K. and P.P. Kundu, A review on aromatic conducting polymers-based catalyst supporting matrices for application in microbial fuel cells. *Polymer Reviews. 54*(3): p. 401-435.

[10]   Oliveira, V.B., et al., Overview on the developments of microbial fuel cells. *Biochemical Engineering Journal. 73*: p. 53-64.

[11]   Ghasemi, M., et al., Nano-structured carbon as electrode material in microbial fuel cells: a comprehensive review. *Journal of Alloys and Compounds. 580*: p. 245-255.

[12]   Mao, L. and W.S. Verwoerd, Selection of organisms for systems biology study of microbial electricity generation: a review. *International Journal of Energy and Environmental Engineering. 4*(1): p. 1-18.

[13]   Peralta-Yahya, P.P., et al., Microbial engineering for the production of advanced biofuels. *Nature. 488*(7411): p. 320-328.

[14]   Kim, B.H., I.S. Chang, and G.M. Gadd, Challenges in microbial fuel cell development and operation. *Applied microbiology and biotechnology,* 2007. *76*(3): p. 485-494.

[15]   Calabrese Barton, S., J. Gallaway, and P. Atanassov, Enzymatic biofuel cells for implantable and microscale devices. *Chemical Reviews,* 2004. *104*(10): p. 4867-4886.

[16]   Pant, D., et al., A review of the substrates used in microbial fuel cells (MFCs) for sustainable energy production. *Bioresource technology. 101*(6): p. 1533-1543.

[17]   Huang, L., J.M. Regan, and X. Quan, Electron transfer mechanisms, new applications, and performance of biocathode microbial fuel cells. *Bioresource technology. 102*(1): p. 316-323.

[18]   Wei, J., P. Liang, and X. Huang, Recent progress in electrodes for microbial fuel cells. *Bioresource technology. 102*(20): p. 9335-9344.

[19] Logan, B.E., Essential data and techniques for conducting microbial fuel cell and other types of bioelectrochemical system experiments. *ChemSusChem.* 5(6): p. 988-994.

[20] Lovley, D.R. and K.P. Nevin, A shift in the current: new applications and concepts for microbe-electrode electron exchange. *Current opinion in biotechnology.* 22(3): p. 441-448.

[21] Wang, H.-Y., et al., Micro-sized microbial fuel cell: a mini-review. *Bioresource technology.* 102(1): p. 235-243.

[22] Ghasemi, M., et al., New generation of carbon nanocomposite proton exchange membranes in microbial fuel cell systems. *Chemical Engineering Journal.* 184: p. 82-89.

[23] Qian, F., et al., A microfluidic microbial fuel cell fabricated by soft lithography. *Bioresource technology.* 102(10): p. 5836-5840.

[24] Ahmad, F., et al., A review of cellulosic microbial fuel cells: performance and challenges. *Biomass and Bioenergy.* 56: p. 179-188.

[25] Malaeb, L., et al., A Hybrid Microbial Fuel Cell Membrane Bioreactor with a Conductive Ultrafiltration Membrane Biocathode for Wastewater Treatment. *Environmental science & technology.* 47(20): p. 11821-11828.

[26] Lovley, D.R. and K.P. Nevin, Electrobiocommodities: powering microbial production of fuels and commodity chemicals from carbon dioxide with electricity. *Current opinion in biotechnology.* 24(3): p. 385-390.

[27] Sevda, S., et al., High strength wastewater treatment accompanied by power generation using air cathode microbial fuel cell. *Applied Energy.* 105: p. 194-206.

[28] Solanki, K., S. Subramanian, and S. Basu, Microbial fuel cells for azo dye treatment with electricity generation: a review. *Bioresource technology.* 131: p. 564-571.

[29] Mekhilef, S., R. Saidur, and A. Safari, Comparative study of different fuel cell technologies. *Renewable and Sustainable Energy Reviews.* 16(1): p. 981-989.

[30] Grimes, P. Historical pathways for fuel cells. The new electric century. in Battery Conference on Applications and Advances, 2000. *The Fifteenth Annual.* 2000: IEEE.

[31] Stambouli, A.B. and E. Traversa, Solid oxide fuel cells (SOFCs): a review of an environmentally clean and efficient source of energy. *Renewable and Sustainable Energy Reviews,* 2002. 6(5): p. 433-455.

[32] Appleby, A.J., From Sir William Grove to today: fuel cells and the future. *Journal of power sources,* 1990. 29(1): p. 3-11.

[33] Stone, C. and A.E. Morrison, From curiosity to power to change the world. *Solid State Ionics,* 2002. 152: p. 1-13.

[34] Selman, J.R., Molten-salt fuel cells-technical and economic challenges. *Journal of power sources,* 2006. 160(2): p. 852-857.

[35] Dicks, A.L., Molten carbonate fuel cells. *Current Opinion in Solid State and Materials Science,* 2004. *8*(5): p. 379-383.

[36] Weissbart, J. and R. Ruka, A solid electrolyte fuel cell. *Journal of the Electrochemical Society,* 1962. *109*(8): p. 723-726.

[37] Wee, J.-H., Applications of proton exchange membrane fuel cell systems. *Renewable and Sustainable Energy Reviews,* 2007. *11*(8): p. 1720-1738.

[38] Cook, B., Introduction to fuel cells and hydrogen technology. *Engineering Science & Education Journal,* 2002. *11*(6): p. 205-216.

[39] Ormerod, R.M., Solid oxide fuel cells. *Chemical Society Reviews,* 2003. *32*(1): p. 17-28.

[40] Chen, F., et al., Investigation of challenges to the utilization of fuel cell buses in the EU vs transition economies. *Renewable and Sustainable Energy Reviews,* 2007. *11*(2): p. 357-364.

[41] Andujar, J.M., F. Segura, and M.J. Vasallo, A suitable model plant for control of the set fuel cell-DC/DC converter. *Renewable Energy,* 2008. *33*(4): p. 813-826.

[42] Miller, A.R., et al., System design of a large fuel cell hybrid locomotive. *Journal of power sources,* 2007. *173*(2): p. 935-942.

[43] Khaligh, A., et al., Digital control of an isolated active hybrid fuel cell/Li-ion battery power supply. *Vehicular Technology, IEEE Transactions on,* 2007. *56*(6): p. 3709-3721.

[44] Segura, F., E. Duran, and J.M. Andujar, Design, building and testing of a stand alone fuel cell hybrid system. Journal of power sources, 2009. *193*(1): p. 276-284.

[45] Stambouli, A.B. and E. Traversa, Fuel cells, an alternative to standard sources of energy. *Renewable and Sustainable Energy Reviews,* 2002. *6*(3): p. 295-304.

[46] Spiegel, R.J. and J.L. Preston, Technical assessment of fuel cell operation on anaerobic digester gas at the Yonkers, NY, wastewater treatment plant. *Waste Management,* 2003. *23*(8): p. 709-717.

[47] Rittmann, B.E., Opportunities for renewable bioenergy using microorganisms. *Biotechnology and bioengineering,* 2008. *100*(2): p. 203-212.

[48] Potter, M.C., Electrical effects accompanying the decomposition of organic compounds. *Proceedings of the Royal Society of London. Series B, Containing Papers of a Biological Character,* 1911: p. 260-276.

[49] Cohen, B., Thirty-second Annual Meeting of the Society of American Bacteriologists. *J Bacteriol,* 1931. *21*: p. 1-60.

[50] Davis, J.B. and H.F. Yarbrough, Preliminary experiments on a microbial fuel cell. *Science,* 1962. *137*(3530): p. 615-616.

[51] Lewis, K., Symposium on bioelectrochemistry of microorganisms. IV. Biochemical fuel cells. *Bacteriological reviews,* 1966. *30*(1): p. 101.

[52] Allen, R.M. and H.P. Bennetto, Microbial fuel-cells. *Applied Biochemistry and Biotechnology,* 1993. *39*(1): p. 27-40.

[53]   Arends, J., et al., Principles and Technology of Microbial Fuel Cells. *Fuel Cell Science and Engineering: Materials, Processes, Systems and Technology:* p. 147-184.

[54]   Yoshikawa, K., et al., Reconstruction and verification of a genome-scale metabolic model for Synechocystis sp. PCC6803. *Applied microbiology and biotechnology.* *92*(2): p. 347-358.

[55]   Delaney, G.M., et al., Electron-transfer coupling in microbial fuel cells. 2. performance of fuel cells containing selected microorganism-mediator-substrate combinations. *Journal of chemical technology and biotechnology. Biotechnology,* 1984. *34*(1): p. 13-27.

[56]   Kim, H.J., et al., A mediator-less microbial fuel cell using a metal reducing bacterium, Shewanella putrefaciens. *Enzyme and Microbial Technology,* 2002. *30*(2): p. 145-152.

[57]   Pham, C.A., et al., A novel electrochemically active and Fe (III)-reducing bacterium phylogenetically related to Aeromonas hydrophila, isolated from a microbial fuel cell. *FEMS Microbiology Letters,* 2003. *223*(1): p. 129-134.

[58]   Strik, D.P., J.F.H. Snel, and C.J.N. Buisman, Green electricity production with living plants and bacteria in a fuel cell. *International Journal of Energy Research,* 2008. *32*(9): p. 870-876.

[59]   Strik, D.P., et al., Microbial solar cells: applying photosynthetic and electrochemically active organisms. *TRENDS in Biotechnology.* *29*(1): p. 41-49.

[60]   Bombelli, P., et al., Quantitative analysis of the factors limiting solar power transduction by Synechocystis sp. PCC 6803 in biological photovoltaic devices. *Energy & Environmental Science.* *4*(11): p. 4690-4698.

[61]   Mahesh, S., T. Desalegn, and M. Alemayehu, Evaluation of Photosynthetic Microbial Fuel Cell for Bioelectricity Production. *Indian Journal of Energy.* *2*(4): p. 116-120.

[62]   Lam, K.B., et al., A MEMS photosynthetic electrochemical cell powered by subcellular plant photosystems. *Microelectromechanical Systems, Journal of,* 2006. *15*(5): p. 1243-1250.

[63]   He, Z., et al., Self-sustained phototrophic microbial fuel cells based on the synergistic cooperation between photosynthetic microorganisms and heterotrophic bacteria. *Environmental science & technology,* 2009. *43*(5): p. 1648-1654.

[64]   Zou, Y., et al., Photosynthetic microbial fuel cells with positive light response. *Biotechnology and bioengineering,* 2009. *104*(5): p. 939-946.

[65]   Nishio, K., K. Hashimoto, and K. Watanabe, Light/electricity conversion by a self-organized photosynthetic biofilm in a single-chamber reactor. *Applied microbiology and biotechnology.* *86*(3): p. 957-964.

[66]   Velasquez-Orta, S.B., T.P. Curtis, and B.E. Logan, Energy from algae using microbial fuel cells. *Biotechnology and bioengineering*, 2009. *103*(6): p. 1068-1076.

[67]   Chiao, M., K.B. Lam, and L. Lin, Micromachined microbial and photosynthetic fuel cells. *Journal of Micromechanics and Microengineering*, 2006. *16*(12): p. 2547.

[68]   Zhou, M., et al., An overview of electrode materials in microbial fuel cells. *Journal of Power Sources. 196*(10): p. 4427-4435.

[69]   Logan, B., et al., Graphite fiber brush anodes for increased power production in air-cathode microbial fuel cells. Environmental science & technology, 2007. *41*(9): p. 3341-3346.

[70]   Aelterman, P., et al., Loading rate and external resistance control the electricity generation of microbial fuel cells with different three-dimensional anodes. *Bioresource technology*, 2008. *99*(18): p. 8895-8902.

[71]   ter Heijne, A., et al., Performance of non-porous graphite and titanium-based anodes in microbial fuel cells. *Electrochimica Acta*, 2008. *53*(18): p. 5697-5703.

[72]   Zhao, F., et al., Challenges and constraints of using oxygen cathodes in microbial fuel cells. *Environmental science & technology*, 2006. *40*(17): p. 5193-5199.

[73]   De Schamphelaire, L., P. Boeckx, and W. Verstraete, Evaluation of biocathodes in freshwater and brackish sediment microbial fuel cells. *Applied microbiology and biotechnology. 87*(5): p. 1675-1687.

[74]   Zhu, N., et al., Improved performance of membrane free single-chamber air-cathode microbial fuel cells with nitric acid and ethylenediamine surface modified activated carbon fiber felt anodes. *Bioresource technology. 102*(1): p. 422-426.

[75]   Rosenbaum, M., et al., Interfacing Electrocatalysis and Biocatalysis with Tungsten Carbide: A High-Performance, Noble-Metal-Free Microbial Fuel Cell. *Angewandte Chemie International Edition*, 2006. *45*(40): p. 6658-6661.

[76]   Zhao, F., et al., Application of pyrolysed iron (II) phthalocyanine and CoTMPP based oxygen reduction catalysts as cathode materials in microbial fuel cells. *Electrochemistry Communications*, 2005. *7*(12): p. 1405-1410.

[77]   Park, D.H. and J.G. Zeikus, Improved fuel cell and electrode designs for producing electricity from microbial degradation. *Biotechnology and bioengineering*, 2003. *81*(3): p. 348-355.

[78]   Aelterman, P., et al., Microbial fuel cells operated with iron-chelated air cathodes. *Electrochimica Acta*, 2009. *54*(24): p. 5754-5760.

[79]   Freguia, S., et al., Non-catalyzed cathodic oxygen reduction at graphite granules in microbial fuel cells. *Electrochimica Acta*, 2007. *53*(2): p. 598-603.

[80]   Cheng, S. and B.E. Logan, Ammonia treatment of carbon cloth anodes to enhance power generation of microbial fuel cells. *Electrochemistry Communications*, 2007. *9*(3): p. 492-496.

[81] Scott, K., et al., Fuel cell power generation from marine sediments: Investigation of cathode materials. *Journal of chemical technology and biotechnology,* 2008. *83*(9): p. 1244-1254.

[82] Selembo, P.A., M.D. Merrill, and B.E. Logan, Hydrogen production with nickel powder cathode catalysts in microbial electrolysis cells. *International Journal of Hydrogen Energy.* *35*(2): p. 428-437.

[83] Huang, Y.-X., et al., A new cathodic electrode deposit with palladium nanoparticles for cost-effective hydrogen production in a microbial electrolysis cell. *International Journal of Hydrogen Energy.* *36*(4): p. 2773-2776.

[84] Hennebel, T., et al., Dehalogenation of environmental pollutants in microbial electrolysis cells with biogenic palladium nanoparticles. *Biotechnology letters.* *33*(1): p. 89-95.

[85] Liu, X.-W., et al., Nano-structured manganese oxide as a cathodic catalyst for enhanced oxygen reduction in a microbial fuel cell fed with a synthetic wastewater. *Water research.* *44*(18): p. 5298-5305.

[86] Fan, Y., et al., Nanoparticle decorated anodes for enhanced current generation in microbial electrochemical cells. *Biosensors and Bioelectronics.* *26*(5): p. 1908-1912.

[87] Clauwaert, P. and W. Verstraete, Methanogenesis in membraneless microbial electrolysis cells. *Applied microbiology and biotechnology,* 2009. *82*(5): p. 829-836.

[88] Call, D. and B.E. Logan, Hydrogen production in a single chamber microbial electrolysis cell lacking a membrane. *Environmental science & technology,* 2008. *42*(9): p. 3401-3406.

[89] Rozendal, R.A., et al., Performance of single chamber biocatalyzed electrolysis with different types of ion exchange membranes. *Water research,* 2007. *41*(9): p. 1984-1994.

[90] Sleutels, T.H.J.A., et al., Ion transport resistance in microbial electrolysis cells with anion and cation exchange membranes. *International Journal of Hydrogen Energy,* 2009. *34*(9): p. 3612-3620.

[91] Zhang, X., et al., Separator characteristics for increasing performance of microbial fuel cells. *Environmental science & technology,* 2009. *43*(21): p. 8456-8461.

[92] Biffinger, J.C., et al., Diversifying biological fuel cell designs by use of nanoporous filters. *Environmental science & technology,* 2007. *41*(4): p. 1444-1449.

[93] Harnisch, F., U. Schröder, and F. Scholz, The suitability of monopolar and bipolar ion exchange membranes as separators for biological fuel cells. *Environmental science & technology,* 2008. *42*(5): p. 1740-1746.

[94] Behera, M., P.S. Jana, and M.M. Ghangrekar, Performance evaluation of low cost microbial fuel cell fabricated using earthen pot with biotic and abiotic cathode. *Bioresource technology. 101*(4): p. 1183-1189.

[95] Rosenbaum, M., M.A. Cotta, and L.T. Angenent, Aerated Shewanella oneidensis in continuously fed bioelectrochemical systems for power and hydrogen production. *Biotechnology and bioengineering. 105*(5): p. 880-888.

[96] Rabaey, K., et al., Tubular microbial fuel cells for efficient electricity generation. *Environmental science & technology,* 2005. *39*(20): p. 8077-8082.

[97] He, Z., et al., An upflow microbial fuel cell with an interior cathode: assessment of the internal resistance by impedance spectroscopy. *Environmental science & technology,* 2006. *40*(17): p. 5212-5217.

[98] Ringeisen, B.R., et al., High power density from a miniature microbial fuel cell using Shewanella oneidensis DSP10. *Environmental science & technology,* 2006. *40*(8): p. 2629-2634.

[99] Hou, H., et al., Microfabricated microbial fuel cell arrays reveal electrochemically active microbes. *PLoS One,* 2009. *4*(8): p. e6570.

[100] Chen, Y.-P., et al., An innovative miniature microbial fuel cell fabricated using photolithography. *Biosensors and Bioelectronics. 26*(6): p. 2841-2846.

[101] Moriuchi, T., et al., Development of a flexible direct photosynthetic/metabolic biofuel cell for mobile use. *International Journal of Precision Engineering and Manufacturing,* 2009. *10*(1): p. 75-78.

[102] Zuo, Y., et al., Isolation of the exoelectrogenic bacterium Ochrobactrum anthropi YZ-1 by using a U-tube microbial fuel cell. *Applied and environmental microbiology,* 2008. *74*(10): p. 3130-3137.

[103] Boon, N., et al., Metabolites produced by Pseudomonas sp. enable a Gram-positive bacterium to achieve extracellular electron transfer. *Applied microbiology and biotechnology,* 2008. *77*(5): p. 1119-1129.

[104] Liu, H. and B.E. Logan, Electricity generation using an air-cathode single chamber microbial fuel cell in the presence and absence of a proton exchange membrane. *Environmental science & technology,* 2004. *38*(14): p. 4040-4046.

[105] Zhuang, L., et al., Comparison of membrane-and cloth-cathode assembly for scalable microbial fuel cells: construction, performance and cost. *Process Biochemistry. 45*(6): p. 929-934.

[106] Ieropoulos, I., J. Greenman, and C. Melhuish, Microbial fuel cells based on carbon veil electrodes: stack configuration and scalability. *International Journal of Energy Research,* 2008. *32*(13): p. 1228-1240.

[107] Aelterman, P., et al., Continuous electricity generation at high voltages and currents using stacked microbial fuel cells. *Environmental science & technology,* 2006. *40*(10): p. 3388-3394.

[108] Oh, S.E. and B.E. Logan, Voltage reversal during microbial fuel cell stack operation. *Journal of Power Sources,* 2007. *167*(1): p. 11-17.

[109] Dekker, A., et al., Analysis and improvement of a scaled-up and stacked microbial fuel cell. *Environmental science & technology,* 2009. *43*(23): p. 9038-9042.

[110] Clauwaert, P., et al., Litre-scale microbial fuel cells operated in a complete loop. *Applied microbiology and biotechnology,* 2009. *83*(2): p. 241-247.

[111] Zhang, F., et al., Effects of anolyte recirculation rates and catholytes on electricity generation in a litre-scale upflow microbial fuel cell. *Energy & Environmental Science. 3*(9): p. 1347-1352.

[112] Rabaey, K., et al., Cathodic oxygen reduction catalyzed by bacteria in microbial fuel cells. *The ISME journal,* 2008. *2*(5): p. 519-527.

[113] Rabaey, K. and R.A. Rozendal, Microbial electrosynthesis-revisiting the electrical route for microbial production. *Nature Reviews Microbiology. 8*(10): p. 706-716.

[114] Thrash, J.C. and J.D. Coates, Review: direct and indirect electrical stimulation of microbial metabolism. *Environmental science & technology,* 2008. *42*(11): p. 3921-3931.

[115] Lowy, D.A., et al., Harvesting energy from the marine sedimentâ€"water interface II: kinetic activity of anode materials. *Biosensors and Bioelectronics,* 2006. *21*(11): p. 2058-2063.

[116] Ter Heijne, A., et al., Cathode potential and mass transfer determine performance of oxygen reducing biocathodes in microbial fuel cells. *Environmental science & technology. 44*(18): p. 7151-7156.

[117] Bird, L.J., V. Bonnefoy, and D.K. Newman, Bioenergetic challenges of microbial iron metabolisms. *Trends in microbiology. 19*(7): p. 330-340.

[118] Shi, L., et al., The roles of outer membrane cytochromes of Shewanella and Geobacter in extracellular electron transfer. *Environmental Microbiology Reports,* 2009. *1*(4): p. 220-227.

[119] Rabaey, K., et al., Microbial fuel cells for sulfide removal. *Environmental science & technology,* 2006. *40*(17): p. 5218-5224.

[120] Chae, K.-J., et al., Effect of different substrates on the performance, bacterial diversity, and bacterial viability in microbial fuel cells. *Bioresource technology,* 2009. *100*(14): p. 3518-3525.

[121] De Schamphelaire, L., et al., Microbial community analysis of anodes from sediment microbial fuel cells powered by rhizodeposits of living rice plants. *Applied and environmental microbiology. 76*(6): p. 2002-2008.

[122] Logan, B.E., Exoelectrogenic bacteria that power microbial fuel cells. *Nature Reviews Microbiology,* 2009. *7*(5): p. 375-381.

[123] Lovley, D.R., Bug juice: harvesting electricity with microorganisms. *Nature Reviews Microbiology,* 2006. *4*(7): p. 497-508.

[124] Lee, H.-S., C.s.I. Torres, and B.E. Rittmann, Effects of substrate diffusion and anode potential on kinetic parameters for anode-respiring bacteria. *Environmental science & technology*, 2009. *43*(19): p. 7571-7577.

[125] Milliken, C.E. and H.D. May, Sustained generation of electricity by the spore-forming, Gram-positive, Desulfitobacterium hafniense strain DCB2. *Applied microbiology and biotechnology*, 2007. *73*(5): p. 1180-1189.

[126] Rabaey, K., et al., Microbial phenazine production enhances electron transfer in biofuel cells. *Environmental science & technology*, 2005. *39*(9): p. 3401-3408.

[127] Wrighton, K.C., et al., A novel ecological role of the Firmicutes identified in thermophilic microbial fuel cells. *The ISME journal*, 2008. *2*(11): p. 1146-1156.

[128] Gorby, Y.A., et al., Electrically conductive bacterial nanowires produced by Shewanella oneidensis strain MR-1 and other microorganisms. *Proceedings of the National Academy of Sciences*, 2006. *103*(30): p. 11358-11363.

[129] Leang, C., et al., Alignment of the c-type cytochrome OmcS along pili of Geobacter sulfurreducens. *Applied and environmental microbiology*. *76*(12): p. 4080-4084.

[130] Inoue, K., et al., Specific localization of the c-type cytochrome OmcZ at the anode surface in current-producing biofilms of Geobacter sulfurreducens. *Environmental Microbiology Reports*. *3*(2): p. 211-217.

[131] Richter, H., et al., Cyclic voltammetry of biofilms of wild type and mutant Geobacter sulfurreducens on fuel cell anodes indicates possible roles of OmcB, OmcZ, type IV pili, and protons in extracellular electron transfer. *Energy & Environmental Science*, 2009. *2*(5): p. 506-516.

[132] Reguera, G., et al., Extracellular electron transfer via microbial nanowires. *Nature*, 2005. *435*(7045): p. 1098-1101.

[133] Morita, M., et al., Potential for direct interspecies electron transfer in methanogenic wastewater digester aggregates. *MBio*. *2*(4): p. e00159-11.

[134] Nielsen, L.P., et al., Electric currents couple spatially separated biogeochemical processes in marine sediment. *Nature*. *463*(7284): p. 1071-1074.

[135] Holmes, D.E., D.R. Bond, and D.R. Lovley, Electron transfer by Desulfobulbus propionicus to Fe (III) and graphite electrodes. *Applied and environmental microbiology*, 2004. *70*(2): p. 1234-1237.

[136] Marsili, E., et al., Shewanella secretes flavins that mediate extracellular electron transfer. *Proceedings of the National Academy of Sciences*, 2008. *105*(10): p. 3968-3973.

[137] Dutta, P.K., et al., Role of sulfur during acetate oxidation in biological anodes. *Environmental science & technology*, 2009. *43*(10): p. 3839-3845.

[138] Schröder, U., J. Nießen, and F. Scholz, A generation of microbial fuel cells with current outputs boosted by more than one order of magnitude. *Angewandte Chemie International Edition*, 2003. *42*(25): p. 2880-2883.

[139] Cheng, S., H. Liu, and B.E. Logan, Power densities using different cathode catalysts (Pt and CoTMPP) and polymer binders (Nafion and PTFE) in single chamber microbial fuel cells. *Environmental science & technology,* 2006. *40*(1): p. 364-369.

[140] Rozendal, R.A., et al., Principle and perspectives of hydrogen production through biocatalyzed electrolysis. *International Journal of Hydrogen Energy,* 2006. *31*(12): p. 1632-1640.

[141] Gregory, K.B., D.R. Bond, and D.R. Lovley, Graphite electrodes as electron donors for anaerobic respiration. *Environmental microbiology,* 2004. *6*(6): p. 596-604.

[142] Rosenbaum, M., et al., Cathodes as electron donors for microbial metabolism: which extracellular electron transfer mechanisms are involved? *Bioresource technology. 102*(1): p. 324-333.

[143] Cheng, K.Y., G. Ho, and R. Cord-Ruwisch, Anodophilic biofilm catalyzes cathodic oxygen reduction. *Environmental science & technology,* 2009. *44*(1): p. 518-525.

[144] Rozendal, R.A., et al., Hydrogen production with a microbial biocathode. *Environmental science & technology,* 2007. *42*(2): p. 629-634.

[145] Freguia, S., S. Tsujimura, and K. Kano, Electron transfer pathways in microbial oxygen biocathodes. *Electrochimica Acta. 55*(3): p. 813-818.

[146] Clauwaert, P., et al., Biological denitrification in microbial fuel cells. *Environmental science & technology,* 2007. *41*(9): p. 3354-3360.

[147] Erable, B., et al., Marine aerobic biofilm as biocathode catalyst. *Bioelectrochemistry. 78*(1): p. 51-56.

[148] Cao, X., et al., A completely anoxic microbial fuel cell using a photo-biocathode for cathodic carbon dioxide reduction. *Energy & Environmental Science,* 2009. *2*(5): p. 498-501.

[149] Butler, C.S., et al., Bioelectrochemical perchlorate reduction in a microbial fuel cell. *Environmental science & technology. 44*(12): p. 4685-4691.

[150] Pham, H.T., et al., High shear enrichment improves the performance of the anodophilic microbial consortium in a microbial fuel cell. *Microbial biotechnology,* 2008. *1*(6): p. 487-496.

[151] Amend, J.P. and E.L. Shock, Energetics of overall metabolic reactions of thermophilic and hyperthermophilic Archaea and Bacteria. *FEMS microbiology reviews,* 2001. *25*(2): p. 175-243.

[152] Liu, H., S. Cheng, and B.E. Logan, Production of electricity from acetate or butyrate using a single-chamber microbial fuel cell. *Environmental science & technology,* 2005. *39*(2): p. 658-662.

[153] Cheng, S., H. Liu, and B.E. Logan, Increased power generation in a continuous flow MFC with advective flow through the porous anode and reduced electrode spacing. *Environmental science & technology,* 2006. *40*(7): p. 2426-2432.

[154] Larminie, J., A. Dicks, and M.S. McDonald, *Fuel cell systems explained.* Vol. 2. 2003: Wiley New York.

[155] Franks, A.E. and K.P. Nevin, Microbial fuel cells, a current review. *Energies.* 3(5): p. 899-919.

[156] Lovley, D.R., Microbial energizers: fuel cells that keep on going. *Microbe-American Society for Microbiology,* 2006. *1*(7): p. 323.

[157] Lovley, D.R., et al. Electricity production with electricigens. in *Bioenergy.* 2008: ASM Press.

[158] Bond, D.R. and D.R. Lovley, Electricity production by Geobacter sulfurreducens attached to electrodes. *Applied and environmental microbiology,* 2003. *69*(3): p. 1548-1555.

[159] Nevin, K.P., et al., Power output and columbic efficiencies from biofilms of Geobacter sulfurreducens comparable to mixed community microbial fuel cells. *Environmental microbiology,* 2008. *10*(10): p. 2505-2514.

[160] Kim, N., et al., Effect of initial carbon sources on the performance of microbial fuel cells containing Proteus vulgaris. *Biotechnology and bioengineering,* 2000. *70*(1): p. 109-114.

[161] Lu, N., et al., Electricity generation from starch processing wastewater using microbial fuel cell technology. *Biochemical Engineering Journal,* 2009. *43*(3): p. 246-251.

[162] Ren, Z., L.M. Steinberg, and J.M. Regan, Electricity production and microbial biofilm characterization in cellulose-fed microbial fuel cells. *Water Science and Technology,* 2008. *58*(3): p. 617.

[163] Zhang, Y., et al., Generation of electricity and analysis of microbial communities in wheat straw biomass-powered microbial fuel cells. *Applied and environmental microbiology,* 2009. *75*(11): p. 3389-3395.

[164] Zhang, C., et al., Pyridine degradation in the microbial fuel cells. *Journal of Hazardous materials,* 2009. *172*(1): p. 465-471.

[165] Luo, H., et al., Phenol degradation in microbial fuel cells. *Chemical Engineering Journal,* 2009. *147*(2): p. 259-264.

[166] Zhu, X. and J. Ni, Simultaneous processes of electricity generation and p-nitrophenol degradation in a microbial fuel cell. *Electrochemistry Communications,* 2009. *11*(2): p. 274-277.

[167] You, S.J., Q.L. Zhao, and J.Q. Jiang, *Biological wastewater treatment and simultaneous generating electricity from organic wastewater by microbial fuel cell.* Huan jing ke xue= Huanjing kexue, 2006. *27*(9): p. 1786-1790.

[168] Feng, Y., et al., Brewery wastewater treatment using air-cathode microbial fuel cells. *Applied microbiology and biotechnology,* 2008. *78*(5): p. 873-880.

[169] Galvez, A., J. Greenman, and I. Ieropoulos, Landfill leachate treatment with microbial fuel cells; scale-up through plurality. *Bioresource technology,* 2009. *100*(21): p. 5085-5091.

[170] Patil, S.A., et al., Electricity generation using chocolate industry wastewater and its treatment in activated sludge based microbial fuel cell and analysis of developed microbial community in the anode chamber. *Bioresource technology,* 2009. *100*(21): p. 5132-5139.

[171] Freguia, S., et al., Microbial fuel cells operating on mixed fatty acids. *Bioresource technology. 101*(4): p. 1233-1238.

[172] Morris, J.M. and S. Jin, Feasibility of using microbial fuel cell technology for bioremediation of hydrocarbons in groundwater. *Journal of environmental science and health. Part A, Toxic/hazardous substances & environmental engineering,* 2008. *43*(1): p. 18-23.

[173] Tender, L.M., et al., The first demonstration of a microbial fuel cell as a viable power supply: powering a meteorological buoy. *Journal of Power Sources,* 2008. *179*(2): p. 571-575.

[174] Chaudhuri, S.K. and D.R. Lovley, Electricity generation by direct oxidation of glucose in mediatorless microbial fuel cells. *Nature biotechnology,* 2003. *21*(10): p. 1229-1232.

[175] Rabaey, K., et al., A microbial fuel cell capable of converting glucose to electricity at high rate and efficiency. *Biotechnology letters,* 2003. *25*(18): p. 1531-1535.

[176] Rosenbaum, M., U. Schröder, and F. Scholz, Investigation of the electrocatalytic oxidation of formate and ethanol at platinum black under microbial fuel cell conditions. *Journal of Solid State Electrochemistry,* 2006. *10*(10): p. 872-878.

[177] Tender, L.M., et al., Harnessing microbially generated power on the seafloor. *Nature biotechnology,* 2002. *20*(8): p. 821-825.

[178] DeLong, E.F. and P. Chandler, Power from the deep. *Nature biotechnology,* 2002. *20*(8): p. 788-789.

[179] Ieropoulos, I., J. Greenman, and C. Melhuish. Imitating metabolism: Energy autonomy in biologically inspired robots. In *Proceedings of the AISB.* 2003.

[180] Wei, J., et al., Carbonization and Activation of Inexpensive Semicokeâ€¢ Packed Electrodes to Enhance Power Generation of Microbial Fuel Cells. *ChemSusChem. 5*(6): p. 1065-1070.

[181] Santoro, C., et al., The effects of wastewater types on power generation and phosphorus removal of microbial fuel cells (MFCs) with activated carbon (AC) cathodes. *International Journal of Hydrogen Energy.*

[182] HaoYu, E., et al., Microbial fuel cell performance with non-Pt cathode catalysts. *Journal of Power Sources,* 2007. *171*(2): p. 275-281.

[183] Santoro, C., et al., Power generation of microbial fuel cells (MFCs) with low cathodic platinum loading. *International Journal of Hydrogen Energy. 38*(1): p. 692-700.

[184] Sun, Y. and J. Cheng, Hydrolysis of lignocellulosic materials for ethanol production: a review. *Bioresource technology, 2002. 83*(1): p. 1-11.

[185] Zuo, Y., P.-C. Maness, and B.E. Logan, Electricity production from steam-exploded corn stover biomass. *Energy & Fuels, 2006. 20*(4): p. 1716-1721.

[186] Catal, T., et al., Utilization of mixed monosaccharides for power generation in microbial fuel cells. *Journal of chemical technology and biotechnology. 86*(4): p. 570-574.

[187] Zhang, Y., et al., Electricity generation and microbial community response to substrate changes in microbial fuel cell. *Bioresource technology. 102*(2): p. 1166-1173.

[188] Oh, S.-E., et al., Evaluation of electricity generation from ultrasonic and heat/alkaline pretreatment of different sludge types using microbial fuel cells. *Bioresource technology. 165*: p. 21-26.

[189] Cheng, S. and B.E. Logan, Increasing power generation for scaling up single-chamber air cathode microbial fuel cells. *Bioresource technology. 102*(6): p. 4468-4473.

[190] Ieropoulos, I., et al., EcoBot-II: An artificial agent with a natural metabolism. *Journal of Advanced Robotic Systems, 2005. 2*(4): p. 295-300.

[191] Shantaram, A., et al., Wireless sensors powered by microbial fuel cells. *Environmental science & technology, 2005. 39*(13): p. 5037-5042.

[192] Chiao, M., et al. A miniaturized microbial fuel cell. in *Technical Digest of the 2002 Solid-State Sensors and Actuators Workshop. 2002.*

[193] Wilkinson, S., "Gastrobots"-benefits and challenges of microbial fuel cells in foodpowered robot applications. *Autonomous Robots, 2000. 9*(2): p. 99-111.

[194] Ieropoulos, I., C. Melhuish, and J. Greenman, Artificial metabolism: towards true energetic autonomy in artificial life. In *Advances in Artificial Life. 2003*, Springer. p. 792-799.

[195] Ieropoulos, I., C. Melhuish, and J. Greenman. *Energetically autonomous robots. in Proceedings of the 8th Intelligent Autonomous Systems Conference. 2004*: Citeseer.

[196] Melhuish, C., et al., Energetically autonomous robots: Food for thought. *Autonomous Robots, 2006. 21*(3): p. 187-198.

[197] Liu, H., S. Grot, and B.E. Logan, Electrochemically assisted microbial production of hydrogen from acetate. *Environmental science & technology, 2005. 39*(11): p. 4317-4320.

[198] Logan, B.E. and S. Grot, A bioelectrochemically assisted microbial reactor (BEAMR) that generates hydrogen gas. *Patent application, 2005. 60*(588): p. 022.

[199] Heilmann, J., *Microbial fuel cells: proteinaceous substrates and hydrogen production using domestic wastewater.* University Park, PA: The Pennsylvania State University, 2005.

[200] Holzman, D.C., Microbe power! *Environmental health perspectives,* 2005. *113*(11): p. A754.

[201] Wang, A., et al., Integrated hydrogen production process from cellulose by combining dark fermentation, microbial fuel cells, and a microbial electrolysis cell. *Bioresource technology. 102*(5): p. 4137-4143.

[202] Habermann, W. and E.H. Pommer, Biological fuel cells with sulphide storage capacity. *Applied microbiology and biotechnology,* 1991. *35*(1): p. 128-133.

[203] Min, B., et al., Electric power generation by a submersible microbial fuel cell equipped with a membrane electrode assembly. *Bioresource technology. 118*: p. 412-417.

[204] Jang, J.K., et al., Construction and operation of a novel mediator-and membrane-less microbial fuel cell. *Process Biochemistry,* 2004. *39*(8): p. 1007-1012.

[205] Moon, H., et al., Residence time distribution in microbial fuel cell and its influence on COD removal with electricity generation. *Biochemical engineering journal,* 2005. *27*(1): p. 59-65.

[206] He, Z., S.D. Minteer, and L.T. Angenent, Electricity generation from artificial wastewater using an upflow microbial fuel cell. *Environmental science & technology,* 2005. *39*(14): p. 5262-5267.

[207] Oh, S., B. Min, and B.E. Logan, Cathode performance as a factor in electricity generation in microbial fuel cells. *Environmental science & technology,* 2004. *38*(18): p. 4900-4904.

[208] Min, B., et al., Electricity generation from swine wastewater using microbial fuel cells. *Water research,* 2005. *39*(20): p. 4961-4968.

[209] Feng, Y., et al., Treatment of biodiesel production wastes with simultaneous electricity generation using a single-chamber microbial fuel cell. *Bioresource technology. 102*(1): p. 411-415.

[210] Kim, J.R., B. Min, and B.E. Logan, Evaluation of procedures to acclimate a microbial fuel cell for electricity production. *Applied microbiology and biotechnology,* 2005. *68*(1): p. 23-30.

[211] Ahn, Y. and B.E. Logan, Effectiveness of domestic wastewater treatment using microbial fuel cells at ambient and mesophilic temperatures. *Bioresource technology. 101*(2): p. 469-475.

[212] Khan, M.R., et al., Design and Fabrication of Membrane Less Microbial Fuel Cell (ML-MFC) using Food Industries Wastewater for Power Generation. *Journal of Chemical Engineering. 27*(2): p. 55-59.

[213] Wang, Z., B. Lim, and C. Choi, Removal of Hg2+as an electron acceptor coupled with power generation using a microbial fuel cell. *Bioresource technology.* *102*(10): p. 6304-6307.

[214] Lefebvre, O., et al., Optimization of a microbial fuel cell for wastewater treatment using recycled scrap metals as a cost-effective cathode material. *Bioresource technology.* *127*: p. 158-164.

[215] Zhuang, L., et al., Scalable microbial fuel cell (MFC) stack for continuous real wastewater treatment. *Bioresource technology.* *106*: p. 82-88.

[216] Wang, Y.-P., et al., A microbial fuel cellâ€"membrane bioreactor integrated system for cost-effective wastewater treatment. *Applied Energy.* *98*: p. 230-235.

[217] Zhang, Y. and I. Angelidaki, Submersible microbial fuel cell sensor for monitoring microbial activity and BOD in groundwater: focusing on impact of anodic biofilm on sensor applicability. *Biotechnology and bioengineering.* *108*(10): p. 2339-2347.

[218] Davila, D., et al., Silicon-based microfabricated microbial fuel cell toxicity sensor. *Biosensors and Bioelectronics.* *26*(5): p. 2426-2430.

[219] Stein, N.E., H.V.M. Hamelers, and C.N.J. Buisman, Stabilizing the baseline current of a microbial fuel cell-based biosensor through overpotential control under non-toxic conditions. *Bioelectrochemistry.* *78*(1): p. 87-91.

[220] Liu, B., Y. Lei, and B. Li, A batch-mode cube microbial fuel cell based shock biosensor for wastewater quality monitoring. *Biosensors and Bioelectronics.* *62*: p. 308-314.

[221] Stein, N.E., et al., On-line detection of toxic components using a microbial fuel cell-based biosensor. *Journal of Process Control.* *22*(9): p. 1755-1761.

[222] Stein, N.E., H.V.M. Hamelers, and C.N.J. Buisman, Influence of membrane type, current and potential on the response to chemical toxicants of a microbial fuel cell based biosensor. *Sensors and Actuators B: Chemical.* *163*(1): p. 1-7.

[223] Peixoto, L., et al., *In situ* microbial fuel cell-based biosensor for organic carbon. *Bioelectrochemistry.* *81*(2): p. 99-103.

[224] Kaur, A., et al., Microbial fuel cell type biosensor for specific volatile fatty acids using acclimated bacterial communities. *Biosensors and Bioelectronics.* *47*: p. 50-55.

[225] King, S.T., et al., Detecting recalcitrant organic chemicals in water with microbial fuel cells and artificial neural networks. *Science of the Total Environment.* *497*: p. 527-533.

[226] Chang, I.S., et al., Continuous determination of biochemical oxygen demand using microbial fuel cell type biosensor. *Biosensors and Bioelectronics,* 2004. *19*(6): p. 607-613.

[227] Chang, I.S., et al., Improvement of a microbial fuel cell performance as a BOD sensor using respiratory inhibitors. *Biosensors and Bioelectronics,* 2005. *20*(9): p. 1856-1859.

[228] Kim, B.H., et al., Novel BOD (biological oxygen demand) sensor using mediator-less microbial fuel cell. *Biotechnology letters,* 2003. *25*(7): p. 541-545.

[229] Moon, H., et al., Improving the dynamic response of a mediator-less microbial fuel cell as a biochemical oxygen demand (BOD) sensor. *Biotechnology letters,* 2004. *26*(22): p. 1717-1721.

[230] Kang, K.H., et al., A microbial fuel cell with improved cathode reaction as a low biochemical oxygen demand sensor. *Biotechnology letters,* 2003. *25*(16): p. 1357-1361.

[231] Stein, N.E., H.V.M. Hamelers, and C.N.J. Buisman, The effect of different control mechanisms on the sensitivity and recovery time of a microbial fuel cell based biosensor. *Sensors and Actuators B: Chemical. 171:* p. 816-821.

[232] Stein, N.E., et al., Kinetic models for detection of toxicity in a microbial fuel cell based biosensor. *Biosensors and Bioelectronics. 26*(7): p. 3115-3120.

[233] Greenman, J., et al., Electricity from landfill leachate using microbial fuel cells: comparison with a biological aerated filter. *Enzyme and Microbial Technology,* 2009. *44*(2): p. 112-119.

[234] Di Lorenzo, M., et al., *A single chamber packed bed microbial fuel cell biosensor for measuring organic content of wastewater.* 2009.

[235] Holmes, D.E., et al., Microbial communities associated with electrodes harvesting electricity from a variety of aquatic sediments. *Microbial ecology,* 2004. *48*(2): p. 178-190.

[236] Bond, D.R., et al., Electrode-reducing microorganisms that harvest energy from marine sediments. *Science,* 2002. *295*(5554): p. 483-485.

[237] Reimers, C.E., et al., Harvesting energy from the marine sediment-water interface. *Environmental science & technology,* 2001. *35*(1): p. 192-195.

[238] Liu, Z., et al., Microbial fuel cell-based biosensor for in situ monitoring of anaerobic digestion process. *Bioresource technology. 102*(22): p. 10221-10229.

[239] Gregory, K.B. and D.R. Lovley, Remediation and recovery of uranium from contaminated subsurface environments with electrodes. *Environmental science & technology,* 2005. *39*(22): p. 8943-8947.

[240] Cheng, S., et al., Efficient recovery of nano-sized iron oxide particles from synthetic acid-mine drainage (AMD) water using fuel cell technologies. *Water research. 45*(1): p. 303-307.

[241] Matsumoto, N., et al., Extension of logarithmic growth of Thiobacillus ferrooxidans by potential controlled electrochemical reduction of Fe (III). *Biotechnology and bioengineering,* 1999. *64*(6): p. 716-721.

[242] Fradler, K.R., et al., Augmenting microbial fuel cell power by coupling with supported liquid membrane permeation for zinc recovery. *Water research. 55*: p. 115-125.

[243] Semiat, R., Energy issues in desalination processes. *Environmental science & technology,* 2008. *42*(22): p. 8193-8201.

[244] Luo, H., et al., Microbial desalination cells for improved performance in wastewater treatment, electricity production, and desalination. *Bioresource technology. 105*: p. 60-66.

[245] Qu, Y., et al., Simultaneous water desalination and electricity generation in a microbial desalination cell with electrolyte recirculation for pH control. *Bioresource technology. 106*: p. 89-94.

[246] Jacobson, K.S., D.M. Drew, and Z. He, Efficient salt removal in a continuously operated upflow microbial desalination cell with an air cathode. *Bioresource technology. 102*(1): p. 376-380.

[247] Yuan, L., et al., Capacitive deionization coupled with microbial fuel cells to desalinate low-concentration salt water. *Bioresource technology. 110*: p. 735-738.

[248] Zhou, M., et al., Recent advances in microbial fuel cells (MFCs) and microbial electrolysis cells (MECs) for wastewater treatment, bioenergy and bioproducts. *Journal of chemical technology and biotechnology. 88*(4): p. 508-518.

[249] Lusk, P. and A.E. Wiselogel, *Methane recovery from animal manures: the current opportunities casebook.* Vol. 3. 1998: National Renewable Energy Laboratory Golden, CO.

[250] Pham, T.H., et al., Microbial fuel cells in relation to conventional anaerobic digestion technology. *Engineering in Life Sciences,* 2006. *6*(3): p. 285-292.

[251] Angenent, L.T., et al., Production of bioenergy and biochemicals from industrial and agricultural wastewater. *TRENDS in Biotechnology,* 2004. *22*(9): p. 477-485.

[252] Rabaey, K. and W. Verstraete, Microbial fuel cells: novel biotechnology for energy generation. *TRENDS in Biotechnology,* 2005. *23*(6): p. 291-298.

[253] Fu, J., J. Reinhold, and N.W. Woodbury, Peptide-modified surfaces for enzyme immobilization. *PloS one. 6*(4): p. e18692.

[254] Mu, Y., et al., Electrochemically active bacteria assisted nitrobenzene removal from wastewater. *Journal of Biotechnology. 150*: p. 147.

[255] Wen, Q., et al., Simultaneous processes of electricity generation and ceftriaxone sodium degradation in an air-cathode single chamber microbial fuel cell. *Journal of Power Sources. 196*(5): p. 2567-2572.

[256] Pham, H., et al., Enhanced removal of 1, 2-dichloroethane by anodophilic microbial consortia. *Water research,* 2009. *43*(11): p. 2936-2946.

[257] Heijne, A.T., et al., Copper recovery combined with electricity production in a microbial fuel cell. *Environmental science & technology. 44*(11): p. 4376-4381.

[258] Williams, K.H., et al., Electrode-based approach for monitoring in situ microbial activity during subsurface bioremediation. *Environmental science & technology,* 2009. *44*(1): p. 47-54.

[259] Qin, B., et al., Nickel ion removal from wastewater using the microbial electrolysis cell. *Bioresource technology. 121*: p. 458-461.

[260] Pant, D., et al., An introduction to the life cycle assessment (LCA) of bioelectrochemical systems (BES) for sustainable energy and product generation: relevance and key aspects. *Renewable and Sustainable Energy Reviews. 15*(2): p. 1305-1313.

[261] Rismani-Yazdi, H., et al., Cathodic limitations in microbial fuel cells: an overview. *Journal of Power Sources,* 2008. *180*(2): p. 683-694.

[262] O'Hayre, R.P., et al., *Fuel cell fundamentals.* 2006.

[263] Gil, G.-C., et al., Operational parameters affecting the performannce of a mediator-less microbial fuel cell. *Biosensors and Bioelectronics,* 2003. *18*(4): p. 327-334.

[264] Nevin, K.P., et al., Anode biofilm transcriptomics reveals outer surface components essential for high density current production in Geobacter sulfurreducens fuel cells. *PLoS One,* 2009. *4*(5): p. e5628.

[265] Logan, B.E. and J.M. Regan, Microbial fuel cells-challenges and applications. *Environmental science & technology,* 2006. *40*(17): p. 5172-5180.

[266] Morrow, W.R., W.M. Griffin, and H.S. Matthews, Modeling switchgrass derived cellulosic ethanol distribution in the United States. *Environmental science & technology,* 2006. *40*(9): p. 2877-2886.

[267] Ragauskas, A.J., et al., The path forward for biofuels and biomaterials. *Science,* 2006. *311*(5760): p. 484-489.

In: Microbial Catalysts. Volume 2
Editors: Shadia M. Abdel-Aziz et al.

ISBN: 978-1-53616-088-8
© 2019 Nova Science Publishers, Inc.

Chapter 5

PECTINASE: AN IMPORTANT MICROBIAL BIOCATALYST AND ITS INDUSTRIAL APPLICATIONS

Fazilath Uzma[1], Lakshmeesha Thimmappa Ramachandrappa[2],
Venkataramana Mudili[3], Mohan Chakrabhavi Dhananjaya[2],
Shobith Rangappa[4], Girish Kesturu Subbaiah[5],
Chandra Nayaka Siddaiah[2,*] and Srinivas Chowdappa[1]

[1]Department of Microbiology and Biotechnology, Bangalore University,
Jnanabharathi Campus Bangalore, Karnataka, India
[2]Department of studies in Biotechnology, University of Mysore,
Mysuru, Karnataka, India
[3]Department of Microbiology, Defence Food Research Laboratory,
Mysore, Karnataka, India
[4]Adichunchanagiri Institute for Molecular Medicine, Nagamangala, Karnataka, India
[5]Department of Studies & Research in Biochemistry,
Tumkur University, Tumkur, India

ABSTRACT

Enzymes are biocatalysts produced by all living cells to bring about specific biochemical reactions as a part of metabolic processes occurring within the cells. The potential of microorganisms for use as biocatalysts has drawn a great deal of attention in a variety of industrial processes. Pectinases are a complex and diverse group of enzymes involved in the degradation of pectic substances by the hydrolysis of pectin, a structural polysaccharide usually present in the primary cell wall and middle lamella in the tissues of higher plants. They are known to be produced from a wide variety of microorganisms including bacteria, fungi and actinomycetes. Protopectinases, esterases and

---

* Corresponding Author's E-mail: moonnayak@gmail.com.

depolymerases are the most widely studied groups of pectinases and each of these finds a significant importance in the biotechnological industry. Microbial pectinases have wide applications in several industries including pulp, textile, food, fruit and juice, etc. proving to be the future enzymes of the commercial sector. The present chapter focuses on the structure, classification, production techniques and applications of microbial pectinase.

Keywords: enzymes, biocatalysts, pectinase

1. INTRODUCTION

Enzymes are biocatalysts which play a key role in all stages of metabolic processes and biochemical reactions. The potential of microorganisms for use as biocatalysts has drawn a great deal of attention in a variety of industrial processes. The biocatalytic uses of enzymes have grown immensely in recent years since they have a high degree specificity/selectivity and can catalyze a wide array of reactions. Microbial enzymes are obtained from different microorganisms and are rapidly gaining popularity in the commercial sector for its widespread applications as organic catalysts in numerous processes on an industrial scale. Several microorganisms including bacteria, fungi and actinomycetes, have been studied for the biosynthesis of different enzymes for use in the commercial production of valuable products (Vermelho et al., 2012). The microbial world has shown to be very heterogeneous in its ability to synthesize different types of pectolytic enzymes with different mechanisms of action and biochemical properties (Torres et al., 2005; Gummadi & Panda, 2003).

Pectic enzymes are produced by both prokaryotic and eukaryotic microorganisms, which primarily synthesize alkaline and acid pectinases respectively. They have been reported in *Bacillus spp., Clostridium spp., Pseudomonas spp., Aspergillus spp., Monilla laxa, Fusarium spp., Verticillium spp., Penicillium spp., Sclerotinia libertiana, Coniothyrium diplodiella*, Thermomyces lanuginosus, Polyporus squamosus (Jayani et al., 2005). Furthermore, the production of these enzymes has also been described in yeast (Theuil et al., 2011).

The filamentous fungi such as *Aspergillus niger, A. awamori, Penicillium restrictum, T. viride, M. piriformis* are used in both submerged as well as solid state fermentation for production of various industrially important products such as citric acid, ethanol, etc. The Food and Drugs Administration (FDA) has approved *A. niger, A. oryzae, P. expansum*, for use in food industry as they are Generally Regarded as Safe (GRAS) and hence are employed in food industry. Apart from fungi, bacteria such as *Bacillus licheniformis, Aeromonascavi, Lactobacillus sp., Saccharomyces, Candida* and *Streptomycetes* are also employed in the production of pectinases. (Saranraj and Stella. 2013)

Pectinases are a complex and diverse group of enzymes involved in the degradation of pectic substances. Pectin and other pectic substances are complex polysaccharides that contribute firmness and structure to plant tissues as a part of the middle lamella. The diverse forms of pectic substances in plant cells may probably account for the existence of various forms of pectinolytic enzymes.

Pectinases are classified according to their mode of secretion as extracellular and intracellular pectinases. An extracellular enzyme is secreted outside the cell into the medium in which that cell is living. Extracellular enzymes usually convert large substrate molecules (i.e., food for the cell or organism) into smaller molecules that can then be more easily transported into the cell, whereas an intracellular enzyme operates within the cell membrane. Both intracellular and extracellular pectinases are classified based on the mode of their action on the galacturonan part of pectin molecules (Singhania et al., 2009).

Pectinases play a significant role in plants as they help in cell wall extension and softening of some plant tissues during maturation and storage (Triparthi et al., 2014). Spoilage of fruits and vegetables by rotting, plant pathogenicity, symbiosis and decomposition of plant deposits are some of the manifestations of pectinolytic enzymes (Raghuwanshi et al., 2013). The microbial pectinase accounts approxiamately for 25% of total global enzyme sales (Nirmaladevi et al., 2014). Most commercial preparations of pectinases are produced from *Aspergillus niger* (Jayani et al., 2005).

Pectin is known to contain neutral sugars which are present in side chains. The most common side chain sugars are xylose, galactose and arabinose. Pectins function as an adhesive which holds the other cell wall polysaccharides like cellulose and hemicellulose (i.e., xyloglugan or glucuronarabinoxylan) and proteins such as hydroxyproline-rich glycoprotein extension, together. In plants, pectins are present in all stages of development. The composition depends not only on the species but also on tissue, stage of growth, maturity and growth conditions. Pectins are heterogeneous with respect to both chemical structure and molecular weight (Shembekar et al., 2009).

Pectinases can be produced by both Submerged Fermentation (SmF) and Solid State Fermentation (SSF). SmF involves cultivation of microorganisms on liquid broth which requires high volumes of water, continuous agitation and generates lot of effluents. SSF utilizes solid substrates for the growth of microorganisms and subsequent product formation with total absence or limited use of water, generally under aerobic conditions (Binod et al., 2008).

Generally, substrates used for SSF include grains such as rice, corn, root, tubers and legumes. Apart from these, peels from fruit and vegetable industry waste are also being used.

Pectins are widely used in textile, food industry (Bakery and Dairy) and have demonstrated their commercial utilization as texturizing, thickening, emulsifyinh, stabilising and clarifying agents. (Liu et al., 2012).

It is also studied for its potential in drug delivery in the pharmaceutical industry (Khule et al., 2012). Pectin is used in making biodegradable films (Raghuwanshi et al., 2013). Despite these applications, pectins are similar to cellulose and hemicelluloses, in converting common waste materials to soluble sugars, ethanol (Turner et al., 2007), and biogas (Hutnan et al., 2000).

2. STRUCTURE OF PECTIN

Pectin is a major component of the primary cell wall of all the terrestrial plants. It is a complex heterogenous structural polysaccharide comprising of a glycan matrix rich in galacturonic acid. Pectin is crosslinked with cellulosolic and hemicellulosolic fibers of the cell wall with side chains consisting of L-rhamnose, arabinose, galactose and xylose. Pectic substances are complex colloidal acid polysaccharides with a backbone of galacturonic acid residues linked by (1 ± 4) linkages. The hydroxyl groups on C2 and C3 may be acetylated. The carboxyl groups of galacturonic acid are partially esterified by methyl groups or they may be completely or partially neutralized by sodium, potassium or ammonium ions (Pasha et al., 2013).

Based on the biochemical studies, the pectinases are comprised of members dominated by a chain of 1,4-linked alpha-D-galactosyluronic acid residues (GalA). There are three types of pectic domains, which include homogalacturonan, rhamnogalacturonan-I, and rhamnogalacturonan-II (Abbott et al., 2007).

Homogalacturonan (HG) is composed of a linear chain of 1,4-linked alpha-D-galactosyluronic acid residues. Demethyl-esterification at the C-6 carboxyl results in large assemblies of homogalacuronan within the cell wall matrix, allowing calcium cross-linkages to form.

Rhamnogalacturonan-I (RG-I) are regions where galacturonic acid units are replaced with the disaccharide repeat $[(1\rightarrow2)\text{-}\alpha\text{-L-rhamnose-}(1\rightarrow4)\text{-}\alpha\text{-D-galacturonic acid}]n$. This replacement causes steric hindrance or a "kink" in the linear backbone and allows the bonding of various sugar side chains, including D-galactose, L-arabinose, and D-xylose.

Rhamnogalacturonan-II (RG-II) is a branched pectic domain containing a homogalacturonan backbone with various complex side chains bonded to the GalA residues (O'Neill et al. 2004). These three polysaccharide domains form covalent linkages throughout the primary cell wall matrix and middle lamellae, and provide considerable potential for structural modulation by a wide range of pectinase enzymes. This matrix forms a crystalline structure that allows it to trap water and other molecules, giving pectin its gelling properties. It has been traditionally assumed that the structure of pectin was primarily composed of a HG backbone interspersed with RG-I and RG-II regions (Figure 1). Pectic regions with high densities of rhamnose are considered "hairy" regions due to their highly branched configuration, while those with less branching are termed "smooth" regions (Figure 2) (Pérez et al. 2000). Most of pectic-polymers are comprised of smooth homogalacturonan and ramified hairy regions. Smooth regions consist of a linear homogalacturonan backbone, while hairy regions consist of rhamnogalacturonan backbone with side-branches of varying length (Aguilar et al., 2008). However, recent alternative structures have been proposed, placing RG-I at the backbone with long side chains of HG, further branching into RG-II (Willats et al. 2006).

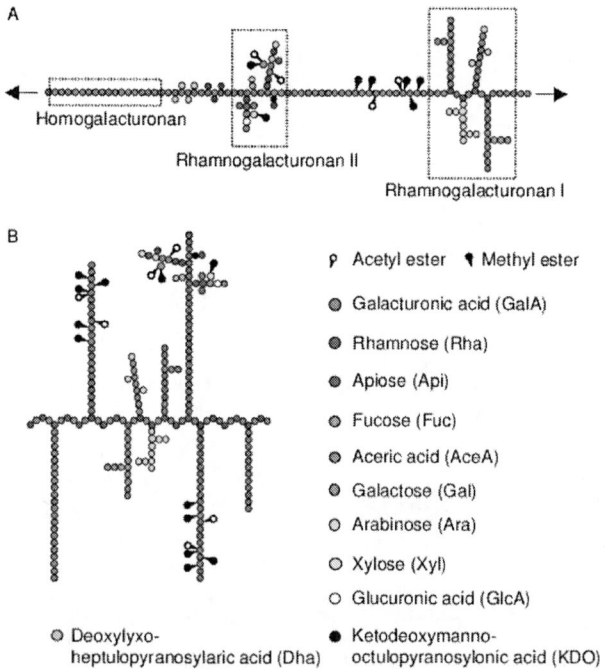

Source: Willats et al., 2006.

Figure 1. The basic structure of pectin. Schematic representations of the conventional (A) Recently proposed alternative (B) Structures of pectin. The polymers shown here are intended only to illustrate the some of the major domains found in most pectins rather than definitive structures.

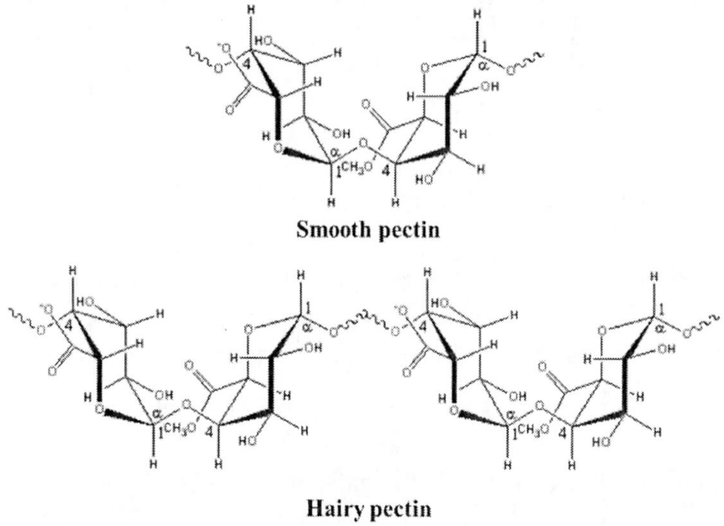

Smooth pectin

Hairy pectin

Figure 2. Structure depicting smooth and hairy pectin.

2.1. Classification

Pectic enzymes have two classes namely, pectin, esterases and pectin depolymerases. Pectin esterase has the ability to de-esterify pectin by the removal of methoxy residues. Pectin depolymerases readily split the main chain and are further classified as polygalacturonase (PG) and pectinlyases (PL) (Haidar and Fazaelipoor, 2010).

Based on its mode of action and substrate preference, these enzymes are classified into three types:

I.   Protopectinases: degrade the insoluble protopectin and give rise to highly polymerized soluble pectin.
II. Depolymerases: catalyze the hydrolytic cleavage of the α-(1- 4)-glycosidic bonds in the D-galacturonic acid moieties of the pectic substances
III. Esterases: catalyze the de-esterification of pectin by the removal of methoxy esters (Table 1)

2.2. Protopectinases

Protopectinases catalyse the solubilisation of protopectin. They catalyse the reaction which converts the insoluble protopectin to soluble pectin.

Two types of PPases have been recognised based on their reaction mechanisms:

(i) A-type PPases which react with the polygalacturonic acid region of protopectin
(ii) B-type PPases which react on the polysaccharide chains that may connect the polygalacturonic acid chain and cell wall constituents (Jayani et al., 2005).

2.3. Depolymerases

Depolymerases act on pectic substances by two different mechanisms:

(i) The depolymerases catalyze the hydrolytic cleavage with the introduction of water across the oxygen bridge by hydrolysis.
(ii) The glycosidic bond is broken down via a trans-elimination lysis by depolymerases which does not involve the participation of water molecules.

Polygalacturonase(PG) and polymethylgalacturonase(PMG) breakdown pectate and pectin, respectively by the mechanism of hydrolysis. However, polygalacturonate lyase and polymethylgalacturonate lyase breakdown pectate and pectin by β elimination.

Depending upon the pattern of action, i.e., random or terminal, these enzymes are termed as Endo or Exo enzymes, respectively (Figure 3)

2.4. Polygalacturonases (PG)

Polygalacturonase catalyzes hydrolysis of α-1,4-glycosidic linkages in polygalacturonic acid producing D-galacturonate. Both groups of hydrolase enzymes (PMG and PG) can act in an endo- or exo- mode. Endo-PG (EC 3.2.1.15) and Endo-PMG catalyze random cleavage of substrate. Exo-PG (EC 3.2.1.67) and Exo-PMG catalyze hydrolytic cleavage at substrate non reducing end producing monogalacturonate or digalacturonate in some cases (Kashyap et al., 2001). Hydrolases are produced mainly by fungi, being more active on acid or neutral medium at temperatures between 40°C and 60°C (Pedrolli et al., 2009).

2.5. Pectate Lyases (PGL)

Pectate lyase cleaves glycosidic linkages preferentially on polygalacturonic acid forming unsaturated product (Δ-4,5-D-galacturonate) through transelimination reaction. PGL has an absolute requirement of Ca^{2+} ions. Hence, it is strongly inhibited by chelating agents as EDTA (Jayani et al., 2005). Pectate lyases are classified as endo-PGL (EC 4.2.2.2) which acts towards substrate in a random way and exo-PGL (EC 4.2.2.9) that catalyze the substrate cleavage from nonreducing end chain (Pedrolli et al., 2009).

2.6. Pectin Lyases (PL)

Pectin lyase catalyzes the random cleavage of pectin, preferentially high esterified pectin, producing unsaturated methyloligogalacturonates through transelimination of glycosidic linkages. PLs do not have an absolute requirement of Ca^{2+} but they are stimulated by this and other cations (Jayani et al., 2005). Most described pectin lyases are endo-PLs (EC 4.2.2.10) (Sinitsyna et al., 2007). Van Alebeek and co-workers (2002) conducted a detailed study of the action mode of pectin lyase A from *Aspergillus niger* which produces mono, di, tri and tetragalacturonates, besides unsaturated di, tri and tetragalacturonates from methyloligogalacturonates. Unsaturated monogalacturonates are not identified in the reaction products during any assay. Both lyase groups are classified into polysaccharides lyase family 1.

Complete degradation of pectin substrate requires enzymes that cleave the rhamnogalacturonan chain.

3. RHAMNOGALACTURONAN RHAMNOHYDROLASES (RG RHAMNOHYDROLASE)

RG rhamnohydrolase, rhamnogalacturonan α-Lrhamnopyranohydrolase or α-L-rhamnosidase (EC 3.2.1.40) catalyzes hydrolytic cleavage of the rhamnogalacturonan chain at nonreducing end producing rhamnose. These enzymes are classified into different glycosyl-hydrolase families (Pedrolli et al., 2009).

3.1. Rhamnogalcturonan Galacturonohydrolases (RG Galacturonohydrolase)

RG galacturonohydrolase (EC 3.2.1) catalyzes hydrolytic cleavage of the rhamnogalacturonan chain at nonreducing end producing monogalacturonate. It is classified into glycosyl-hydrolase family (Pedrolli et al., 2009).

3.2. Rhamnogalacturonan Hydrolases (RG Hydrolases)

RG hydrolase randomly hydrolyses the rhamnogalacturonan chain producing oligogalacturonates (Pedrolli et al., 2009).

3.3. Rhamnogalacturonan Lyases (RG Lyases)

RG lyase (EC 4.2.2.-) catalyzes the random transelimination of the rhamnose-galcturonate linkage from rhamnogalacturonan chain, producing an unsaturated galacturonate at nonreducing end of one oligomer and a second oligomer containing a rhamnose as a reducing end residue These enzymes are classified into polysaccharides-lyase family (Pedrolli et al., 2009).

3.4. Rhamnogalacturonan Acetylesterases (RG Acetylesterases)

Rhamnogalacturonan acetylesterase (EC 3.1.1) catalyzes hydrolytic cleavage of acetyl groups from rhamnogalacturonan chain. It is classified into carbohydrate esterase family (Pedrolli et al., 2009).

4. Xylogalacturonan Hydrolase

Xylogalacturonase (EC 3.2.1) catalyzes hydrolytic cleavage of glycosidic linkages between two galacturonate residues in xylose-substituted rhamnogalacturonan chain, producing xylose-galacturonate dimmers. These enzymes are classified into glycosyl-hydrolase family (Pedrolli et al., 2009).

5. Pectin Esterase

Pectin esterase (PE, E.C. 3.1.1.11), is a carboxylic acid esterase and belongs to the hydrolase group of enzymes and catalyzes the deesterification of methyl ester linkages of galacturonan backbone of pectic substances to release acidic pectins and methanol. The resulting pectin is then acted upon by polygalacturonases and lyases. The mode of action of PE varies depending on its origin. Fungal PEs reacts differently from plant PEs. Fungal PEs act by a multi-chain mechanism, removing the methyl groups at random whereas plants PEs tend to act either at the non-reducing end or next to a free carboxyl group, and proceed along the molecule by a single chain mechanism. The reaction catalyzed by PE can be represented as follows (Jayani et al., 2005).

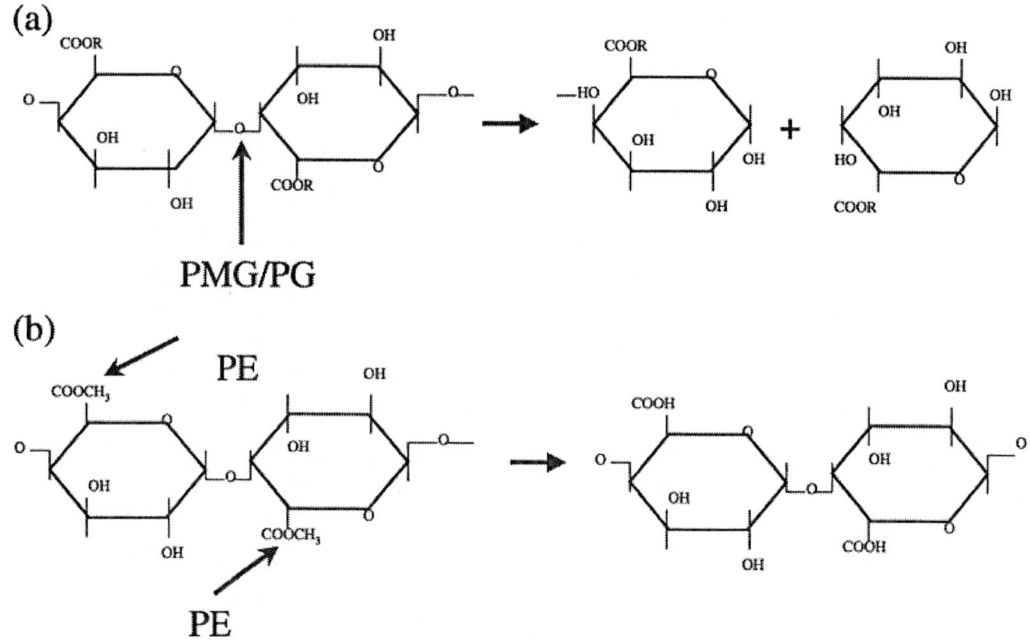

Figure 3. Mode of action of pectinases: (a) R = H for PG and CH3 for PMG; (b) PE; and (c) R = H for PGL and CH3 for PL. The arrow indicates the place where the pectinase reacts with the pectic substances. PMG, polymethylgalacturonases; PG, polygalacturonases (EC 3.2.1.15); PE, pectinesterase (EC 3.1.1.11); PL, pectin lyase (EC-4.2.2.10), Source: Pedrolli et al., 2009.

Table 1. Classification of pectinolytic enzymes

Enzyme	E.C. No.	Modified EC systematic name	Action mechanism	Action pattern	Primary substrate	Product
Esterase						
1. Pectin methyl esterase	3.1.1.11		Hydrolysis	Random	Pectin	Pectic acid + methanol
Depolymerizing enzymes						
(a) Hydrolases						
1. Protopectinases	3.2.1	Poly-(1-4)-a-D-galactosiduronate glycanohydrolase	Hydrolysis	Random	Protopectin	Pectin
2. Endopolygalacturonase	3.2.1	67 Poly-(1-4)-a-D-galactosiduronate glycanohydrolase	Hydrolysis	Random	Pectic acid	Oligogalacturonates
3. Exopolygalacturonase	3.2.1.82		Hydrolysis	Terminal	Pectic acid	Monogalacturonates
4. Exopolygalacturonan-digalacturono hydrolase	4.2.2.2	Poly-(1-4)-a-D-galactosiduronate digalacturonohydrolase	Hydrolysis	Penultimate bonds	Pectic acid	Digalacturonates
5. Oligogalacturonate hydrolase	4.2.2.9	Poly-(1-4)-a-D-galactosiduronate lyase	Hydrolysis	Terminal	Trigalacturonate	Monogalacturonates
6. D4:5 Unsaturated oligogalacturonate hydrolases	4.2.2.6	Poly-(1-4)-a-D-galactosiduronate exolyase	Hydrolysis	Terminal	D4:5(Galacturonate)n	Unsaturated monogalacturonates & saturated (n-1)
7. Endopolymethyl-galacturonases	4.2.2.10	Oligo-D-galactosiduronate lyase	Hydrolysis	Random	Highly esterified pectin	Oligomethylgalacturonates
8. Endopolymethyl-galacturonase		Poly(methyl galactosiduronate) lyase	Trans-elimination	Terminal	Highly esterified pectin	Oligogalacturonates
Lyases			Trans-elimination	Random	Pectic acid	Unsaturated oligogalacturonates
1. Endopolygalacturonase lyase			Trans-elimination	Penultimate bond	Pectic acid	Unsaturated digalacturonates
2. Exopolygalacturonase lyase			Trans-elimination	Terminal	Unsaturated digalacturonates	Unsaturated monogalacturonates
3. Oligo-D-galactosiduronate lyase			Trans-elimination	Random	Unsaturated poly-(methyl-digalacturonates)	Unsaturated methyloligogalacturonates
4. Endopolymethyl-D-galactosiduronate Lyase				Terminal	Unsaturated poly-(methyl-digalacturonates)	Unsaturated methylmonogalacturonates
5. xopolymethyl-D-galactosiduronate						

Source: Jayani et al., 2005.

The PE reaction is represented below:

$$\text{Pectin} + n\text{H}_2\text{O} \xrightarrow{\text{PE}} \text{Pectate} + \text{C}_2\text{H}_5\text{OH}$$

PE activity is implicated in cell wall metabolism including cell growth, fruit ripening, abscission, senescence and pathogenesis. PE has been reported in plant pathogenic bacteria and fungi such as *Rhodotorula sp., Phytophthora infestans, Erwinia chrysanthemi* B341, *Saccharomyces cerevisiae, Lachnospira pectinoschiza, Pseudomonas solanacearum, Aspergillus niger, Lactobacillus lactis subsp. Cremoris, Penicillium frequentans , E. chrysanthemi 3604 , Penicillium occitanis* and *A. japonicus* (Hasunama et al., 2003; Jayani et al., 2005).

6. ROLE OF PGASES IN PLANT PATHOGENESIS

Polygalacturonases play an important role in virulence of some fungi and bacteria. Most of the pathogenic microorganisms such as *Aspergillus flavus, Alternaria citri, Claviceps purpurea, Agrobacterium tumefaciens* and *Ralstonia solanacearum* produce a large variety of plant cell wall enzymes and polygalacturonases that are known to play an important role in phytopathogenicity (Kars et al., 2005). The pectic substances are rich in methyl-esterified galacturonic acid. The distribution of esterified residues and the esterification level in the pectin molecule changes according to the plant life cycle and between different species. The ability of some phytopathogenic microorganisms to produce a variety of pectinolytic enzymes that differ in their substrate specifity provides them with more efficacy in cell wall pectin degradation and facilitates in plant infection (Pedrolli et al., 2009). The pathogenic microorganisms attack plant by virtue of the pectic substances as they are most accessible than the other fibres in plant tissue (Hoondal et al., 2002).

Endo-PGases are widely distributed among fungi, bacteria and have been reported in *Aureobasidium pullulans, Rhizoctonia solani Kuhn, Fusarium moniliforme, Neurospora crassa, Rhizopus stolonifer, Aspergillus sps, Thermomyces lanuginosus and Peacilomyces clavisporus*. Endo- PGases have also been cloned and genetically studied in a large number of microbial species (Fernández-González et al., 2004). In contrast, exo-PGases occur less frequently. They have been reported in *Agrobacterium tumefaciens, Bacteroides thetaiotamicron, E. chrysanthemi, Alternaria mali, Fusarium oxysporum, Ralstonia solanacearum, Bacillus sp*, very few strains of Bacillus sps. produce both pectate lyase (PL) and polygalacturonase (PGL) (Soriano et al., 2005). Previously, Pectate lyase (PL) activity has been detected in supernatants of phytopathogenic bacteria and has been described as the most important cause of soft rot disease. Pectin lyase and pectin methyl esterase are also cited as pectinolytic enzymes produced by microorganisms. *P. marginalis* has been described as bacteria producing PL, PNL and small amounts of PME (Koboyashi et al., 2001).

7. PECTIC SUBSTANCES

Pectic substances are polysaccharides of high molecular weight, with a negative charge, appearing mostly in the middle lamella and the primary cell wall of higher plants, found in the form of calcium pectate and magnesium pectate. They are formed by a central chain containing a variable amount although in high proportion of galacturonic acid residues linked through α (1- 4) glycosidic bonds partially esterified with methyl groups (Figure 4). This molecule is known as pectin, while the demethylated molecule is known as polygalacturonic acid or pectic acid. Two types of pectins- Homogalacturonic pectins(HG), consisting of D-galacturonic acids and Rhamnogalacturonic pectins in which the galacturonic acid chains are discontinued by rhamnose residues joined together by α(1-2) bond with side chains of galactose and arabinose. The D-galacturonic acid residues of both types of pectins may be esterified with methanol and pectins are further categorised as high methoxyl or low methoxyl depending on the degree of methylation (Fernández-González et al., 2004).

Several L-rhamnopyranosyl residues may be attached to the main chain through its C-1 and C-2 atoms. In addition, galacturonate residue may be acetylated at the C-2 and C-3 positions, and side chains of residues of neutral sugars may be linked to the galacturonic acid or to the C-4 of the rhamnose residue in the main chain (Mohnen, 2008; Caffall & Mohnen, 2009).

Pectic substances tend to form a gel structure when portions of HG are cross-linked forming a three dimensional crystalline network in which water and solutes are trapped (Figure 5). Various factors determine gelling properties including temperature, pectin type, esterification degree, acetylation degree, pH, sugar and other solutes, and mainly the interaction between calcium ions and pectin unesterified carboxyl groups.

In high-ester pectins, the junction zones are formed by the cross-linking of HG by hydrogen bridges and hydrophobic forces between methoxyl groups, both promoted by high sugar concentration and low pH. Two antiparallel pectin chains can be condensed in the cell wall by cross-linking with Ca^{2+} ions to form 'junction zones' or the multiple 'eggbox'. In some species, pectins may be cross-linked to other pectins or non-cellulosic polysaccharides by ester linkages with dihydroxycinnamic acids such as diferulic acid (Shembekar et al., 2009). Pectic polysaccharides have been used as bioactive food ingredients and as detoxifying agents. It is an adequade infant food supplement (Gummadi et al., 2003).

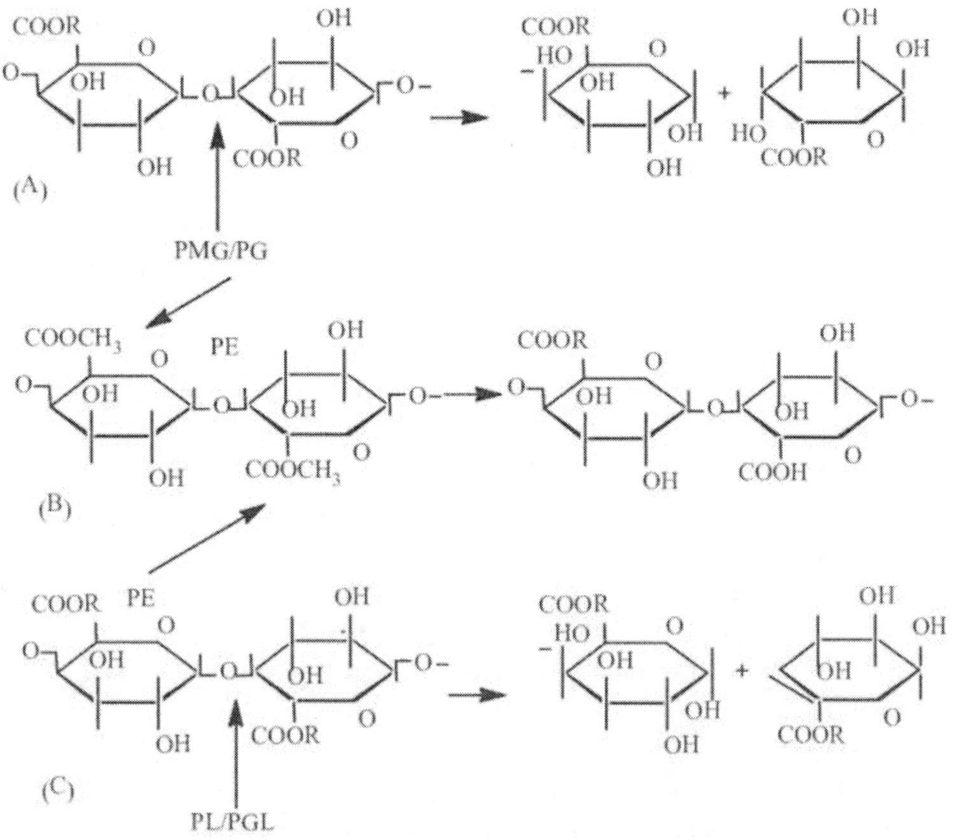

Figure 4. Mechanism of action of Pectinases 147. PMG, polymethylg-alacturonases; PG, polygalacturonases; PE, pectinesterase; PL, pectin lyase. (Kantharaj, et al., 2017).

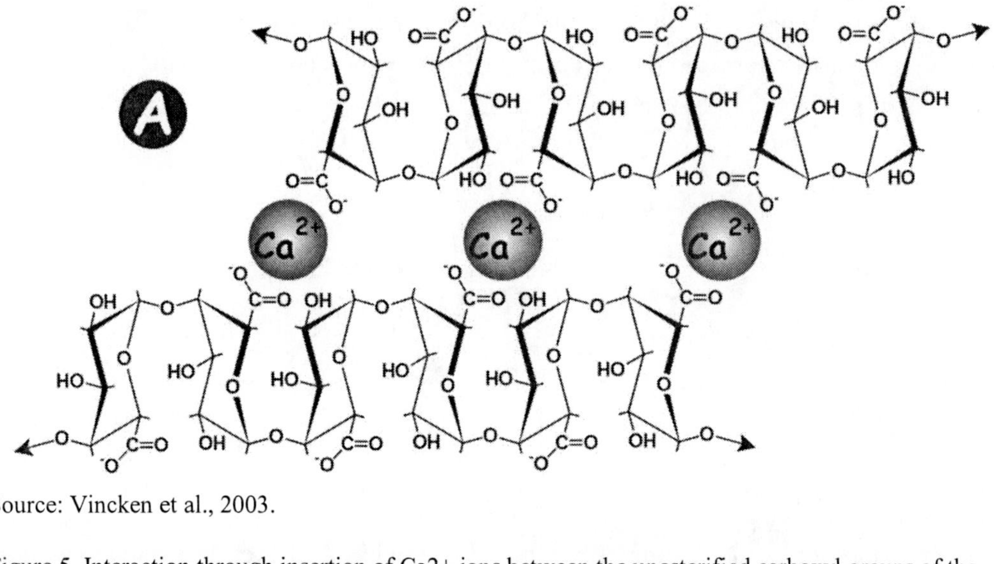

Source: Vincken et al., 2003.

Figure 5. Interaction through insertion of Ca2+ ions between the unesterified carboxyl groups of the galacturonosyl residues of two HG chains.

In unripe fruit, pectin is found as a water insoluble pectic substance, the protopectin, bounded to cellulose microfibrils conferring rigidity on cell walls. During ripening the fruit enzymes alter the pectin structure by breaking the pectin backbone or side chains, resulting in a more soluble molecule (Kashyap et al., 2001).

Pectic substances are mainly classified into the following four types:

I.   **Protopectin:** is the water insoluble pectic substance present in intact tissue. Protopectin on restricted hydrolysis yields pectin or pectic acids.
II. **Pectic acid:** is the soluble polymer of galacturonans that contains negligible amount of methoxyl groups. Normal or acid salts of pectic acid are called pectates.
III. **Pectinic acids:** is the polygalacturonan chain that contains > 0 and < 75% methylated galacturonate units. Normal or acid salts of pectinic acid are referred to as pectinates.
IV. **Pectin (Polymethyl galacturonate):** is the polymeric material in which, at least, 75% of the carboxyl groups of the galacturonate units are esterified with methanol. It confers rigidity on cell wall when it is bound to cellulose in the cell wall (Figure 6)

Pectic substances represent between 0.5 - 4% of fresh weight plant material (Jayani et al., 2005) (Table 2). In addition to their role as cementing and lubricating agents in the cell walls of higher plants, they are responsible for the texture of fruits and vegetables during growth, maturation and their storage (Mohnen 2008; Caffall & Mohnen, 2009).

Furthermore, pectic substances are involved in the phytopathogenicity between plant hosts and their pathogens (Kashyap et al., 2001). This indicates that the pectic substances are present in various forms in plant cells and this is the probable reason for the existence of various forms of pectinolytic enzymes.

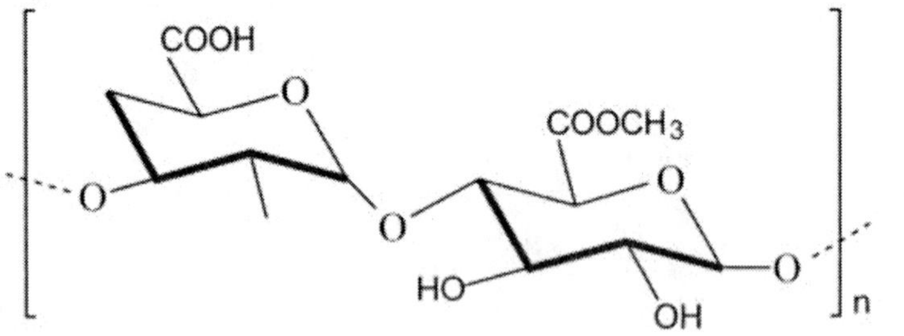

Figure 6. Structure of pectin molecule.

Pectic substances are commonly amorphous with a degree of polymerization. Compared with young actively growing tissues, lignified tissues have a low content of pectic substances. The content of the pectic substances is very low in higher plants. They are mainly found in fruits and vegetables, constitute a large part of some algal biomass (up to 30%) and occur in low concentration in forestry or agricultural residues. Polysaccharides from cell walls of ripe pears were reported to contain 11.5% pectic substances, 16.1% lignin, 21.4% glucosan, 3.5% galactan, 1.1% mannan, 21% xylan and 10% arabinan. Contrary to the proteins, lipids and nucleic acids, pectic substances do not have a defined molecular weight. The relative molecular masses of pectic substances range from 25 to 360 kDa.

Table 2. Composition of pectin in different fruits and vegetables

Fruit/vegetable	Pectic substance (%)
Apple	0.5 – 1.6
Banana	0.7 – 1.2
Peaches	0.1 – 0.9
Strawberries	0.6 – 0.7
Cherries	0.2 – 0.5
Peas	0.9 – 1.4
Carrots	6.9 – 18.6
Orange pulp	12.4 – 28.0
Potatoes	1.8 – 3.3
Tomatoes	2.4 – 4.6
Sugar beet	10.0 – 30.0

8. ROLE OF MICROBES IN PECTINASE PRODUCTION

Enzymes are the biocatalysts that regulate many chemical changes in living tissues (Prathyusha and Suneetha, 2011). Pectinases constitute a group of at least seven enzymes that catalyse the breakdown of pectin. Pectolysis is one of the most important processes for plant as it plays a role in cell elongation and growth as well as fruit ripening. Microbial pectinases are important in plant pathogenesis, symbiosis, decomposition of plant deposits (Lang and Dornenberg, 2000). The microorganisms that abundantly produce pectinolytic enzymes are saprophytic fungi which include species of Mucor, Aspergillus, Paecilomyces, Rhizopus, etc. (Gummadi and Panda 2003; Guimaraes et al., 2008). Thus, by breaking down pectin polymer for nutritional purposes, microbial pectolytic enzymes play an important role in nature (Yadav et al., 2009). These enzymes are inducible, produced only when needed and they contribute to the natural carbon cycle (Hoondal et al., 2002).

Several studies on microbial enzymes have depicted the production of multiple pectinase forms which differ on molecular mass and kinetic properties which in turn improves the ability of the microorganism to adapt to different environmental changes (Naidu and Panda 2003; Kars et al., 2005).

9. METHODS OF PECTINASE PRODUCTION

Microbial enzymes are commercially produced either through submerged Fermentation (SmF) or Solid Substrate Fermentation (SSF) techniques. The SmF techniques for enzyme production are generally conducted in stirred tank reactors under aerobic conditions using batch or fed batch systems. SmF is cultivation of microorganisms on liquid broth which requires high volumes of water, continuous agitation and generates lot of effluents. High capital investment, energy costs and the need for infrastructural requirements in large scale production make the application of SmF techniques in enzyme production more impractical. On the contrary, SSF incorporates microbial growth and product formation on or within particles of a solid substrate under aerobic conditions, in the absence or near absence of free water (Aguilar et al., 2008)

Generally, pectinases are produced by SmF and SSF systems (Kaur et al., 2004; Martin et al., 2010) Studies have been conducted on comparative production of pectinases by SmF and SSF processes (Mienda et al., 2011).

10. SUBMERGED FERMENTATION

Submerged fermentation is the cultivation of microorganisms in liquid nutrient broth. This involves growing carefully selected micro organisms in vessels containing desired nutrients and agitated. As the microorganisms break down the nutrients, they release the enzymes into broth. With the expansion of large-scale fermentation technologies, the production of microbial enzymes accounts for a noteworthy proportion of the biotechnology industrial output. Fermentation takes place in large vessels (fermenter) with volumes of up to 1,000 cubic metres. Submerged fermentation is a well developed system and has the advantages of better opportunities for process control, analysis and aims to increase fermentation yield through the use of optimized medium (Martin et al., 2010) (Table 3). SmF is used to produce a wide variety of microbial bioactive metabolites. Pectinolytic enzymes on a large scale are produced from *Aspergillus* and *Penicillium sps*. For the industrial production of pectinolytic enzymes, it is important to optimise the cultural conditions to yield better production on inexpensive carbon sources (Saranraj and Naidu, 2014). The price of commercially available enzymes that are produced mostly by SmF is too high for applications (Fawzi et al., 2003).

Batch-fed and continuous fermentation processes are common types of Submerged Fermentation. In the batch-fed process, sterilised nutrients are supplemented to the fermenter during microbial growth. In the continuous process, sterilised liquid nutrients are fed into the fermenter at the same flow rate as the fermentation broth leaving the system. This will attain a steady-state production. Parameters like temperature, pH, oxygen consumption and carbon dioxide formation are automatically controlled to optimise the fermentation process (Renge et al., 2012)

Table 3. Advantages and disadvantages of SMF process

Advantages	Disadvantages
• Measure of process parameters is easier than with solid-state fermentation. • Bacterial and yeast cells are evenly distributed throughout the medium. • There is a high water content which is ideal for bacteria.	• High costs due to the expensive media • Large reactors are needed and the behaviour of the organism cannot be predicted at times. • There is also a risk of contamination

Post production, removal of insoluble products, e.g., microbial cells is normally done by centrifugation. Since most industrial enzymes are extracellular secreted by cells into the external environment, they persist in the fermented broth after the biomass is removed. The enzymes in the remaining broth are then concentrated by evaporation, membrane filtration or crystallization depending on their intended application. Pure enzyme preparations are prepared by gel or ion exchange chromatography. Applications where commercial enzyme products are needed, the crude powder enzymes are made into granules for convenient use or liquid formulations are preferred because they are easier to handle (Singhania et al., 2009).

11. SOLID STATE FERMENTATION (SSF)

Solid state fermentation (SSF) is generally defined as the growth of microorganisms on solid materials in the absence of free water. It has tremendous potential for the production of enzymes. Studies on the production of enzymes by SSF are increasing because of the potential advantages such as simplicity, high productivity and concentrated products over submerged fermentations (Table 4) (Mienda et al., 2011).

Substrates that are employed in the production of enzyme should be solid, as solid substrate can encourage good growth of microbes. Substrates provide all needed nutrients to the microorganisms for their growth. Generally agro-industrial wastes are employed for the pectinase production (Pandey et al., 2009). Production of enzymes from agro wastes is very essential as they contain large amounts of cellulose, hemicellulose and pectin, which could serve as inducers for the production of cellulase, xylanase, and pectinases, respectively (Gholifar et al., 2010).

Microorganisms are widely accepted as the best sources for the production of enzymes from agro wastes. Bacteria and fungi are known to produce wide array of industrial enzymes. (Sumantha et al., 2005). Pectinase production has been reported from bacteria including actinomycetes (Phutela et al., 2005), yeast and fungi (Niturea et al., 2008).

Fungi produce several extracellular enzymes that result in the decomposition of organic matter with the pectinolytic enzymes being one among them. Fungal pectinases break down the middle lamella in plants so that it can extract nutrients from the plant.

Various substrates that are presently being used for pectinase production include sugarcane bagasse, wheat bran, rice bran, wheat straw, rice straw, sorghum stems, saw dust, corn cobs, sun flower heads, coconut coir pith, banana waste, tea waste, sugar beet pulp, apple pomace, orange peel, soya bean pulp powder, lemon peel, coffee pulp, wheat husk, wheat bran, soya bran, banana peel, pineapple peels, etc. (Okafer et al., 2010; Durairajan and SivaSankari 2014)

According to Patil and Chaudhari (2010), it was found that orange bagasse as the best substrate for PG production by *Penicillium sp.* Similar studies were also carried out by Silva et al., (2002) where orange bagasse and wheat bran gave higher yields of PG when inoculated with culture *P. viridicatum* RFC3. Orange bagasse used in the study was obtained after the extraction of juice from fruit processing industries in high quantities

Table 4. Advantages and disadvantages of the SSF process

Advantages	Disadvantages
The culture media are simple. Some substrates can be used directly as a solid media or enriched with nutrients	The used microorganisms are limited those that grow in reduced levels of humidity
The product of interest is concentrated, that which facilitates its purification	The determination of parameters such as humidity, pH, free oxygen and dioxide of carbon, constitute a problem due to the lack of monitoring devices
The used inoculum is the natural flora of the substrates, spores or cells	The scale up of SSF processes has not been worked out to the industrial level.
The low humidity content and the great inoculums used in a SSF reduce vastly the possibility of a microbial contamination	Static condition is mostly preferred as agitation most often proved to be very difficult.
The quantity of waste generated is smaller than the SmF	Difficulties are usually encountered in biomass determination.
The enzymes are low sensitive to catabolic repression or induction	There is possibility of contamination with unwanted fungal species.
The main advantage of SSF over SmF is the higher concentration of products and less effluent generation	Aeration may be difficult sometimes due to high solid concentrations

Source: Mienda et al., 2011.

The effect of carbon and nitrogen sources on the productivity of pectinases has been studied (Almeida et al., 2003; Patil and Dayanand, 2006a). Other factors like particle size, moisture levels, concentration of nutrients, pH, temperature, nature of solvent, etc. is also taken into consideration during pilot scale production (Martin et al., 2004). Thermal stability plays an important role especially when dealing with enzymes, which in turn is a function of the exposure time (Martin et al., 2010). A comparative account of Smf and SSF has been listed in Table 5.

There is a need to highlight recent developments on several aspects related to pectinase production due to their wide applications. The pectinolytic enzymes from microbes have generally focused on the induction of enzyme production under various conditions, fermentation process, various substrate purification and characterization and use of this enzyme for different industrial processes. In order to use enzyme from the isolates for commercial application, it must have desirable biochemical and physiochemical characteristics combined with low cost of production.

Table 5. Comparison between liquid and solid substrate fermentations

Factor	SmF	SSF
Substrates	SmF utilizes free flowing liquid substrates, such as molasses and broths.	SSF utilizes solid substrates, like bran, bagasse, and paper Pulp
Utilization of substrates	The substrates are utilized quite rapidly and need to be constantly supplemented with nutrients.	The substrates are utilized very slowly and steadily, so the same substrate can be used for long fermentation periods.
Aseptic conditions	Heat sterilisation and aseptic Control	Vapour treatment, non sterile Conditions
Water	High volumes of water consumed and effluents discarded	Limited consumption of water, low effluent
Metabolic Heating	Easy control of temperature	Low heat transfer capacity Easy aeration and high surface exchange air/substrate
pH control	Easy pH control	Buffered solid substrates
Mechanical agitation	Good homogenization	Static conditions preferred
Scale up	Industrial equipments Available	Need for Engineering & New
Inoculation	Easy inoculation	Spore inoculation
Contamination	Risk of contamination	Risk of contamination for low rate growth fungi
Energetic consideration	High energy consuming	Low energy consuming
Volume of Equipment	High volumes and high cost	Low volumes &low costs of Equipments
Effluent & pollution	High volumes of polluting effluents	No effluents, less pollution
Concentration of Products	30 - 80 g/l	100/300g/l

Source: Maghsoodi and Yaghmaei, 2010.

12. APPLICATION OF PECTINOLYTIC ENZYMES

The applications of microbial pectinases in industries are vast. Pectins have numerous and important applications in the food and pharmaceutical industries. In the food sector, it is primarily used as a gelling agent, replacing sugars and fats in low-calorie food and as nutritional fiber (Gummadi et al., 2003). The pharmaceutical industry offers them as preparations to reduce cholesterol or to act as a lubricant in the intestines thus promoting normal peristaltic movement without causing irritation. In addition, these polysaccharides are used as drug delivery systems, which can also reduce the toxicity of these and make their activity longer lasting without altering their therapeutic effects (Morris et al., 2010).

12.1. Textile Processing and Bioscouring of Cotton Fibers

Bioscouring is a novel process for removal of non-cellulosic material present on the surface of the cotton. Raw cotton contains about 90% of cellulose and various noncellulosics such as waxes, pectins, proteins, fats, lignin-containinig impurities and colouring matter. Pectinases have been used in conjunction with amylases, lipases, cellulases and hemicellulases to remove sizing agents from cotton in a safe and ecofriendly manner (Hoondal et al., 2002). The pectin content of cotton fibre can be decreased by about 30% in the pectinase treatment. Removal of pectin results in lower amounts of waxes on the cotton surface and improves the water absorbency of the fabric. The enzymatic treatment has no effect on the tensile strength of the fabric (Agrawal et al., 2008)

12.2. Extraction of Fruit Juices

Pectic enzymes are used in the fruit juice industries and wine making often come from fungal sources. Pectinases (acidic) are widely used in extraction, clarification, and removal of pectin in fruit juices and in winemaking, are often produced by *Aspergillus niger*. Pectins contribute to fruit juice viscosity and turbidity. A mixture of pectinases and amylases are used to clarify fruit juices by decreasing the filtration time upto 50%. Pectinases have been used to increase the pressing efficiency of the fruits for juice extraction along with other enzymes such as cellulases, arabinases and xylanases (Gailing et al., 2000). The crushing of pectin-rich fruits results in high viscosity juice which remains with the fruit pulp in a gelatinous structure, thereby hindering the juice extraction process. The addition of Pectinase to the extraction process improves the fruit juice yield by decreasing the juice viscosity, thereby improving the juice concentration capacity (Jayani et al., 2005).

Pectinases used in fruit juice industries are commonly employed for producing sparkling clear juices wherein enzymes are added in order to increase the juice yield during pressing and straining of the juice and to remove suspended matter to give sparkling clear juices (free of haze). Apples, pear, strawberry, raspberry, blackberry, pear, grape juice and wine are few examples where pectic enzymes are used to produce haze free juices. Pectin enzymes containing high levels of polygalacturonase activity are added to fruit juices such as oranges, guava, apricot, papaya, pineapple and banana to stabilize the cloud of citrus juices, purees and nectars. Pectinases are used to soften the peel of citrus fruits for removal by vacuum infusion technique which can be widely used for the production of canned products. The exploitation of pectinases, mainly exo-PG have been well established in variety of fruit juice and wine processing industries to increase the juice yield, clarification, promoting antioxidant formation and juice concentrate production. The addition of such exogenous enzyme also allows more specific degradation which is necessary to give a characteristic smooth texture, colour and increases level of reducing sugar (Pasha et al., 2013).

12.3. Retting of Plant Fibres

Retting and Pectinolytic enzymes are involved in the retting and degumming of jute, flax, and coir from coconut husks. Retting is a fermentation process in which certain bacteria (e.g., Clostridium, Bacillus) and certain fungi (e.g., Aspergillus, Penicillium) decompose the pectin of the bark and release fiber (Hoondal et al., 2002).

12.4. Degumming of Fiber Crops

Bast fibers are the soft fibers formed in groups outside the xylem, phloem or pericycle, e.g., Ramie and sun hemp. Ramie fibers are an excellent natural textile, but decorticated ramie fibers contain 20 ± 35% ramie gum consisting of pectin and hemicelluloses making it essential to degum the fibres for textile manufacturing (Kashyap et al., 2001). In a classical degumming process, this gum is removed by treatment of decorticated fibers with hot alkaline solution (12 ± 20% NaOH solution) with or without application of pressure. In addition to the high consumption of energy, this treatment is polluting, toxic and non-biodegradable. Biotechnological degumming using pectinases along with xylanases presents an eco-friendly and economic alternative for the production of degummed fibres (Kapoor and Kuhad., 2002). Pectinolytic enzymes from actinomycetes have demonstrated good correlation between the pectate lyase activity and the degumming effects, resulting in good separation of the fiber (Kashyap et al., 2001).

12.5. Pretreatment of Pectic Waste Waters

The wastewater from the citrus-processing industry contains pectinaceous materials that are barely decomposed by microbes during the activated-sludge treatment (Pasha et al., 2013). The treatment of waste water from citrus processing industries containing pectic substances include various procedures which include physical dewatering, spray irrigation, chemical coagulation, direct activated sludge treatment and chemical hydrolysis resulting in the formation of methane (Hoondal et al., 2002). These have several disadvantages, such as the high cost of treatment and longer treatment times in addition to environmental pollution from the use of chemicals. Thus, an alternative, cost effective, and environmentally friendly method is the use of pectinases from bacteria, which selectively remove pectic substances from the waste water. The pretreatment of pectic wastewater from vegetable/food processing industries with alkaline pectinase and alkalophilic pectinolytic microbes facilitates removal of pertinacious material and renders it suitable for decomposition by activated sludge treatment (Raghuvanshi et al., 2013).

12.6. Paper and Pulp Industry

The use of microbial enzymes for paper making has substantially increased over the years. The alkaline peroxide bleaching of mechanical pulp solubilizes acidic polysaccharides (pectins) which are interfering substances during paper making process. Pectinases can depolymerize polymers of galacturonic acid, and consequently lower the cationic demand of pectin solutions and the filtrates obtained from peroxide bleaching (Pedrolli et al., 2009).

12.7. Coffee and Tea Fermentation

Pectinase treatment destroys the foam forming property of instant tea powders and aides in tea fermentation. Mucilaginous coat from coffee beans during Coffee fermentation are removed by the addition of pectinases. Pectinases are sometimes added to remove the pulpy bean layer consisting of pectic substances (Jayani et al., 2005; Raghuwanshi et al., 2013)

12.8. Animal Feed

Pectinases are used for the production of animal feeds. Intensive research in to the use of varies enzymes in animal and poultry feeds started in the early 1980s. The first commercial success was addition of β-glucanase in to barley-based feed diets. Usually a feed enzyme preparation is a multi enzyme cocktail containing glutanases, xylanases, proteinases, pectinase and amylases. Enzyme addition reduces the feed viscosity, which increases absorption of nutrients, liberates nutrients, either by hydrolysis of non-biodegradable fibers or by liberating nutrients blocked by these fibers, and reduces the amount of faeces (Jayani et al., 2005).

12.9. Oil Extraction

Citrus oil such as lemon oil can be extracted with pectinase as this enzyme destroys the emulsifying properties of pectin that interfere with the collection of oils from citrus peel extracts. Plant cell wall degrading enzyme preparation has begun to be used in olive oil preparation. The enzyme is added during the process of grinding of olives which aids in easy removal of oil and is accomplished by subsequent separation procedures (Raghuwanshi et al., 2013).

12.10. Improvement of Chromaticity and Stability of Red Wines

Pectinases improve the colour, turbidity and enhanced stability of red wines when added to macerated fruits before the addition of yeast during wine making when compared to non- enzyme treated wines (control).

These wines also showed greater stability as compared to the control (Jayani et al., 2005). The addition of pectolytic enzymes leads to increased levels of methanol in wine (Raghuwanshi et al., 2013) due to the activity of pectin methyl esterase. The maximum concentration of methanol in wine is regulated as it is toxic. Therefore, pectin methyl esterase activity should be at low concentrations in commercial mixtures. The functions of pectic enzymes in the winemaking process are to maximize juice yield, facilitate filtration and intensify the flavour and colour of wine so obtained (Sieiro et al., 2003).

12.11. Purification of Plant Viruses

In cases where the virus particle is restricted to phloem, alkaline pectinases are used to obtain pure viral preparations if the virus particle is restricted to the phloem tissue. (Jayani et al., 2005)

CONCLUSION

Industrial uses of enzymes have greatly increased during the past few years. Continued and increased usage of presently available enzymes with novel applications and further development of newer enzymes systems in various industries are of foremost importance in the evergrowing commercial sector. Pectinases are the enzymes that hydrolyze pectic substances and have been used to increase yields and clarity of fruit juices and wine. Most of the studies have concentrated with the screening, isolation, production, purification, characterization and applications of pectinolytic enzymes in increasing the fruit juice yield and its clarification. Newer applications of pectinolytic enzymes in other industries such as textiles, paper and pulp, oil, animal feed, etc. are being investigated. Much research is underway by various enzyme manufacturers to find new and improved sources of pectinase for better yield and stability. Screening a large number of microorganisms for pectinolytic enzymes combined with protein engineering, direct evolution and metagenome approaches can lead to development of novel strains. The study on genetic machinery on enzyme system used by microbes for complete breakdown of pectin is the most important tool for developing an economical and ecofriendly approach. Future studies on pectic enzymes should be devoted to the understanding of the regulatory mechanism of the enzyme secretion at the molecular level and generation of recombinant strains with higher activity.

REFERENCES

Abbott, D. W., Hrynuik, S. & Boraston, A. B. (2007). Identification and characterization of a novel periplasmic polygalacturonic acid binding protein from Yersinia enterocolitica. *J. Mol. Biol.*, *367*, 1023–1033.

Agrawal, P. B., Nierstrasz, V. A., Bouwhuis, G. H. & Warmoeskerken, M. M. C. G. (2008). Cutinase and pectinase in cotton bioscouring: an innovative and fast bioscouring process. *Biocatalysis and Biotransformation.*, *26*(5), 412-421.

Aguilar, C. N., Sánchez, G. G., Barragán, P. A., Herrera, R. R., Martínez-Hernandez, J. L. & Contreras-Esquivel, J. C. (2008). Perspectives of Solid State Fermentation for Production of Food Enzymes. *American J Biochem Biotechnol.*, *4* (4), 354-366.

Alimardani-Theuil, P., Gainvors-Claisse, A. & Duchiron, F. (2011). Yeasts: An attractive source of pectinases- From gene expression to potential application: A review. *Process Biochem.*, *46*, 1525-1537.

Almeida, C., Brányik, T., Moradas-Ferreira, P. & José. (2003). Teixeira Continuous Production of Pectinase by Immobilized Yeast Cells on Spent Grains. *J Bioscience Bioengineering.*, *96*(6), 513-518.

Binod, P., Singhania, R. R., Soccol, C. R. & Pandey, A. (2008). "Industrial enzymes," in *Advances in Fermentation Technology*, A. Pandey, C. Larroche, C. R. Soccol, and C.-G. Dussap, Eds Asiatech Publishers, New Delhi, India. pp. 291–320.

Caffall, K. H. & Mohnen, D. (2009). The structure, function, and biosynthesis of plant cell wall pectic polysaccharides. *Carbohydr Res*, *344*, 1879–1900.

Durairajan, B. & Siva Sankari, P. (2014). Optimization of Solid State Fermentation Conditions for the production of Pectinases by Aspergillus niger. *J. Pharm. BioSci.*, 250-57.

Favela-Torres, E., Volke-Sepulveda, T. & Viniegra-Gonzalez, G. (2006). Production of hydrolytic depolymerising pectinases. *Food Technol and Biotechnol*, *44*(2), 221-227.

Fawzi, E. M. (2003). Production and Purification of β-glucosidase and Protease By Fusarium Proliferatum NRRL 26517 grown on Ficus Nitida wastes. *Annals Of Microbiology.*, *53* (4), 463-476.

Fernández-González, M., Úbeda, J. F., Vasudevan, T. G., Otero, R. R. C. & Briones, A. I. (2004). Evaluation of polygalacturonase activity in Saccharomyces cerevisiae wine strains. *FEMS Microbiol Lett.*, *237*, 261-267.

Gholifar, E., Asadi, A., Akbari, M. & Atashi, M. P. (2010). Effective Factors in Agricultural Apple Waste in Islamic Republic of Iran: A Comparative Study. *J Hum Ecol*, *32*(1), 47-53.

Guimarães, P. G., Rico-Gray, J. V., Reis, S. F. & Thompson, J. N. (2008). Asymmetries in specialization in ant–plant mutualistic. *Proc. R. Soc. B*, 2006, *273.*, doi: 10.1098/rspb.2006.3548.

Gummadi, S. N. & Panda, T. (2003). Purification biochemical properties of microbial pectinases: A review. *Process Biochem.*, *38*, 987-996.

Haider, A. & Fazelipoor, M. H. (2010). Pectinase production in a defined medium using surface culture fermentation. *Ind. J Ind Chem.*, *1*(1), 5-10.

Hannan, A., Bajwa, R. & Latif, Z. (2009). Status of Aspergillus niger strains for pectinase production potential. *Pak J of Phytopathol.*, *21* (1), 77-82.

Hasunuma, T., Fukusaki, E. I. & Kobayashi, A. (2003). Methanol production is enhanced by expression of an Aspergillus niger pectin methylesterase in tobacco cells. *J. Biotechnol.*, *106*, 45–52.

Hoondal, G. S., Tiwari, R. P., Tewari, R., Dahiya, N. & Beg, Q. K. (2002). Microbial alkaline pectinases and their industrial applications: a review. *Appl. Microbiol. Biotechnol.*, *59*, 409-18.

Hutnan, M., Drtil, M. & Mrafkova, L. (2000). Anaerobic biodegradation of sugar beet pulp. *Biodegradation.*, *11*, 203-211.

Jayani, R. S., Saxena, S. & Gupta, R. (2005). Microbial pectinolytic enzymes: A review. *Process Biochemistry.*, *40*(9), 2931-2944.

Kapoor, M. & Kuhad, R. C. (2002). Improved polygalacturonase production from Bacillus sp. MG-cp-2 under submerged (SmF) and solid state (SSF) fermentation. *Lett Appl Microbiol*, *34*(3), 317-322.

Kars, I., Krooshof, G., Wagemakers, L., Joosten, R., Benen, J. A. E. & Kan, J. A. L. (2005). Necrotizing activity of five Botrytis cinerea endopolygalacturonases produced in Pichia pastoris. *The Plant Journal.*, *43*, 213-225.

Kashyap, Vohra D. R., Chopra, P. K. S. & Tewari, R. (2001). Applications of pectinases in the commercial sector: A review. *Bioresource Technol.*, *77*(3), 215- 227.

Kaur, G., Kumar, S. & Satyanarayana, T. (2004). Production, characterization and application of a thermostable polygalacturonase of a thermophilic mould Sporotrichum thermophile. Apinis. *Bioresource Technol. 94*, 239–243.

Khule, N. R., Mahale, N. B., Shelar, D. S., Rokade, M. M. & Chaudhari, S. R. (2012). Extraction of pectin from citrus fruit peel and use as natural binder in paracetamol tablet. *Der Pharmacia Lettre.*, *4* (2), 558-564.

Koboyashi, T., Higaki, N., Yajima, N., Suzumatsu, A., Haghihara, H., Kawai, S., et al. (2001). Purification and properties of a galacturonic acid-releasing exopolygalacturonase from a strain of Bacillus. *Biosci Biotechnol Biochem.*, *65*, 842–7.

Lang, C. & Dornenburg, H. (2000). Perspectives in the biological function and the technological application of polygalacturonases. *Applied Microbial. Biotechnol.*, *53*, 366-375.

Liu, K., Zhao, G., He, B., Chen, L. & Huang, L. (2012). Immobilization of pectinases and lipase on macroporous resin coated with chitosan for treatment of whitewater from papermaking. *Bioresour. Technol.*, *123*, 616-619.

Maghsoodi, V. & Yaghmaei, S. (2010). Comparison of Solid Substrate and Submerged Fermentation for Chitosan Production by Aspergillus niger. *Transactions C: Chemistry and Chemical Engineering.*, *17*(2), 153-157.

Martin, N., Guez, M. A. U., Sette, L. D., Da Silva, R. & Gomes, E. (2010). Pectinase production by a Brazilian thermophilic fungus Thermomucor indicae-seudaticae N31 in solid-state and submerged fermentation. *Microbiology.*, *79*(3), 306-313.

Martin, N., Souza, S. R., Silva, R. & Gomes, E. (2004). Pectinase production by fungal strains in solid-state fermentation using agro-industrial bioproduct. *Braz. arch. biol. technol. [online].* *47*(5), 813-819.

Mienda, S. B., Idi, A. & Umar, A. (2011). Microbiological features of solid state fermentation and its applications- An overview. *Res. Biotechnol.*, *2*(6), 21-26.

Mohnen, D. (2008). Pectin structure and biosynthesis. *Curr Opin Plant Biol.*, Jun, *11*(3), 266-77. doi: 10.1016/j.pbi.2008.03.006.

Morris, G., Kök, S., Harding, S. & Adams, G. (2010). Polysaccharide drug delivery systems based on pectin and chitosan. *Biotechnol & Genetic Engineering Rev.*, Vol. *27*, 257-284.

Naidu, G. S. N. & Panda, T. (2003). Studies on pH and thermal deactivation of pectolytic enzymes from Aspergillus niger. *Biochem. Eng. J.*, *16*, 57–67.

Nirmaladevi, D., Anilkumar, M. & Srinivas, C. (2014). Production and Characterization of Exopolygalacturonase from Fusarium Oxysporum f. Sp. Lycopersici. *Int J Pharm Bio Sci.*, *5*(1), (B), 666 – 675.

Niturea, S. K., Kumarb, A. K., Parabc, P. B. & Panta, A. (2008). Inactivation of polygalacturonase and pectate lyase produced by pH tolerant fungus Fusarium moniliforme NCIM 1276 in a liquid medium and in the host tissue. *Microbiological Research.*, *163*, 51—62.

Okafor, U. A., Okachi, V. I., Chinedu, S. N., Ebuehi, O. A. T. & Onygeme-Okerenta, B. M. (2010). Pectinolytic activity of wild-type filamentus fungi fermented on agro-wastes. *Afr J Microbiol Res.*, *4*(24), 2729-2734.

ONeill, M. A., Ishi, T., Albersheim, P. & Darvill, A. G. (2004). Rhamnogalacturonan II: structure and function of a borate cross-linked cell wall pectic polysaccharide. *Annu. Rev. Plant Physiol. Plant Mol. Biol.*, *55*, 109–139.

Pandey, A., Selvakumar, P., Soccol, C. R., et al., (2009). Solid state fermentation for the production of Industrial enzymes. *Curr. Sci.*, *77*, 149-162.

Pasha, K. M., Anuradha, P. & Subbarao, D. (2013). Applications of Pectinases in Industrial Sector. *Int. J. Pure Appl. Sci. Technol.*, *16*(1), 89-95.

Patil, N. P. & Chaudhari, B. L. (2010). Production and Purification of Pectinase by Soil isolate Penicillium Sp and search for better Agro-Residue for its SSF. *Recent Research in Science and Technology*, *2*(7), 36-42

Patil, S. R. & Dayanand, A. (2006). Optimization of process for the production of fungal pectinases from deseeded sunflower head in submerged and solid-state conditions. *Bioresource Technol.*, *97*(18), 2340-2344.

Pedrolli, D. B., Monteiro, A. C., Gomes, E. & Carmona, E. C. (2009). Pectin and Pectinases: Production, Characterization and Industrial Application of Microbial Pectinolytic Enzymes. *The Open Biotechnology Journal.*, *3*, 9-18.

Pérez, S., Mazeau, K. & Hervé du Penhoat, C. (2000). The three-dimensional structures of the pectic polysaccharides. *Plant Physiol Biochem.*, *38*, 37–55.

Perinbam, Kantharaj., Bharath, Boobalan., Seeni, Sooriamuthu. & Ravikumar, Mani. (2017). "Lignocellulose Degrading Enzymes from Fungi and Their Industrial Applications." *Int J Cur Res Rev|*, Vol *9*, no. 21, 1.

Phutela, U., Dhuna, V., Sandhu, S. & Chadha, B. S. (2005). Pectinase and polygalacturonase production by thermophilic Aspergillus fumigates isolated from decomposting orange peels, *Brazilian journal of Microbiology*, *36*(1), 63-69.

Raghuwanshi, S., Gupta, A., Srivastava, A., Singh, D., Nema, R., Kumar, A. J., Pratap, B., Verma, S., Shrivastva, H., Dadse, D., Binjhade, D., Ghidode, S. & Thakre, S. (2013). Pectolytic enzyme and its global applications: A Review. *CMBT J Sci Tech.*, *1*(1), 1-9.

Renge, V. C., Khedkar, S. V. & Nandurkar, N. R. (2012). Enzyme Synthesis By Fermentation Method: A Review. *Sci. Revs. Chem. Commun.*, *2*(4), 585-590.

Saranraj, P. & Naidu, M. A. (2014). Microbial Pectinases: A Review. *Global J Trad Med Sys.*, *3*(1), 1-9.

Saranraj, P. & Stella, D. (2013). Fungal Amylase: A Review. *Int J of Microbiol Res.*, *4*(2), 203 - 211.

Shembekar, V. S. & Dhotre, A. (2009). Studies of pectin degrading microorganisms from soil. *J microbial world.*, *11*(2), 216-222.

Sieiro, C., Poza, M., Vilanova, M. & Villa, T. G. (2003). Heterologous expression of the Saccharomyces cerevisiae PGU1 gene in Schizosaccharomyces pombe yields an enzyme with more desirable properties for the food industry. *Appl Environ Microbiol.*, *69*, 1861-1865.

Silva, D., Martins, E. S., Silva, R. & Gomes, E. (2002). Pectinase production from Penicillium viridicatum RFC3 by solid state fermentation using agricultural residues and agro-industrial by-product. *Braz. J. Microbiol.*, *33*, 318-324.

Singhania, R. R., Patel, A. K., Soccol, C. R. & Pandey, A. (2009). Recent advances in solid-state fermentation. Biochem. *Eng. J.*, *44*, 13–18.

Sinitsyna, O. A., Fedorova, E. A., Semenova, M. V., et al. (2007). Isolation and characterization of extracellular pectin lyase from Penicillium canescens. *Biochem* (Moscow), *72*(5), 565-71.

Soriano, M., Diaz, P. & Pastor, F. I. J. (2005). Pectinolytic systems of two aerobic sporogenous bacterial strains with high activity on pectin. *Curr. Microbio.*, *50*, 114–118.

Sumantha, A., Sandhya, C., Szakacs, G., Soccol, C. R. & Pandey, A. (2005). Production and partial purification of a neutral metalloprotease by fungal mixed substrate fermentation, *Food Technol Biotechnol.*, *43* (4), 313-319.

Tripathi, G. D., Javed, Z. Sushma. & Singh, A. K. (2014). Pectinase production and purification from Bacillus subtilis isolated from soil. *Adv Appl Sci Res.*, *5*(1), 103-105.

Turner, P., Mamo, G. & Karlsson, E. N. (2007). Potential and utilization of thermophiles and thermostable enzymes in biorefining. *Microbial Cell Factories.*, *6*, 9 doi:10.1186/1475-2859-6-9.

Van Alebeek, G. J. W. M., Christensen, T. M. I. E., Schols, H. E., Mikkelsen, J. D. & Voragen, A. G. J. (2002). Mode of action of pectin lyase A of Aspergillus niger on differently C6-substituted oligogalacturonides. *J Biol Chem*, *277*(29), 25929-36.

Vermelho, A. B., Supuran, C. T. & Guisan, J. M. (2012). Microbial Enzyme: Applications in Industry and in Bioremediation. Hindawi Publishing Corporation Enzyme Research, Article ID 980681. doi:10.1155/2012/980681.

Vincken, J. P., Schols, H. A., Oomen, R. J. F. J., et al. (2003). If homogalacturonan were a side chain of rhamnogalacturonan I: implications for cell wall architecture. *Plant Physiol*, *132*, 1781-9.

Viniegra-González, G., Favela-Torres, E., Aguilar, C. N., Rómero-Gomez, S. J., Díaz-Godínez, G. & Augur, C. (2003). Advantages of fungal enzyme production in solid state over liquid fermentation systems. *Biochem. Eng. J.*, *13*, 157–167.

Willats, W. G. T., Knox, P. & Mikkelsen, J. D. (2006). Pectin: new insights into an old polymer are starting to gel. *Trends Food Sci Technol.*, *17*, 97-104.

Yadav, S., Yadav, P. K., Yadav, D. & Yadav, K. D. S. (2008). Purification and characterization of an alkaline pectin lyase from Aspergillus flavus. *Process Biochem.*, *43*, 547-502.

In: Microbial Catalysts. Vol. 2
Editors: Shadia M. Abdel-Aziz et al.

ISBN: 978-1-53616-088-8
© 2019 Nova Science Publishers, Inc.

Chapter 6

MICROBIAL ENZYME: AN EFFECTIVE REPLACEMENT OF INDUSTRIAL CATALYST

M. Nuruzzaman Khan, PhD, Ismat Z. Luna, Taslim Ur Rashid, Khandoker S. Salem, Md. Minhajul Islam, Asaduz Zaman, Sadia Sharmeen, Papia Haque, and Mohammed Mizanur Rahman,[*] PhD

Department of Applied Chemistry and Chemical Engineering, Faculty of Engineering and Technology, University of Dhaka, Dhaka, Bangladesh

ABSTRACT

In this chapter the activity of microbial enzyme in daily life to industrial production are discussed. Since the dawn of mankind, enzymes have been used in cheese production and indirectly via yeasts and bacteria in food manufacturing. Owing to improved production technologies, engineered enzyme properties and new applications based industrial enzyme business is steadily growing. The major part of enzymes is produced with GRAS-status. Usually the production organism and often also the individual enzyme have been genetically engineered for maximal productivity and optimized enzyme properties. Isolated enzymes have found several applications in food, starch, detergents, textile, pulp, paper, lather and fine chemical industry. Enzymes are used in production of chirally pure amino acids and rare sugars. They are also used in production of fructose and penicillin derivatives as well as several other chemicals. Enzymes are now considered as a part of a rapidly growing biocatalyst industry also involving genetically enhanced living cells as chemical production factories.

[*] Corresponding Author's E-mail: mizanur.rahman@du.ac.bd.

1. INTRODUCTION

Most of the reactions in living organisms are catalyzed by protein molecules called enzymes. Enzymes are complex protein molecules. They are produced by living organisms to catalyze the biochemical reactions required for life. Although enzymes are formed within living cells, they can continue to function *in vitro* (in the test-tube) and their ability to perform very specific chemical transformations is making them increasingly useful in industrial processes. Enzymes can rightly be called the catalytic machinery of living systems [1, 2].

Man has indirectly used enzymes almost since the beginning of human history. Enzymes are responsible for the biocatalytic fermentation of sugar to ethanol by yeasts, a reaction that forms the bases of beer and wine manufacturing.

Enzymes oxidize ethanol to acetic acid. This reaction has been used in vinegar production for thousands of years. Similar microbial enzymetic reactions of acid forming bacteria and yeasts are responsible for aroma forming activities in bread production and in preserving activities in sauerkraut preparation[3]. The fermentative activity of microorganisms was discovered only in 18th century and finally proved by the French scientist Louis Pasteur. The study of enzymes is a fairly recent activity. Over the century the research on microorganisms lead the further improvement and exciting discovery of specific enzymes for applications in foods, animal feeds, pulp, paper, lather, detergents and fine chemicals production [4].

This chapter concentrates on the selection and production of industrially suitable enzymes in large scale and their applications in technical industrial production and fine chemical industries as well as pharmaceuticals manufacturing. The use of microorganisms as biocatalysts in chemical production is, however, an interesting and growing field. The techniques of genetic, protein and pathway engineering are making chemical production by living cells an interesting green alternative to replace traditional chemical processes.

2. HISTORY OF ENZYME TECHNOLOGY

Enzyme technology may be rather a new arena of scientific exploration but the application of enzyme has been around since down of mankind. Man has indirectly used enzymes almost since the beginning of human history in the production of beer, wine, cheese, vinegar and yoghurt. The secret knowledge of alcoholic brews production from barley was known to Egyptians, Sumerians and Babylonians peoples [1]. The Greek epic poems "The Iliad" and "The Odyssey" written around 700 BC, both refer the uses of calfs and kids stomachs (sources of rennet) for the production of cheese and the early Christian and Sanskrit writings describe fermented dairy products. Since 800 BC, sour dough bread had been produced in Europe.

The first purified chemical to be produced by biotechnology was ethanol, which was being manufactured to fortify wines and beers by the 14th century. The biocatalytic activity of enzyme was employed for fermentation of sugar to ethanol by yeasts, a reaction that forms the bases of beer and wine manufacturing. Throughout the centuries, nobody understood the underlying chemistry or even that living organisms were involved. The fermentative activity of microorganisms was discovered only in 18th century and finally proved by the French scientist Louis Pasteur. He discovered that yeast converts the sugars to ethanol. In 1878 German physiologist William Kühne named the agents responsible for catalyzing the reactions "Enzymes". The term "Enzyme" comes from Latin word, which literally mean "in yeast". Eduard Buchner after a series of experiments at the University of Berlin by 1897 proved that, yeast extract could convert glucose to ethanol and CO_2 without the need for any living organisms. Such reactions were called unorganized ferments and named the enzyme "Zymase". The first application of cell free enzymes was the use of rennin isolated from calf or lamb stomach in cheese making. Rennin is an aspartic protease which coagulates milk protein and has been used for hundreds of years by cheese makers. Röhm in Germany prepared the first commercial trypsin enzyme in 1914 [5, 6]. This enzyme isolated from animals degraded proteins and was used as a detergent. It proved to be so powerful compared to traditional washing powders. The real breakthrough of enzymes occurred with the introduction of microbial proteases into washing powders. The first commercial bacterial Bacillus protease was marketed in 1959 and became big business when Novozymes in Denmark started to manufacture it and major detergent manufactures started to use it around 1965.

In food industry, besides cheese manufacturing enzymes were used already in 1930 in fruit juice manufacturing. These enzymes clarify juice. They are called pectinases, which contain numerous different enzyme activities. The major usage of microbial enzymes in food industry started in 1960s in starch industry [7]. The traditional acid hydrolysis of starch was completely replaced by alpha-amylases and glucoamylases, which could convert starch with over 95%, yield to glucose. Starch industry became the second largest user of enzymes after detergent industry.

3. Enzymes Classification

The sequence information of a growing number of organisms opens the possibility to characterize all the enzymes of an organism on a genomic level. More than 2000 different enzyme activities have been isolated and characterized till now. The smallest known organism, *Mycoplasma genitalium*, contains 470 genes of which 145 are related to gene replication and transcription. Baker's yeast has 7000 genes coding for about 3000 enzymes. Thousands of different variants of the natural enzymes are known till now. The number of reported 3-dimensional enzyme structures is rapidly increasing [8, 9]. In the year 2000 the structure of about 1300 different proteins were known. The enzymes are classified into six major categories (Table 1) based on the nature of the chemical reactions they catalyze.

Table 1. Enzyme classes and types of reactions

Enzyme Commission number	Class of enzyme	Reaction profile
EC 1	Oxidoreductases	Oxidation reactions involve the transfer of electrons from one molecule to another. In biological systems we usually see the removal of hydrogen from the substrate. Typical enzymes in this class are called dehydrogenases. For example, alcohol dehydrogenase catalyzes reactions of the type $R–CH_2\ OH \longrightarrow A$ $R–CHO + H_2\ A$, where A is an acceptor molecule. If A is oxygen, the relevant enzymes are called oxidases or laccases; if A is hydrogen peroxide, the relevant enzymes are called peroxidases.
EC 2	Transferases	This class of enzymes catalyzes the transfer of groups of atoms from one molecule to another. Aminotransferases or transaminases promote the transfer of an amino group from an amino acid to an alpha-oxoacid.
EC 3	Hydrolases	Hydrolases catalyze hydrolysis, the cleavage of substrates by water. The reactions include the cleavage of peptide bonds in proteins, glycosidic bonds in carbohydrates, and ester bonds in lipids. In general, larger molecules are broken down to smaller fragments by hydrolases.
EC 4	Lyases	Lyases catalyze the addition of groups to double bonds or the formation of double bonds through the removal of groups. Thus bonds are cleaved using a principle different from hydrolysis. Pectatelyases, for example, split the glycosidic linkages by beta-elimination.
EC 5	Isomerases	Isomerases catalyze the transfer of groups from one position to another in the same molecule. In other words, these enzymes change the structure of a substrate by rearranging its atoms.
EC 6	Ligases	Ligases join molecules to gether with covalent bonds. These enzymes participate in biosynthetic reactions where new groups of bonds are formed. Such reactions require the input of energy in the form of cofactors such as ATP.

A limited fraction of all the known enzymes are commercially available and even smaller amount is used in large quantities. More than 75% of industrial enzymes are hydrolases. Protein-degrading enzymes constitute about 40% of all enzyme sales. Proteinases have found new applications but their use in detergents is the major market.

More than fifty commercial industrial enzymes are available and their number increases steadily. Some enzymes with potential applications in industries are listed in Table 2.

Table 2. A selection of enzymes used in industrial processes

Class	Industrial enzymes
Oxidoreductases	Catalases Glucose oxidases Laccases
Transferases	Fructosyl transferases Glucosyl transferases
Hydrolases	Amylases Cellulases Lipases Mannanases Pectinases Phytases Proteases Pullulanases Xylanases
Lyases	Pectatelyases Alpha-acetolactate Decarboxylases
Isomerases	Glucose isomerases Epimerases Mutases Lyases Topoisomerases
Ligases	Argininosuccinate Glutathione synthase

4. GLOBAL MARKET

Modern industries have begun to explore the advantages of enzymes. Microorganisms used in the past have now been replaced by purified enzymes.

Table 3. The estimated value of the worldwide use of industrial enzymes

	2002	2003	2004	2009
Technical enzymes	978.2	1009.2	1040.0	1222.0
Food enzyme	701.0	720.0	740.0	863.0
Animal feed enzyme	210.8	215.6	220.0	267.0
Total	1890.0	1945.0	2000.0	2352.0

This has led to the growing market for industrial enzymes. In Table 3, F Hasan *et al*[10] showed the estimated value of the worldwide use of industrial enzymes and the market segmentation. The technical industries, dominated by the detergent, starch, textile, paper, fuel and alcohol industries, accounts for the major consumption of industrial enzymes.

5. ENZYME PRODUCTION

Other than plant and animal source most of the enzymes are, however, produced by microorganisms by submerged cultures in large reactors called fermenters. The enzyme production process can be divided into following phases: a) enzyme selection, b) selection of a production strain, c) construction of an overproducing strain by genetic engineering, d) optimization of culture medium and production conditions, e) optimization of recovery process (and purification if needed) and f) formulation of a stable enzyme product.

5.1. Selection of Microbial Production Strains

Criteria used in the selection of an industrial enzyme include specificity, reaction rate, pH and temperature, optima and stability, effect of inhibitors and affinity to substrates. Enzymes used in paper industry should not contain cellulose-degrading activity as a side activity, because this activity would damage the cellulose fibers. Enzymes used in animal feed industry must be thermo tolerant to survive in the hot extrusion process used in animal feed manufacturing. The same enzymes must have maximal activity at the body temperature of the animal. Enzymes used in industrial applications must usually be tolerant against various heavy metals and have no need for cofactors [1, 11]. They should be maximally active already in the presence of low substrate concentration, so that the desired reaction proceeds to completion in a realistic time frame.

In choosing the production strain several aspects have to be considered. Ideally the enzyme is secreted from the cell. This makes the recovery and purification process much simpler compared to production of intracellular enzymes, which must be purified from thousands of different cell proteins and other components. Secondly, the production host should have a GRAS-status, which means that it is Generally Regarded as Safe. This is especially important when the enzyme produced by the organism is used in food processes [12]. Thirdly, the organism should be able to produce high amount of the desired enzyme in a reasonable time frame. The industrial strains typically produce over 50-g/l extracellular enzyme proteins. Most of the industrial enzymes are produced by a relatively few microbial hosts like Aspergillus and Trichoderma fungi, Streptomyces fungi imperfecti and Bacillus bacteria. Yeasts are not good producer of extracellular enzymes and are rarely used for this purpose. Most of the industrially used microorganisms have been genetically modified to overproduce the desired activity and not to produce undesired side activities.

5.2. Enzyme Production by Microbial Fermentation

Once the biological production of organism has been genetically engineered to overproduce the desired products, a production process has to be developed. The optimization of a fermentation process includes media composition, cultivation type and process conditions. This is a demanding task and often involves as much effort as the intracellular engineering of the cell. The bioprocess engineer asks questions like: is the organism in question safe or are extra precautions needed, what kind of nutrients the organism needs and what is their optimal/ economical concentration, how the nutrients should be sterilized, what kind of a reactor is needed (mass transfer, aeration, cooling, foam control, sampling), what needs to be measured and how is the process controlled, how is the organism cultivated (batch, fed-batch or continuous cultivation), what are the optimal growth conditions, what is the specific growth and product formation rate, what is the yield and volumetric productivity, how to maximize cell concentration in the reactor, is the product secreted out from the cells, how to degrade the cell if the product is intracellular, does some of the raw materials or products inhibit the organism and finally, how to recover, purify and preserve the product. A typical enzyme production scheme is shown in Figure 1.

ENZYME PRODUCTION FLOW SHEET

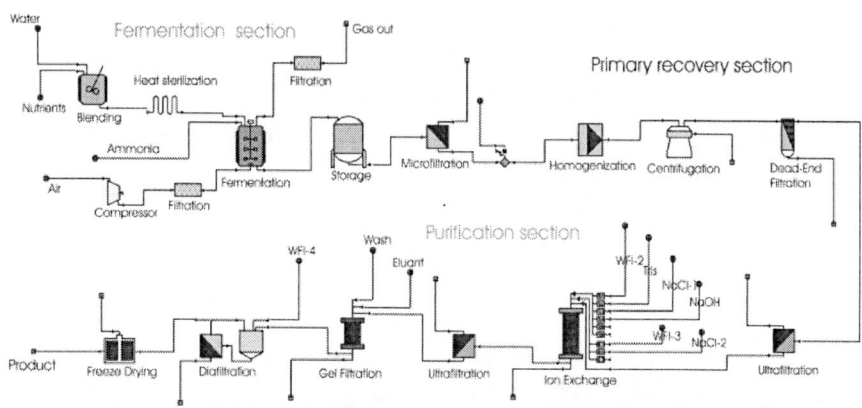

Figure 1. A typical enzyme production scheme. Large volume industrial enzymes are usually not purified. Their recovery is often finalized by an ultrafiltration step. Specialty enzymes need more purification.

The large volume industrial enzymes are produced in 50 – 500 m³ fermenters. The extracellular enzymes are often recovered after cell removal (by vacuum drum filtration, separators or microfiltration) by ultrafiltration. If needed the purification is carried out by ion exchange or gel filtration. The final product is either a concentrated liquid with necessary preservatives like salts or polyols or alternatively granulated to a non-dusty dry product. Enzymes are proteins, which like any protein can cause and have caused in the past allergic reactions. Therefore protective measures are necessary in their production and application.

Table 4. Large scale enzyme applications (adapted from www.novozymes.com)

Industry	Enzyme	Effect
Detergent	Amylases	Remove starch based stains
	Proteinase	Protein degradation
	Lipase	Fat removal
	Cellulose	Color brightening
Textile	Cellulase	Microfibril removal
	Laccase	Color brightening
Pulp and paper	Xylanase	Biobleaching
	Lipases	Reduces 'pitch' which causes paper to stick to rollers
Food industry		
Fruit juice	Pectinase	Juice clarification,
	Cellulase,	Juice extraction
	Xylanase	
Baking industry	Xylanase	Dough conditioning
	Alpha-amylase	Loaf volume; shelf-life
	Glucose oxidase	Dough quality
Dairy industry	Rennin	Protein coagulation
	Lactase	Lactose hydrolysis
	Glucanase	Filter aid
	Papain	Haze control
Animal feed	Xylanase	Fiber solubility
	Phytase	Release of phosphate
Leather Industry	Proteinase (trypsin).	Bating treatment of leather
	Proteases	Hydrolysis of non-collagenous constituents.
	Lipases	Grease removal

6. Enzyme Technology

The field of enzyme technology deals with the application approach of an enzyme in practical processes. Enzymes accelerate different chemical reactions with high specificity and are not permanently modified by their participation in reactions. The simplest way to use enzymes is to add them into a process stream, where they catalyze the desired reaction and are gradually inactivated during the process. This happens in many bulk enzyme applications like liquefaction of starch with amylases, bleaching of cellulose pulp with xylanases or use of enzymes in animal feed. In these applications the price of the enzymes must be kept low to make their use economical. Extracellularly produced bulk enzyme concentrates cost only US$ 10-20/kg protein.

But enzymes are costlier than chemical catalysts, in general, and cost effectiveness of enzyme based processes could be reached by the repeated use of enzymes. Since enzymes remain in solution with products, it is not possible to recover them easily from the reaction mixture. Immobilization of enzyme on solid support is an alternative approach to reuse the stock. The largest application of an immobilized enzyme is the conversion of glucose syrup to high fructose syrup for food applications. In the early applications the glucose isomerase enzyme containing cells were permeabilised and immobilized on a solid support. The enzyme containing support material was packed into a column through which the glucose solution was passed. Finnish Sugar Company developed in early 80s an alternative method where the intracellular glucose isomerase from Streptomyces rubiginosus was purified by crystallization and the pure enzyme was bound to an anion exchange resin, which can be regenerated with fresh enzyme after the previous one is inactivated. Another way to immobilize enzymes is to use ultrafiltration membranes in the reactor system. The large enzyme molecules cannot pass the membrane but the small molecular reaction products can. Therefore enzymes are retained in a reaction system and the products leave the system continuously. This method has been used in production of chirally pure amino acids from racemic mixtures of amino acid derivatives. Enzymes have also been immobilized on membranes for analytical purposes. The best-known example is glucose oxidase, which is used to measure glucose concentrations in biological samples. Many different laboratory methods for enzyme immobilization based on chemical reaction, entrapment, specific binding or absorption have been developed [13, 14].

A novel approach to use enzymes was introduced by Finnish Sugar Company in the late 80s. It was based on the use of cross-linked crystalline glucose isomerase. Enzyme crystals contain usually 30-80% free water and the enzyme is active even in the cross-linked insoluble form. The dimensions of an enzyme reactor, packed with this kind of a material, are considerably smaller compared to traditional immobilized systems because the carrier matrix can be completely omitted. The concept, originally developed in Finland, was later applied to other enzymes by Altus Ltd in USA, which has developed novel applications for the CLECs, which is the trademar for Cross-Linked Enzyme Crystals. These applications include chiral separations, controlled release of chemicals, specific separations and recently even cofactor entrapment into the crystal structure. All this is possible because an enzyme crystal contains water, pores, active center, hydrophobic areas and ionic properties.

7. INDUSTRIAL APPLICATIONS

Enzymes have long been used as alternatives to chemicals to improve the efficiency and cost-effectiveness of a wide range of industrial systems and processes. They are currently used in basic and applied arenas of research as well as in a wide range of product design and manufacturing processes, such as those pertaining to the food, beverage, pharmaceutical, detergent, leather processing, and peptide synthesis. They are widely distributed in nature and play a vital role in life processes. Table 5 summarizes major large-scale enzyme applications. Each of them is discussed in the text in some detail. Industrial Enzymology is recommended as a good resource text for those who need a more comprehensive treatment of an individual subject.

7.1. Food Industry

Food and beverage enzymes constitute the largest segment of industrial enzymes with revenues of nearly $5.1 billion in 2011 which is expected to grow to $7 billion by 2017, at a composed annual growth rate of 6.3% [15]. In terms of government regulation, enzymes used in food can be divided into food additives and processing aids. Most food enzymes are considered as processing aids, with only a few used as additives, such as lysozyme and invertase [16]. The processing aids are used during the manufacturing process of foodstuffs, and do not have a technological function in the final food. All these materials are expected to be safe, under the guidance of good manufacturing practice (GMP). The key issue in evaluating safety of enzyme preparations is the safety assessment of the production strain. Only about nine recombinant microorganisms are considered, generally recognized As Safe (GRAS) based on FDA regulations. These are from a relatively small number of bacterial and fungal species primarily *A. oryzae, A. niger, B. subtilis*and *B. licheniformis* for food enzyme production from a security point of view [17].

7.1.1. Baking Industry
The α-amylases have been used in the baking industry widely. These enzymes are generally added to the dough of bread in order to degrade the starch into smaller dextrins, which are further fermented by the yeast. The α-amylase enhances the fermentation rate and the reduction of the viscosity of dough, which results in improvements in the volume and texture of the product [18]. The baking industry uses amylases to delay the staling of bread and other baked products [19]. Proteases are used on a large commercial scale in the production of bread, baked goods, crackers and waffles [20]. These enzymes can be added to reduce mixing time, to decrease dough consistency, to assure dough uniformity, to regulate gluten strength in bread, to control bread texture and to improve flavor. In

addition, proteases have largely replaced bisulfite, which was previously used to control consistency through reduction of gluten protein disulfide bonds, while proteolysis breaks down peptide bonds. In both cases, the final effect is a similar weakening of the gluten network [21]. Xylanases combined with amylases, lipases and many oxidoreductases to attain specific effects on the rheological properties of dough and organoleptic properties of bread. Both xylanases and proteases enzymes have also been used to improve the quality of biscuits, cakes and other baked products [6]. Glucose oxidase has been used successfully to remove residual glucose and oxygen in foods and beverages aiming to increase their shelf life. The hydrogen peroxide generated by this enzyme presents antimicrobial properties, and is easily removed by catalase utilization.

7.1.2. Fruit Juice

Juices extracted from ripe fruit contain a significant amount of pectin that imparts a cloudy appearance. Pectins are colloidal in nature, making solutions viscous and holding other materials in suspension. Pectinesterase removes methyl groups from the pectin molecules exposing carboxyl groups, which in the presence of bi- or multivalent cations, such as calcium, form insoluble salts which can readily be removed. At the same time, polygalacturonase, a type of pectolitic enzyme, degrades macromolecular pectin, causing reduction in viscosity and destroying the protective colloidal action so that suspended materials will settle out [1]. It results in a crystal clear juice. Cellulase is another enzyme that is used in production of fruit juice. Cellulose exists in fruit pulp in high amount which is not digestible by human but using cellulase enzymes this can be hydrolyzed into monosaccharide such as glucose which is easily digestible. Xylanase can be also used as juice clarifying agent along with pectinase [22].

7.1.3. Dairy Industry

In dairy industries some enzymes are required for the production of cheeses, yogurt and other dairy products, while others are required for improving texture or flavor. The major application of proteases in the dairy industry is for the manufacture of cheese. Calf rennin had been preferred in cheese making due to its high specificity, but microbial proteases produced by GRAS microorganisms like *Mucormiehei*, *Bacillus subtilis*, *Mucorpusillus Lindt* and *Endothiaparasitica*are gradually replacing it. These four recombinant proteases have been approved by FDA for cheese production [7]. The primary function of these enzymes in cheese making is to hydrolyze the specific peptide bond (Phe105-Met106) that generates para-k-casein and macropeptides [23] resulting in the formation of a cheese curd. Production of calf rennin (chymosin) in recombinant *A. nigervarawamori* amounted to about 1 g/L after nitrosoguanidine mutagenesis and selection for 2-deoxyglucose resistance [24]. Further improvement was done by para sexual recombination resulting in a strain producing 1.5 g/L from parents producing 1.2 g/L [25]. Lactose in milk is not tolerable to 70% of the global adult population. Lactase enzyme is used in dairy industries for preventing crystallization of lactose in dairy products by converting it into glucose and galactose [1]. Unpasteurized milk contains lipoprotein lipases which contribute to a piquant flavor in aged cheese by breaking down milk fat to free acids. Lipases enzymes serve important roles in human practices as ancient as yogurt and cheese fermentation. However, lipases are also being exploited as cheap and versatile catalysts to degrade lipids in more modern applications. External lipase may be used during production to enhance flavor by synthesis of esters of short chain fatty acids and alcohols, which are known flavor and fragrance compounds [26].

7.1.4. Confectioneries

Soft candy cookies and other treats made with sugar have short life because sugar sucrose in the product begins to crystallize soon after the confection is produces. An enzyme named as invertase converts this sucrose into two type of monosaccharaides, glucose and fructose, and thus prevent the formation of sugar crystals [1].

7.1.5. Manufacture of Sweet Syrup

An extremely important use for fungal amylases is in conversion of partially acid hydrolyzed starch to sweet syrups. Acid hydrolysis is a random action, whereas enzymic hydrolysis is a patterned one. By proper control of the type and proportion of enzymes used (a-amylase, amyloglucosidase, maltase) syrups of almost any desired proportions of glucose, maltose, and dextrins may be produced.

7.1.6. Manufacture of Soy Products

Soybeans serve as a rich source of food, due to their high content of good-quality protein. Proteases have been used from ancient times to prepare soy sauce and other soy products[23]. The alkaline and neutral proteases of fungal origin play an important role in the processing of soy sauce. Proteolytic modification of soy proteins helps to improve their functional properties. Treatment of soy proteins with alcalase at pH 8 results in soluble hydrolysates with high solubility, good protein yield, and low bitterness. The hydrolysate is used in protein-fortified soft drinks and in the formulation of dietetic feeds.

7.1.7. Debittering of Protein Hydrolysates

Protein hydrolysates have several applications, e.g., as constituents of dietetic and health products, in infant formulae [18] and clinical nutrition supplements, and as flavoring agents. The bitter taste of protein hydrolysates is a major barrier to their use in food and health care products. The intensity of the bitterness is proportional to the number of hydrophobic amino acids in the hydrolysate. The presence of a proline residue in the center of the peptide also contributes to the bitterness. The peptidases that can cleave hydrophobic amino acids and proline are valuable in debittering protein hydrolysates. Aminopeptidases from lactic acid bacteria are available under the trade name Debitrase. Carboxypeptidase A has a high specificity for hydrophobic amino acids and hence has a great potential for debittering. A careful combination of an endoprotease for the primary hydrolysis and an aminopeptidase for the secondary hydrolysis is required for the production of a functional hydrolysate with reduced bitterness.

7.1.8. Meat Tenderizing

Myofibrillar proteins and connective tissue protein cause tenderer of some meat slices. Protease enzymes are used like papin and bromelain to tenderize tougher cuts of meat for many year [27]. But due to very fine line between tender and mushy, the process is difficult to control and more specific proteases have been used to make this tenderizing process more vigorous [28].

7.2. Animal Feed

The use of enzymes in animal feed is of great importance. Consistent increase in the price of feed ingredients becomes a major constraint in most of the developing countries. As a consequence cheaper and nonconventional feed ingredients are used which contain higher percentage of non-starch polysaccharides along with starch which possess chemical cross linking among them therefore, are not well digested by poultry [29, 30]. Developments of heat-stable enzymes, improved specific activity, some new non-starch polysaccharide-degrading enzymes, and rapid, economical and reliable assays for measuring enzyme activity have always been the main concern and have been intensified recently [12].

The global market for feed enzymes is a promising segment in the enzyme industry [31].The use of enzymes as feed additives is restricted in many countries by local regulatory authorities and its applications may therefore vary from country to country [12].

Commercial enzymes used in the animal feed industry are produced by microbial fermentation. Feed enzymes are produced by a batch fermentation process, starting with a seed culture and growth media [32]. After the completion of fermentation, the enzyme protein is separated from the fermentation residues and source organism. Although the source organisms are, in many cases, similar among enzyme products, the types and activity of enzymes produced can vary widely depending on the strain selected, the growth of substrate and culture conditions used [33-35]. Compared to the fermentation extract, these enzyme products are relatively concentrated and purified, possessing specific, controlled enzyme activities. They do not contain live cells. Enzyme products for animal diets are of fungal (mostly *Trichoderma longibrachiatum*, *Aspergillusniger*, *A. oryzae*) and bacterial (mostly *Bacillus* spp.) origin [36].

The use of exogenous fiber-degrading enzyme additives for ruminants was first examined in the 1960s, as reviewed by Beauchemin and Rode (1996) [37]. Several fibrolytic enzyme products evaluated as feed additives in animal diets were originally developed as silage additives [11].

Benefits of using feed enzymes to poultry diets include [5, 38-57]

a. Reduction in digesta viscosity,
b. Enhanced digestion and absorption of nutrients especially fat and protein,
c. Improved Apparent Metabolizable Energy (AME) value of the diet,
d. Increased feed intake, weight gain, and feed-grain ratio,
e. Reduced beak impaction and vent plugging,
f. Decreased size of gastrointestinal tract,
g. Altered population of microorganisms in gastrointestinal tract, reduced water intake,
h. Reduced water content of excreta,
i. Reduced production of ammonia from excreta,
j. Reduced output of excreta, including reduced N and P
k. Degrade unacceptable components in feed, which are otherwise harmful or of little or no value

7.2.1. Different Feed Enzymes

Feed enzymes commercially available are phytases, proteases, α-galactosidases, glucanases, xylanases, α-amylases, and poly-galacturonases, mainly used for swine and poultry [58]. The protein is abundantly available as a by-product from keratinous wastes, representing a valuable source of proteins and amino acids which is useful for animal feeds [59]. Animal feed grains contains phosphoruos which is bound to phytic acid. Animals need phosphorous for bone growth and other necessary biochemical processes. A specific enzyme, phytase releases the bound phosphorous, making it digestable to the chicken or hog [60, 61]. Addition of xylanase into a rye-based diet of broiler chickens results in reduced intensity of viscosity, thus improving both weight gain of chicks and their feed conversion efficiency [62, 63].

7.2.2. Direct-Fed Microbials and Enzymes

Research in the area of bacterial DFM (Direct-fed microbials) was mainly centered around the concept of feeding beneficial organisms to stressed animals with the general assumption that they will decrease or prevent intestinal establishment of pathogenic microorganisms [64]. Lack of organism specificity, proper dose, survival and the difficulty in defining when animals are the reasons for these findings [65].

Many microorganisms are used in DFM formulations. Some of the most common microbes used include *Lactobacillus acidophilus, L. casei, Enterococcus diacetylactis,* and *Bacillus subtilis.* The most common bacterial organisms in DFM products for ruminants' diets are lactobacilli. These organisms have little effect on ruminal fermentation [66] and the suggested mode of action from these organisms appears to be in the lower gut. *Saccharomyces cerevisiae* (*SC*, a yeast) and fungal fermentation extracts from *Aspergillusoryzae*(*AO*) plays an important role in animal feeds [67, 68].

7.3. Enzymes in Production of Fine Chemicals

Successful application of enzymatic processes in the chemical industry depends mainly on cost competitiveness with the existing and well-established chemical methods [69]. Lower energy demand, increased product titer, increased catalyst efficiency, less catalyst waste and byproducts, as well as lower volumes of wastewater streams, are the main advantages that biotechnological processes have as compared to well-established chemical processes. There are estimated to be only around 150 biocatalytic processes currently applied in industry [70]. However, new scientific developments in genomics, as well as in protein engineering, facilitate the tailoring of enzyme properties to increase that number significantly [8, 71]. Detail procedures for production of chemicals by enzyme are given in next section.

7.3.1. Chiral Compounds

Biocatalysts can be used in production and biotransformation of single enantiomers of chiral compounds. Preparation of chiral medicines, i.e., the synthesis of complex chiral pharmaceutical intermediates, is one of the most important applications in biocatalysis. Esterases, lipases, proteases and ketoreductases are widely applied in the preparation of carboxylic acids, chiral alcohols, amines or epoxides [4, 72, 73]. Kinetic resolution of racemic amines is a much known method used in the synthesis of chiral amines. Acylation of a primary amine moiety by a lipase [74-78] is used by BASF for the resolution of chiral primary amines in a large scale [79].

7.3.2. 4-Hydroxybenzoic Acid

Phenol carboxylase can be used to produce 4-hydroxybenzoic acid in an enzymatic process. A biological carboxylation route gives higher yields and lower waste generation [80].

7.3.3. Beta-Lactam Antibiotics

Enzymes are useful for preparation of beta-lactam antibiotics such as semi-synthetic penicillins and cephalosporins. Beta-lactam antibiotics constitute 60%–65% of the total antibiotic market. Semi-synthetic penicillins and cephalosporins are derived from 6-

aminopenicillanic acid (6-APA) and 7-aminocephalosporanic acid (7-ACA), respectively [74-78, 81].

7.3.4. Asymmetric Synthesis

Novel asymmetric reactions catalyzed by improved microbial enzymes provide a 100% yield [82]. Asymmetric reduction of tetrahydrothiophene-3-one with a wild-type reductase gave the desired alcohol ((R)-tetrahydrothiophene-3-ol) which is a key component in sulopenem, a potent antibacterial developed by Pfizer, but only in 80%–90% enantiomeric excess [83]. Recently, asymmetric synthesis from the corresponding chiral ketones, using transaminases becomes important [73]. Some (R)-selective transaminases have been recently developed using *in silico* strategies for a sequence-based prediction of substrate specificity and enantio-preference [84].

7.3.5. Enzymatic Oligosaccharide Synthesis

Recently isomalto-oligosaccharides, a new class of sugars are being produced using glucosyltransferases. They have potential commercial applications in food industry as non-digestible carbohydrate bulking agent. They are also used to suppress tooth decay associated with consumption of conventional carbohydrates and prevent baked goods going stale [85].

7.3.6 Production of Acrylamide

In 1985, Mitsubishi Rayon Co., Ltd. commenced production of acrylonitrile from acrylamide using immobilized bacterial enzyme nitrile hydratase [86]. This process is low cost, highly qualified and environmentally friendly [87]. About 100,000 tons of acrylamide is produced by this process annually [86]. Pseudomonas chlororaphfs B23 [88] and Rhodococcussp. N-774 [88-91] are used as efficient catalysts for the production of acrylamide. The recovery of unreacted acrylonitrile is not necessary in the biochemical process because the conversion of the latter is more than 99.99% [92, 93].

7.3.7. Production of 2-Methyl Pentanol

An effective enzymatic process using enzyme evolution was developed by the biotechnology company Codexis, in cooperation with Pfizer, to produce 2-methyl pentanol which is an important intermediate for manufacture of pharmaceuticals and liquid crystals [94].

7.3.8. Production of Optically Active Carboxylic Acids

Optically active carboxylic acids can be synthesized through different enzymatic routes catalyzed by lipases, nitrilases or hydroxynitrilelyases. Recent advances have improved the efficiency of these procedures. 2-arylpropanoic acids (e.g., ketoprofen, ibuprofen and naproxen) are synthesized mainly through the kinetic resolution of racemic substrates by lipases from *Candida antarctica*or *Pseudomonas* sp. Using (R,S)-N-profenylazoles, instead of their correspondent esters, proved to be more efficient [95].

7.3.9. Production of Formaldehyde and Formic Acid

A methanol oxidation system with alcohol oxidase and catalase from methylotrophic yeasts can be used for the production of formaldehyde from methanol [13, 96]. Formaldehyde dismutase from *Pseudomonas putfda* catalyzes the dismutation of various aldehydes (including formaldehyde) which leads to the formation of equimolar amounts of the corresponding alcohols and acids [97].

7.3.10. Benzene-1,2,3-Triol

Benzene-1,2,3-triol is polyphenol used in industry. It is industrially produced by autoclaving 3,4,5-trihydroxybenzoic acid obtained from tannins, under strong acidic conditions. A bacterial strain *Cftrobacter*sp. showed an inducible trihydroxybenzoic acid decarboxylase activity in producing benzene- 1,2,3-triol from trihydroxybenzoic acid [98].

7.3.11. Indigo

Indigo is a blue pigment used as a dye in textile industry. Indigo production occurred naturally only in plants. The chemical process was developed by Bayer which involves quite difficult and relatively severe procedures. Ensley *et. al.*, (1983) discovered that expression of the Pseudomonas naphthalene dioxygenase system in *Escherichia coli* can synthesize indigo [99]. The naphthalene dioxygenase system catalyzes the oxidation of indole to produce indigo. Indole is produced from tryptophan in E. *coli* by the action of tryptophanase enzyme. Such combinations of microbial reactions lead to the development of useful applications in the chemical industry [100].

7.3.12. Production of Propene Oxide

The Cetus process was the first enzymatic process for the production of propene oxide [101]. In this process, propene halohydrinis synthesized from propene, where *Caldariornycesfurnago* -chloroperoxidase is used in the presence of hydrogen peroxide [102]. Halohydrinis are further converted to propene oxide using *Flavobucterfurn-*halohydrinepoxidase [103].

7.3.13. Acrylic Acid and Products of Nitriles

R. rhodochrous K22 and *Alcallgenesfaecalls* JM3 produce novel nitrilases. They act on aliphatic nitriles [104] and arylacetonitriles respectively [100, 105]. Stalker *et al.* (1987) reported that Klebsfellaozaernaenitrilase possesses high specificity for bromoxynil as a substrate [106]. Using *Breubacterfum* R3 12 amidase, a continuous immobilized cell reactor was designed to produce acrylic acid via hydrolysis of acrylamide [107]. The enzymatic hydrolysis of a dinitrile, for instance, may give rise to five different products, i.e., monoamide, mononitrile, monoacid mononitrile, diamide, monoamide monoacid, and diacids, whereas acid- or base-catalyzed hydrolysis gives only diamide and diacid. Many other useful compounds may be derived from a nitrile in connection with other or enzymatic reactions [108].

7.3.14. Production of Ethanol

Some bacteria, such as *E. coli, Klebsiella, Erwinia, Lactobacillus, Bacillus,* and *Clostridis* utilize mixed sugars but produce no or only a limited quantity of ethanol. Some microorganisms have been genetically engineered to over produce ethanol from mixed sugar substrates. Various recombinant strains, e.g., *E. coli* K011, *E. coli* SL40, *E. coli* FBR3, *Zymomonas* CP4 (Pzb4), and *Saccharomyces* 1400 (Plnh32) are selected for the fermentation of mixed sugar substrates [109].

7.3.15. Cyclohexanol to Adipic Acid

Cyclohexanone and cyclohexanol are metabolised through a series of sequential oxidation reactions involving several enzymes. A large number of microorganisms can oxidize cyclohexanol to adipic acid efficiently [110].

7.4. Textile Industry

Using enzymes in textile industries has a long tradition. The current application of enzymes in textile industries involves mainly hydrolases and now some extent in oxidoreductase [111]. The enzymatic desizing of cotton with α-amylases is state-of-the-art since many decades [112]. Moreover, cellulases, pectinases, hemicellulases, lipases and catalases are used in different cotton pre-treatment and finishing processes [113]. Other natural fibers are also treated with enzymes. Examples are the enzymatic degumming of silk with sericinases [114], the felt-free-finishing of wool with proteases [115], the softening of jute with cellulases and xylanases [116], fading of denim with cellulase [111]. Enzymes not only enhance the quality of finishing products but also it influences the development of environmentally friendly technologies. The use of enzymes in the chemical processes of the textile industries can save 70,000-90,000 liters of water for every ton of knitwear produces. With approximately 9 million tons of knitwear being produced annually, the world can save 630 billion liters of water every year if all knitwear produced using enzymes [3].

7.4.1. Desizing

Desizing means removing starch from fabrics which was used in sizing operation. Before the discovery of amylases for desizing, process used to be carried out by treating the fabric with acid, alkali or oxidizing agents at high temperatures. The chemical treatment was not totally effective in removing the starch, leading to imperfections in dyeing, and also resulted in a degradation of the cotton fiber destroying the natural, soft feel of the cotton [117]. In the textile industry amylases are used to remove starch-based size for improved and uniform wet processing. Amylase is a hydrolytic enzyme which catalyzes the breakdown of dietary starch to short chain sugars, dextrin and maltose. The advantage of these enzymes is that they are specific for starch, removing it without damaging to the support fabric [111]. An amylase enzyme can be used for desizing processes at low-temperature (30-60°C) and optimum pH is 5,5-6,5 [118]. Use of amylases has decreased the use of harsh chemicals hence results low environmental pollution, discharge of waste chemicals. It also improved the safety of workplace for workers [119].

7.4.2. Scouring

Untreated cotton contains various noncellulosic impurities, such as waxes, pectins, hemicelluloses and mineral salts, present in the cuticle and primary cell wall of the fiber. Before cotton yarn or fabric can be dyed, it needs to be pretreated to remove materials that inhibit dye binding. This step is called scouring. Enzymatic or bioscouring, leaves the cellulose structure almost intact, preventing cellulose weight and strength loss. Several types of enzyme, including pectinases, cellulases, proteases, and lipases/ cutinases, alone or combined have been studied for cotton bioscouring, with pectinases being the most effective. In this pectinase destroy the cotton cuticle structure by digesting the pectin and removing the connection between the cuticle and the body of cotton fiber whereas cellulase can destroy cuticle structure by digesting the primary wall cellulose immediately under the cuticle of cotton. Biological Oxygen Demand (BOD) and Chemical Oxygen Demand (COD) of enzymatic scouring process are 20-45 % as compared to alkaline scouring (100 %). Total Dissolved Solid (TDS) of enzymatic scouring process is 20-50% as compared to alkaline scouring (100%). Handle is very soft in enzymatic scouring compared to harsh feel in alkaline scouring process. Enzymatic scouring makes it possible to effectively scour fabric without negatively affecting the fabric or the environment. It also minimises health risks hence operators are not exposed to aggressive chemicals [111, 120]. Fabric is softer and fluffier than conventional scouring, ideal for terry towel/ knitted goods [119].

7.4.3. Bleaching

The purpose of cotton bleaching is to decolourise natural pigments and to confer a pure white appearance to the fiber. The most common industrial bleaching agent is hydrogen peroxide. Conventional preparation of cotton requires high amounts of alkaline chemicals and consequently, huge quantities of rinse water are generated. However, radical reactions of bleaching agents with the fiber can lead to a decrease in the degree of polymerisation and, thus, to severe damage. Therefore, replacement of hydrogen peroxide by an enzymatic bleaching system would not only lead to better product quality due to less fibre damage but also to substantial savings on washing water needed for the removal of hydrogen peroxide. An alternative to this process is to use a combination of suitable enzyme systems. Amyloglucosidases, pectinases, and glucose oxidases are selected that are compatible concerning their active pH and temperature range [111]. Tzanov *et al.* (2003) reported for the first time the enhancement of the bleaching effect achieved on cotton fabrics using laccases in low concentrations. In addition, the short time of the enzymatic pre-treatment sufficient to enhance fabric whiteness makes this bio-process suitable for continuous operations. Also, Pereira *et al.* (2005) showed that a laccase from a newly isolated strain of *T. hirsuta*was responsible for whiteness improvement of cotton most likely due to oxidation of flavonoids. More recently, Basto *et al.* (2006) proposed a combined ultrasound-laccase treatment for cotton bleaching. They found that the supply of low ultrasound energy (7W) enhanced the bleaching efficiency of laccase on cotton fabrics. Enzymes break down hydrogen peroxide liquor into water and less reactive gaseous oxygen resulting cleaner waste water and reduction of energy and time [111].

7.4.4. Biopolishing

Biopolishing is a finishing process that improves fabric quality by mainly reducing fuzziness and pilling property of cellulosic fibre. It was first adopted by the Danish Firm, Novo Nordisk for the finishing treatment of cellulosic fabrics with cellulose enzymes [119]. The objective of the process is elimination of micro fibrils of cotton [111]. Cellulaseshydrolyse the microfibrils (hairs or fuzz) protruding from the surface of yarn because they are most susceptible to enzymatic attack. This weakens the microfibrils, which tend to break off from the main body of the fibre and leave a smoother yarn surface. It results in softer feeling, cleaner surface, slight improvement in absorbency [119].

7.4.5. Enzymetic Treatment of Denim

Denim is heavy grade cotton. Dye is mainly adsorbed on the surface of the fiber that is why fading can be done without considerable loss of strength. In traditional process sodium hypochlorite or potassium permanganate was used called as pumice stones. Fading was done by the abrasive action of pumice stone. But this pumice stone causes large amount of back-stanning, considerable wears and tear of machine also required in very large amount. Cellulase enzyme is used in denim washing. It works by loosening the indigo dye on the denim. A small dose of enzyme can replace several kilograms of pumice stones. The use of less pumice stones results in less damage to garment, machine and less pumice dust in the laundry environment [111].

7.4.6. Degumming of Silk

Silk degumming process is a fundamental finishing process for silk yarn and silk fabric. The objective of degumming is remove the substrate such as silk gum (sericin), wax and some impurities from silk fibers. The principle of degumming process is breaking the peptide linkage of amino acid in sericin structure into a small molecule, which is soluble in water. The methods used for the degumming are the hydrolysis reaction performed by acid and alkali, but these methods are not eco-friendly and same time they have a big problem on the surface of silk. The thermostable protease basically forms Geo-bacillus genus has been used for the enzymatic degumming of silk which are quite resistant to various chemicals and temperature [121].

7.5. Pulp and Paper

The technology for pulp manufacture is highly diverse, and numerous opportunities exist for the application of microbial enzymes. Historically, enzymes have found some uses in the paper industry, but these have been mainly confined to areas such as modifications of raw starch. However, a wide range of applications in the pulp and paper industry have now been identified. The use of enzymes in the pulp and paper industry has grown rapidly since the mid-1980s. While many applications of enzymes in the pulp and paper industry are still in the research and development stage, several applications have found their way into the mills in an unprecedented short period of time.

7.5.1. Prebleaching of Pulp

Currently the most important application of enzymes is the prebleaching of kraft pulp[9], where lignin is separated from pulp to improve the property and quality of paper. Xylanase in pre-bleaching, which would lower the amount of chlorine compounds used by this process up to 30%, so that a 15–20% reduction in organo-chlorines in the effluents could be achieved. The utilization of xylanases could lead to the replacement of 5–7 kg of chlorine dioxide per ton of Kraft pulp [122]. The enzyme used for the purpose of bio-bleaching of wood pulp should be active in the conditions of alkaline pH, high temperature and free from cellulosic activity. Beside bleaching, it provides brightness and improve viscosity of the pulp [19, 123]. This is probably caused by the selective removal of xylan, as determined by the pentosan values. Xylan, with lower DP than cellulose, can be expected to lower the average viscosity of kraft hemicellulose [9].

7.5.2. Pitch Control

Pitch control is an important aspect in pulp and paper manufacture, and the first example where microbial biotechnology provided successful solutions in this industrial sector [124]. Lipophilic extractives causes' so-called pitch deposits along with pulp and paper manufacturing processes. This lipophilic extractives includes Pitch, that is composed of triglycerides, fatty acids, resin acids, sterols, glycerol esters of fatty acids, other fats, and waxes [9, 124]. Lipase treatments have been shown to reduce pulp triglycerides successfully, by converting triglycerides into fatty acids which is more soluble in water and does not deposit as much. But in addition to triglycerides other compounds such as free and esterified sterols, resin acids, fatty alcohols, alkanes, etc. are also responsible of pitch problems as mentioned before. Sterol esterase which hydrolyze fatty acid esters of sterols, have also been suggested for pitch control. In addition it is found that, lipase also catalyze the hydrolysis of sterol esters. Enzymatic pitch control helps to reduce pitch-related problems to a satisfactory level. At the same time, it also offers other advantages, such as ecofriendly and nontoxic technology, improved pulp and paper quality, reduction in bleaching chemical consumption, reduction of effluent load, and space and cost saving in a mill wood yard by using unseasoned logs. By reducing the outside storage time of logs, this method reduces wood discoloration, wood yield loss, and the natural wood degradation which occurs over longer storage time [9].

7.5.3. Paper Sizing

The use of amylases in the pulp and paper industry is for the modification of starch of coated paper, i.e., for the production of low-viscosity, high molecular weight starch. The coating treatment serves to make the surface of paper sufficiently smooth and strong, to improve the writing quality of the paper. In this application, the viscosity of the natural starch is too high for paper sizing and this can be altered by partially degrading the polymer with amylases in a batch or continuous processes. Starch is a good sizing agent for the finishing of paper, improving the quality and erasebility, besides being a good coating for the paper [125, 126]. The size enhances the stiffness and strength in paper [125].

7.5.4. Deinking

Deinking is another place where microbial enzymes show its usefulness. In this process inks and stickies attach to fibers at surface of the fibers and stuck to microfibrils. Many patent applications have been filed or granted concerning the use of enzymes in deinking. Several patents specify the use of cellulases, particularly alkaline cellulases, for deinking. Few patents claim that esterases can be used, while others specify the use of lipases or pectinases. One patent application employs laccase from white rot fungi. Most of the published literature on deinking deals with cellulases and hemicelluase [9]. Among them esterases help to breakdown ink particles and also break ester bonds in polymers used in toners and adhesives. It improves paper cleanliness, causes less deposits and can be used as substitute for talc or solvent based dispersant.

7.5.5. Elimination of Slime

Slime is the generic name for deposits of microbial origin in a paper mill. It is impractical to run a paper mill as a sterile system. As a result, vast arrays of microbes contaminate the mills, and many of the resulting slime compounds have not been characterized. In some cases, however, specific slime compounds have been characterized, and efficient strategies for their removal can be used. One such case is with levan, which is a â-2,6-linked polymer of fructose that forms a slime film. This compound is secreted by several species of Bacillus and Pseudomonas bacteria that can grow in the recirculated water around the paper machine, especially for fine paper, where the level of inhibiting compounds is low. The enzyme levan hydrolase can hydrolyze this polymer to low-molecular-weight polymers that are water soluble, thereby cleaning the slime out of the system. The enzyme is effective at pH 4-8 and runs best at pH 5.0.

7.5.6. Shives Removal

Shives are small bundles of fibers that have not been separated into individual fibers during the pulping process. They appear as splinters that are darker than the pulp. Shivex, can be used to increase the efficiency of shive removal by bleaching. Shivex is a multicomponent mixture of proteins, some of which are xylanases, but the degree of shive removal by the enzyme is not directly related to the enzyme's xylanase activity or bleach boosting effectiveness by treating brownstock with shivex mills can increase shive removal in subsequent bleaching upto 55% [9].

7.5.7. Debarking of Wood

Removal of the bark is one of the most important steps in all processing of wood. This step consumes substantial amounts of energy. Extensive debarking is needed for high-quality mechanical and chemical pulp, because even small amounts of bark residues cause darkening of the product. Pectinases are found to be key enzymes in the process. One of the major difficulties with enzymatic debarking is the poor infiltration of enzymes in the cambium of whole logs [127].

7.6. Enzyme in Therapeutic Applications

Therapeutic enzymes have a wide variety of specific uses such as oncolytics, thrombolytics, or anticoagulants and as replacements for metabolic deficiencies. Proteolytic enzymes serve as good anti-inflammatory agents. The list of enzymes which have the potential to become important therapeutic agents and its microbial sources are shown in Table 5.

Table 5. Some important enzymes and their therapeutic importance [126-135]

Enzyme	Reaction	Use	Sources
Asparaginase	L-Asparagine $H_2O\rightarrow$L-aspartate + NH_3	Leukaemia	*E. coli*
Collagenase	Collagen hydrolysis	Skin ulcers	*C. perfringens*
Glutaminase	L-Glutamine $H_2O\rightarrow$L-glutamate + NH_3	Leukaemia	*E. coli SFL-1*
Lysozyme	Bacterial cell wall hydrolysis	Antibiotic	*Homo sapiens*
Ribonuclease	RNA hydrolysis	Antiviral	*Yeast and bacteriophages*
Streptokinase	Plasminogen→plasmin	Blood clots	*Streptococci sp.*
Trypsin	Protein hydrolysis	Inflammation	*Homosapiens and* other vertebrates
Uricase	Urate + O_2 →allantoin	Gout	*A. flavus*
Urokinase	Plasminogen→plasmin	Blood clots	Bacillus subtilis
β-Lactamase	β-Lactam ring hydrolysis	Antibiotic resistance	Citrobacterfreundii, Serratiamarcescens,and Klebsiella pneumonia
Penicillin acylase	Binding the rings of benzylpenicillin (penicillin G) and phenoxymethylpenicillin (penicillin V)	Penicillin production/broad spectrum antibiotic production	Penicillium sp.

As compared to the industrial use of enzymes, therapeutically useful enzymes are required in relatively less amounts, but the degree of purity and specificity should be generally high. The cost of these enzymes is high but comparable to those of therapeutic agents or treatments. One of the major applications of therapeutic enzymes is in the treatment of cancer and various other diseases are discussed here. A higher than normal concentration of amylases may predict one of several medical conditions, including acute inflammation of the pancreas, perforated peptic ulcer, strangulation ileus, torsion of an ovarian cyst, macroamylasemia, and mumps. In other body fluids also amylase can be measured, including urine and peritoneal fluid. In various human body fluids the level α-amylase activity is of clinical importance, for example, in diabetes, pancreatitis, and cancer research [128].

7.6.1. Treatment of Damaged Tissue

A large number of proteolytic enzymes of plant and bacterial origin have been studied for the removal of dead skin of burns. Various enzymes of higher quality and purity are now in clinical trials. Debrase gel dressing, containing a mixture of several enzymes extracted from pineapple, received clearance in 2002 from the US FDA for a Phase II clinical trial for the treatment of partial-thickness and full-thickness burns. A proteolytic enzyme (Vibrilase TM) obtained from Vibrio proteolyticus is found to be effective against denatured proteins such as those found in burned skin. The regeneration of injured spinal cord have been demonstrated using chondroitinases, where this enzyme acts by removing the glial scar and thereby accumulating chondroitin sulfate that stops axon growth [129].

7.6.2. Treatment of Cancer

The cancer research has some good instances of the use of enzyme therapeutics. Recent studies have proved that arginine-degrading enzyme (PEGylated arginine deaminase) can inhibit human melanoma and hepatocellular carcinomas [130]. Currently, another PEGylatedengyme, Oncasparl (pegaspargase), has shown good results for the treatment of children newly diagnosed with acute lymphoblastic leukemia and are already in use in the clinic. The normal cells are able to synthesize asparagine but the cancerous cells cannot and thus, die in the presence of asparagine degrading enzyme. Asparaginase and PEG-asparaginase are effective adjuncts for standard chemotherapy. Another important feature of oncogenesis is proliferation. It has been proved that the removal of chondroitin sulfate proteoglycans by chondroitinase AC and, to a lesser extent, by chondroitinase B, stops tumor growth, metastasis and neovascularization [131].

The further application of enzymes as therapeutic agents in cancer is described by antibody-directed enzyme prodrug therapy (ADEPT). A monoclonal antibody carries an enzyme specific to cancer cells, where the enzyme activates a prodrug and destroys cancer cells but not normal cells. This approach is being utilized for the discovery and development of cancer therapeutics based on tumor-targeted enzymes that activate prodrugs. The targeted enzyme prodrug therapy (TEPT) platform, involving enzymes with antibody-like targeting domains, can also be used in this effort [132].

7.6.3. Treatment of Infectious Diseases

Lysozymeisa is naturally occurring antibacterial agent. It used in many foods and consumer products, as it is able to breakdown carbohydrate chains in bacterial cell wall. Lysozyme has also been found to have activity against HIV, as RNase A and urinary RNaseU present selectively degrade viral RNA [133] showing possibilities for the treatment of HIV infection. Chitinases is another naturally occurring antimicrobial agent. The cell wall of various pathogenic organisms, including fungi, protozoa, and helminths is made up of chitin and is a good target for antimicrobials [134]. The lytic enzyme derived from bacteriophage is used to target the cell walls of Streptococcus pneumonia, Bacillus anthracis, and Clostridium perfringens [135]. The application of lytic bacteriophages can be used for the treatment of several infections and could be useful against new drug-resistant bacterial strains.

7.7. Detergents

Detergents were the first large scale application for microbial enzymes. Bacterial proteinases are still the most important detergent enzymes. Some products have been genetically engineered to be more stable in the hostile environment of washing machines with several different chemicals present [19]. These hostile agents include anionic detergents, oxidising agents and high pH. Late 80s lipid degrading enzymes were introduced in powder and liquid detergents. Lipases decompose fats into more water-soluble compounds by hydrolysing the ester bonds between the glycerol backbone and fatty acid. The most important lipase in the market was originally obtained from Humicola lanuginose. It is produced in large scale by Aspergillusoryzae host after cloning the Humicola gene into this organism[18].

Amylases are used in detergents to remove starch based stains. Amylase hydrolyzes gelatinised starch, which tends to stick on textile fibers and bind other stain components. Cellulases have been part of detergents since early 90s. Cellulase is actually an enzyme complex capable of degrading crystalline cellulose to glucose. In textile washing cellulases remove cellulose microfibrils, which are formed during washing and the use of cotton based cloths. This can be seen as colour brightening and softening of the material. Alkaline cellulases are produced by Bacillus strains and neutral and acidic cellulases by Trichoderma and Humicola fungi [34].

7.8. Leather

Leather industry uses proteolytic and lipolytic enzymes in leather processing. The use of these enzymes is associated with the structure of animal skin as a raw material. Enzymes are used to remove unwanted parts from skin. Proteases are used for selective hydrolysis of non-collagenous constituents of the skin and to remove nonfibrillar proteins such as albumins and globulins. Currently, microbial alkaline proteases are used to ensure faster absorption of water and to reduce time required for soaking. The use of proteases as alternatives to hazardous chemicals such as sodium sulphide has proved successful in improving leather quality and in reducing environmental pollution. Alkaline proteases with hydrated lime and sodium chloride are used for dehairing, resulting in significant reduction in wastewater [3]. Trypsin in combination with other proteases of Bacillus and Aspergillus origin is used for bating. The selection of the enzyme depends on its specificity for matrix proteins such as elastin and keratin and the amount of enzyme depends on the 15 type of leather (soft or hard) desired to be produced. Usage of enzymes for dehairing and bating not only prevents pollution problems but is also effective in saving energy. Novo Nordisk manufactures three different proteases viz. Aquaderm TM, NUE, and Pyrase for use in soaking, dehairing and bating respectively.

In dehairing and dewooling phases enzymes are used to assist the alkaline chemical process. This results in a more environmentally friendly process and improves the quality of the leather (cleaner and stronger surface, softer leather, less spots). The used enzymes are typically alkaline bacterial proteases. Lipases are used in this phase or in bating phase to specifically remove grease [3]. The use of lipases is a fairly new development in leather industry.

The next phase is bating which aims at deliming and deswelling of collagen. In this phase the protein is partly degraded to make the leather soft and easier to dye. Pancreatic trypsins were originally used but they are being partly replaced by bacterial and fungal enzymes.

8. FUTURE PROSPECT OF INDUSTRIAL ENZYMOLOGY

Microorganisms provide an impressive amount of catalysts with a wide range of applications across several industries such as, food, animal feed, technical industries, fine chemicals and pharma. There are many factors influencing the growing interest in biocatalysis which include enzyme promiscuity, robust computational methods combined with directed evolution and screening technologies to improve enzyme properties to meet process prospects, the application of one-pot multistep reactions using multifunctional catalysts and the de novo design and selection of catalytic proteins catalyzing any desired chemical reaction. The unique properties of enzymes such as high specificity, fast action and biodegradability allow enzyme-assisted processes in industry to run under milder reaction conditions, with improved yields and reduce waste generation. However, naturally occurring enzymes are often not suitable for such biocatalytic processes without further tailoring or redesign of the enzyme itself in order to fine tune substrate specificity activity or other key catalytic properties.

Recent advances in genomics, metagenomics, proteomics, efficient expression systems and emerging recombinant DNA techniques have facilitated the discovery of new microbial enzymes from nature (through genome and metagenome) or by creating (or evolving) enzymes with improved catalytic properties. The ongoing progress and interest in enzymes provides further success in areas of industrial biocatalysis. The next years should see a lot of exciting developments in the area of bio-transformations. Many future investigations will use combinations of engineered and de novo designed enzymes coupled with chemistry to generate more (and most likely new) chemicals and materials from cheaper (and renewable) resources, which will consequently contribute to establishing a bio-based economy and achieving low carbon green growth.

CONCLUSION

Since the ancient human civilization, enzymes are being known to mankind. But since the 18th century it has been technically known to us as enzymes. Many scientists had tried to study the use of enzymes, and from their pioneer work, we have come to know about its power and utility in our daily life. Enzymes industry is one among the major industries of the world, and there exists a great market for further improvement in this field. To date different types of enzymes are being manufactured by many giant companies and being sold for their important role in different industries like food, dairy, detergent, and chemical as well as for their important lifesaving therapeutically application. Due to advancement of modern biotechnology and protein engineering a new area of enzyme engineering, has evolved which mainly deals with the purification and stability of these important enzymes.

CONFLICTS OF INTEREST

The authors declare no conflict of interest.

REFERENCES

[1]   Underkofler, L. A., Barton, R. R. and Rennert, S. S. *Production of Microbial Enzymes and Their Applications,* 1957, Takamine Laboratory, Division of Miles Laboratories, Inc.: Clifton, New Jersey.

[2]   Adrio, J. L. and A. L. Demain, Microbial enzymes: tools for biotechnological processes. *Biomolecules,* 2014. **4**(1): p. 117-39.

[3]   Pandey, A. and S. R. R, Production and application of industrial enzymes. *Chem Indus Digest,* 2008. **21**: p. 82-91.

[4]   Kirk, O., T. V. Borchert, and C. C. Fuglsang, Industrial enzyme applications. *Current opinion in biotechnology,* 2002. **13**(4): p. 345-351.

[5]   Choct, M., Enzymes for the feed industry: past, present and future. *World's Poultry Science Journal,* 2006. **62**(01): p. 5-16.

[6]   Melim Miguel, A. S., et al., *Enzymes in Bakery: Current and Future Trends.* INTECH, 2013: p. 288-321.

[7]   Pariza, M. W. and E. A. Johnson, Evaluating the safety of microbial enzyme preparations used in food processing: Update for a new century. *Regul. Toxicol. Pharmacol,* 2001. **33**: p. 173-186.

[8]   Jäckel, C. and D. Hilvert, Biocatalysts by evolution. *Current opinion in biotechnology,* 2010. **21**(6): p. 753-759.

[9] Bajpai, P., Application of enzymes in the pulp and paper industry. *Biotechnol Prog,* 1999. **15**(2): p. 147-57.

[10] Hasan, F., A. A. Shah, and A. Hameed, Industrial applications of microbial lipases. *Enzyme and Microbial Technology,* 2006. **39**(2): p. 235-251.

[11] Feng, P., et al., Effect of enzyme preparations on in situ and in vitro degradation and in vivo digestive characteristics of mature cool-season grass forage in beef steers. *Journal of Animal Science,* 1996. **74**(6): p. 1349-1357.

[12] Pariza, M. W. and M. Cook, Determining the safety of enzymes used in animal feed. *Regulatory Toxicology and Pharmacology,* 2010. **56**(3): p. 332-342.

[13] Baratti, J., et al., Preparation and properties of immobilized methanol oxidase. *Biotechnology and Bioengineering, 1*978. **20**(3): p. 333-348.

[14] Walker, J. M., Microbial Enzymes and Biotransformations, in *Methods in Biotechnology,* J. L. Barredo, Editor 2005, Humana press: Totowa, New Jersey.

[15] *World and Enzymes,* 2014: Cleveland, OH, USA. p. 338.

[16] *Collection of Information on Enzymes,* A.F.E. Agency, Editor 2002: Vienna, Austria.

[17] Olempska-Beer, Z. S., R. I. D. Merker, and M. J. DiNovi, Food-processing enzymes from recombinant microorganisms. *Regul. Toxicol. Pharmcol,* 2006. **45**: p. 144-158.

[18] Gurung, N., et al., A broader view: microbial enzymes and their relevance in industries, medicine, and beyond. *Biomed Res Int,* 2013. **2013**: p. 329121.

[19] Nigam, P. S., Microbial enzymes with special characteristics for biotechnological applications. *Biomolecules,* 2013. **3**(3): p. 597-611.

[20] Bombara N., Anon M. C., and P. AMR, Functional Properties of Protease Modified Wheat Flours. *LWT Food Science and Technology, 1*997. **30(5)**: p. 479-486.

[21] Linko Y-Y, Javanainen P, and L. S, Biotechnology of bread baking. *Trends in Food Science and Technology,* 1997. **8(10)**: p. 339-344.

[22] Biely, P., Microbial xylanolytic systems. *Trends Biotechnol,* 1985. **3**: p. 286-290.

[23] Rao, M. B., et al., Molecular and biotechnological aspects of microbial proteases. Microbiol. *Mol. Biol. Rev.,* 1998. **62**(597-635).

[24] Dunn-Coleman, N. S., et al., Commercial levels of chymosin production by Aspergillus. *Biotechnology,* 1991. **9**: p. 976-981.

[25] Bodie, E. A., G. L. Armstrong, and N. S. Dunn-Coleman, Strain improvement of chymosin-producing strains of Aspergillus niger var awamori using parasexual recombination. *Enzyme Microb. Technol,* 1994. **16**: p. 376-382.

[26] Macedo G. A., Lozano. M.M.S., and Pastur. G.M, Enzymatic synthesis of short chain citronellyl esters by a new lipase from Rhizopus sp. Electron. *J. Biotechnol,* 2003. **6(1)**: p. 72-75.

[27]   S, S., et al., Alpha amylase from microbial sources-an overview on recent developments. *Food Technol Biotechnol,* 2006. **44**: p. 173-184.

[28]   Singhania, R. R., et al., Advancement and comparative profiles in the production technologies using solid-state and submerged fermentation for microbial cellulases. *Enzyme and Microbial Technology,* 2010. **46**(7): p. 541-549.

[29]   Adams, E. A. and R. Pough, Non-starch polysaccharides and their digestion in poultry. *Feed Compounder* 1993. **13**: p. 19-21.

[30]   Annison, G., The role of wheat non-starch polysaccharides in broiler nutrition. *Crop and Pasture Science,* 1993. **44**(3): p. 405-422.

[31]   Sullivan, F., *Feed Enzymes: The Global Scenario,* 2007: London, UK.

[32]   Cowan, W. Factors affecting the manufacture, distribution, application and overall quality of enzymes in poultry feeds. in *Joint Proc. 2nd Int. Roundtable on Anim. Feed Biotechnol.—Probiotics, and Workshop on Anim. Feed Enzymes,* Ottawa. 1994.

[33]   Considine, P. and M. Coughlan, Production of carbohydrate-hydrolyzing enzyme blends by solid-state fermentation. *Enzyme System for Lignocellulose Degradation.* Elsevier Applied Science, London, 1989: p. 273-281.

[34]   Gashe, B., Cellulase production and activity by Trichoderma sp. A-001. *Journal of applied bacteriology,* 1992. **73**(1): p. 79-82.

[35]   Lee, B., et al., Media evaluation for the production of microbial enzymes. *Journal of Agricultural and Food Chemistry,* 1998. **46**(11): p. 4775-4778.

[36]   Pendleton, B., The regulatory environment, in *Direct- Fed Microbial, Enzyme and Forage Additive Compendium,* M. S. Muirhea, MN., Editor 2000, The Miller Publishing Co. p. 49.

[37]   Beauchemin, K., et al. Use of feed enzymes in ruminant nutrition. in *Animal Science Research and Development-Meeting Future Challenge*s (Rode, LM, Ed.). 1996.

[38]   Annison, G. and M. Choct, Anti-nutritive activities of cereal non-starch polysaccharides in broiler diets and strategies minimizing their effects. *World's Poultry Science Journal,* 1991. **47**(03): p. 232-242.

[39]   Bedford, M., Mechanism of action and potential environmental benefits from the use of feed enzymes. *Animal Feed Science and Technology,* 1995. **53**(2): p. 145-155.

[40]   Bedford, M., H. Classen, and G. Campbell, The effect of pelleting, salt, and pentosanase on the viscosity of intestinal contents and the performance of broilers fed rye. *Poultry Science,* 1991. **70**(7): p. 1571-1577.

[41]   Benabdeljelil, K. Improvement of barley utilization for layers: effects on hen performance and egg quality. In *Proceedings, 19th World's Poultry Congress.* 1992.

[42] Campbell, G., et al., Genotypic and environmental differences in extract viscosity of barley and their relationship to its nutritive value for broiler chickens. *Animal Feed Science and Technology,* 1989. **26**(3): p. 221-230.

[43] Choct, M., et al., Non-starch polysaccharide-degrading enzymes increase the performance of broiler chickens fed wheat of low apparent metabolizable energy. *The Journal of nutrition,* 1995. **125**(3): p. 485-492.

[44] Classen, H., et al. The relationship of chemical and physical measurements to the apparent metabolisable energy (AME) of wheat when fed to broiler chickens with and without a wheat enzyme source. in *Proc. of 2nd European Symposium on Feed Enzymes.* 1995.

[45] Dunn, N., Combating the pentosans in cereals. *World Poultry,* 1996. **12**(1): p. 24-25.

[46] Esteve-Garcia, E., J. Brufau., A. Perez-Vendrell, A. Miouei, K. Buven. Bioefficiency of enzymes preparation containing ß-glucanases and xylanase activities in broiler diets based on barley or wheat in combination with Flavomycin. *Poultry Science* 1997. **76**: p. 1728-1737.

[47] Gracia, M., et al., Alpha-amylase supplementation of broiler diets based on corn. *Poultry Science,* 2003. **82**(3): p. 436-442.

[48] Jansson, L., et al. Test of the efficacy of virginiamycin and dietary enzyme supplementation against necrotic enteritis disease in broilers. in *Proceedings, 8th European Poultry Conference.* 1990.

[49] Jeroch, H. and S. Dänicke, *Barley in poultry feeding. Proceedings 9th European Symposium on Poultry Nutrition,* 1993, World's Poultry Science Association: Jelenia Gora, Poland. p. 38-46.

[50] Leeson, S. and J. Proulx, Enzymes and barley metabolizable energy. *The Journal of Applied Poultry Research,* 1994. **3**(1): p. 66-68.

[51] Marquardt, R., et al., The nutritive value of barley, rye, wheat and corn for young chicks as affected by use of a *Trichoderma reesei* enzyme preparation. *Animal Feed Science and Technology,* 1994. **45**(3): p. 363-378.

[52] Marquardt, R. R., et al., Use of enzymes to improve nutrient availability in poultry feedstuffs. *Animal Feed Science and Technology, 1996.* **60**(3): p. 321-330.

[53] Odetallah, N., et al., Versazyme supplementation of broiler diets improves market growth performance. *Poultry Science,* 2005. **84**(6): p. 858-864.

[54] Odetallah, N. H., *Enzymes in corn-soybean diets, in Proceedings of the 29th Annual Carolina Poultry Nutrition Conference,* R. North Carolina State University, NC, Editor 2002.

[55] Ouhida, I., J. F. Perez., J. Gasa., F. Puchal., Enzymes (ß-glucanase and arabinoxylanase and/or Sepiolite supplementation and nutritive value of maize-barley-wheat based diets for broiler chickens. *British Poultry Science* 2000. **41**: p. 617-624.

[56]   Saleh, F., et al., Effect of enzymes of microbial origin on in vitro digestibilities of dry matter and crude protein in maize. *The journal of poultry science,* 2003. **40**(4): p. 274-281.

[57]   Wang, Z., et al., Effects of enzyme supplementation on performance, nutrient digestibility, gastrointestinal morphology, and volatile fatty acid profiles in the hindgut of broilers fed wheat-based diets. *Poultry Science,* 2005. **84**(6): p. 875-881.

[58]   Selle, P. H. and V. Ravindran, Microbial phytase in poultry nutrition. *Animal Feed Science and Technology,* 2007. **135**(1): p. 1-41.

[59]   Gushterova, A., et al., Keratinase production by newly isolated Antarctic actinomycete strains. *World Journal of Microbiology and Biotechnology,* 2005. **21**(6-7): p. 831-834.

[60]   Pandey, A., et al., Production, purification and properties of microbial phytases. *Bioresource technology,* 2001. **77**(3): p. 203-214.

[61]   Ushasree, M. V., H. B. S. Sumayya, and A. Pandey, Adopting structural elements from intrinsically stable phytase--A promising strategy towards thermostable phytases. *Indian Journal of Biotechnology,* 2011. **10**(4): p. 458-467.

[62]   Bedford, M. and H. Classen, *influence of dietary xylanase on intestinal viscosity and molecular weight distribution of carbohydrates in rye-fed broiler chicks.* Progress in biotechnology, 1991.

[63]   Van Paridon, P., et al., *application of fungal endoxylanase in poultry diets.* Progress in biotechnology, 1991.

[64]   Vandevoorde, L., H. Christiaens, and W. Verstraete, In vitro appraisal of the probiotic value of intestinal lactobacilli. *World Journal of Microbiology and Biotechnology,* 1991. **7**(6): p. 587-592.

[65]   Fuller, R., Probiotics in man and animals. *J. Appl. Bact,* 1989. **66**: p. 365.

[66]   Ware, D. R., P. L. Read, E. T. Manfredi, Lactation performance of two large dairy herds fed Lactobacillus acidophilus strain BT 1386. *J. Dairy Sci.,* 1988. **71**(Suppl. 1): p. 219 (Abstract).

[67]   Martin, S. A. and D. J. Nisbet, Effect of direct-fed microbials on rumen microbial fermentation. *Journal of Dairy Science,* 1992. **75**(6): p. 1736-1744.

[68]   Savage, D., Microorganisms associated with epithelial surfaces and stability of the indigenous gastrointestinal microflora. Food/Nahrung, 1987. **31**(5-6): p. 383-395.

[69]   Tufvesson, P.r., et al., Guidelines and cost analysis for catalyst production in biocatalytic processes. *Organic Process Research & Development,* 2010. **15**(1): p. 266-274.

[70]   Panke, S. and M. Wubbolts, Advances in biocatalytic synthesis of pharmaceutical intermediates. *Current opinion in chemical biology,* 2005. **9**(2): p. 188-194.

[71]   Lutz, S., Reengineering enzymes. *Science* (New York, NY), 2010. **329**(5989): p. 285.

[72] Kirchner, G., M. P. Scollar, and A. M. Klibanov, Resolution of racemic mixtures via lipase catalysis in organic solvents. *Journal of the American Chemical Society,* 1985. **107**(24): p. 7072-7076.

[73] Zheng, G.-W. and J.-H. Xu, New opportunities for biocatalysis: driving the synthesis of chiral chemicals. *Current opinion in biotechnology,* 2011. **22**(6): p. 784-792.

[74] Brakhage, A. A., *Molecular biotechnology of fungal beta-lactam antibiotics and related peptide synthetases.* Vol. 88. 2004: Springer.

[75] Doo, H. and D. D. RYut, Enzymatic synthesis of phenoxymethylpenicillin using Erwinia aroideae enzyme. *Synthesis,* 1984. **37**(10).

[76] Gavrilescu, M. and Y. Chisti, Biotechnology – a sustainable alternative for. *Biotechnology advances,* 2005. **23**: p. 471-499.

[77] Parmar, A., et al., Advances in enzymatic transformation of penicillins to 6-aminopenicillanic acid (6-APA). *Biotechnology advances,* 2000. **18**(4): p. 289-301.

[78] Torres-Bacete, J., et al., Optimization of 6-aminopenicillanic acid (6-APA) production by using a new immobilized penicillin acylase. *Biotechnology and applied biochemistry,* 2000. **32**(3): p. 173-177.

[79] Sheldon, R. A., E factors, green chemistry and catalysis: an odyssey. *Chemical Communications,* 2008(29): p. 3352-3365.

[80] Aresta, M., et al., Enzymatic synthesis of 4-OH-benzoic acid from phenol and CO< sub> 2</sub>: the first example of a biotechnological application of a Carboxylase enzyme. *Tetrahedron,* 1998. **54**(30): p. 8841-8846.

[81] Volpato, G., R. C. Rodrigues, and R. Fernandez-Lafuente, Use of enzymes in the production of semi-synthetic penicillins and cephalosporins: drawbacks and perspectives. *Current medicinal chemistry,* 2010. **17**(32): p. 3855-3873.

[82] Wohlgemuth, R., Asymmetric biocatalysis with microbial enzymes and cells. *Current opinion in microbiology,* 2010. **13**(3): p. 283-292.

[83] Liang, J., et al., Highly enantioselective reduction of a small heterocyclic ketone: biocatalytic reduction of tetrahydrothiophene-3-one to the corresponding (R)-alcohol. *Organic Process Research & Development,* 2009. **14**(1): p. 188-192.

[84] Höhne, M., et al., Rational assignment of key motifs for function guides in silico enzyme identification. *Nature chemical biology,* 2010. **6**(11): p. 807-813.

[85] Eichler, J., Biotechnological uses of archaeal extremozymes. *Biotechnology advances,* 2001. **19**(4): p. 261-278.

[86] Vandamme, E. and W. Soetaert, *Industrial biotechnology and sustainable chemistry.* 2004.

[87] OCED, *The application of biotechnology to industrial sustainability.* Paris7 OECD, 2001b.

[88] Asano, Y., et al., A new enzymatic method of acrylamide production. *Agricultural and Biological Chemistry,* 1982. **46**(5): p. 1183-1189.

[89]   Watanabe, I., Satoh, Y., Kouno, T., 1979: *Japanese Patent no. 129,*190.

[90]   Watanabe, I., Okumura, M., 1986: *Japanese Patent no. 162,*193.

[91]   Watanabe, I., Y. Satoh, and K. Enomoto, Screening, Isolation and Taxonomical Properties of Microorganisms Having Acrylonitrile-hydrating Activity (Microbiology & Fermentation Industry). *Agricultural and Biological Chemistry,* 1987. **51**(12): p. 3193-3199.

[92]   Nagasawa, T. and H. Yamada, Microbial transformations of nitriles. *Trends in Biotechnology,* 1989. **7**(6): p. 153-158.

[93]   Nakai, K., et al., Development of acrylamide manufacturing process using microorganisms. *Journal of the Agricultural Chemical Society of Japan* (Japan), 1988.

[94]   Gooding, O. W., et al., Development of a practical biocatalytic process for (R)-2-methylpentanol. *Organic Process Research & Development,* 2009. **14**(1): p. 119-126.

[95]   Adrio, J. L. and A. L. Demain, Microbial Enzymes: Tools for Biotechnological Processes. *Biomolecules,* 2014. **4**(1): p. 117-139.

[96]   Sakai Yasuyoshi, Yoshiki Tani. Formaldehyde production bv cells of a mutant of *Candlda bofdlnfl S2* grown in methanol-limited chemostat culture. *Agric. Biol. Chem.,* 1986. **50**: p. 2615-2620.

[97]   Kato, N., et al., Formate production from methanol by formaldehyde dismutase coupled with a methanol oxidation system. *Applied microbiology and biotechnology,* 1988. **27**(5-6): p. 567-571.

[98]   Yoshida, H., Y. Tani, and H. Yamada, Isolation and identification of a pyrogallol producing bacterium from soil. *Agricultural and Biological Chemistry*, 1982. **46**(10): p. 2539-2546.

[99]   Ensley, B. D., Ratzkin, B. J., Osslund, T. D., Simon, M. J., Wackett, L. P., Gibon, D. T., Expression of naphthalene oxidation genes in Escherfchia coli results in the biosynthesis of indigo. *Science* (New York, NY), 1983. **222**: p. 167- 169.

[100] Nagasawa, T., J. Mauger, and H. Yamada, A novel nitrilase, arylacetonitrilase, of Alcaligenes faecalis JM3. *European journal of biochemistry,* 1990. **194**(3): p. 765-772.

[101] Neidleman, D. L., Amon, Jr. W. F., Geigert, J., *US. Patent,* 1981.

[102] Geigert, J., et al., Haloperoxidases: enzymatic synthesis of α, β-halohydrins from gaseous alkenes. *Applied and environmental microbiology,* 1983. **45**(2): p. 366-374.

[103] Geigert, J., et al., Production of Epoxides from α, β-Halohydrins by Flavobacterium sp. *Applied and environmental microbiology,* 1983. **45**(3): p. 1148.

[104] Kobayashi, M., et al., Purification and characterization of a novel nitrilase of Rhodococcus rhodochrous K22 that acts on aliphatic nitriles. *Journal of bacteriology,* 1990. **172**(9): p. 4807-4815.

[105] Mauger, J., T. Nagasawa, and H. Yamada, Occurrence of a novel nitrilase, arylacetonitrilase in Alcaligenes faecalis JM3. *Archives of microbiology,* 1990. **155**(1): p. 1-6.

[106] Stalker, D. M., McBridge, K. E., Cloning and expression in Escherfchfa coll of a Klebslellaozaenae plasmic-borne gene encoding a nitrilase specific for the herbicide bromoxynil. *J. Bacteriol.,* 1987. **169**: p. 955-960.

[107] Arnaud, A., Galzy, P., Commeyras, A., Jallageas. J. C., 1979: *French Patent* no. 73 33613.

[108] Godfredsen S. E., I., K., Yde, B., Andersen, O., Biocatalysis in Organic Syntheses (*Studies on Organic Chemistry,* 22) 1985.

[109] Sun, Y. and J. Cheng, Hydrolysis of lignocellulosic materials for ethanol production: a review. *Bioresource technology,* 2002. **83**(1): p. 1-11.

[110] Brzostowicz, P. C., et al., Simultaneous identification of two cyclohexanone oxidation genes from an environmental Brevibacterium isolate using mRNA differential display. *Journal of bacteriology,* 2000. **182**(15): p. 4241-4248.

[111] Mojsov, K., Application of Enzymes in the Textile Industry: A Review. *International Congress "Engineering, Ecology and Materials in the Processing Industry,* 2011: p. 230-239.

[112] Marcher, D. and H. A. Hagen, *ITB Veredlung.* 1993. **39**: p. 20-32.

[113] Meyer-Stork, L. S., *Maschen-Industrie,* 2002. **52**: p. 32-40.

[114] Gulrajani, M. L., *Rev. Prog. Coloration,* 1992. **22**: p. 79-89.

[115] Forneil, S., *Melliand Textilber,* 1994. **75**: p. 120-125.

[116] Kundu, A. B., et al., *Textile Res. J,* 1991. **61**: p. 720-723.

[117] Araújo, R., M. Casal, and A. Cavaco-Paulo, Application of enzymes for textile fibres processing. *Biocatalysis and Biotransformation,* 2008. **26**(5): p. 332-349.

[118] Cavaco-Paulo, A. and G. M. Gübitz, *Textile Processing with Enzymes,* 2003, Woodhead Publishing Ltd: England.

[119] Vigneswaran, C., N. Anubumani, and M. Anathasubramanian, Biovision in Textile Wet processing Industry- Technological Chellanges. *JTATM,* 2011. **7**(1): p. 1-13.

[120] Pawar, S. B., H. D. Shah, and G. R. Andhorika, *Man-Made Textiles in India.* 2002. **45**(4): p. 133.

[121] Rajasekhar, A., et al., Thermostable Bacterial Protease - A New Way for Quality Silk Production. *International Journal of Bio-Science and Bio-Technology*, 2011. **3**(4): p. 43-56.

[122] Polizeli, M. L., et al., Xylanases from fungi: properties and industrial applications. *Appl Microbiol Biotechnol,* 2005. **67**(5): p. 577-91.

[123] Bajpai, P. ; Bhardwaj, N. K.; Maheshwari, S.; Bajpai, P. K.; *Appita,* 1993. **46**(**4**): p. 274-276.

[124] Gutierrez, A., J. C. del Rio, and A. T. Martinez, Microbial and enzymatic control of pitch in the pulp and paper industry. *Appl Microbiol Biotechnol,* 2009. **82**(6): p. 1005-18.

[125] Gupta, R., et al., Microbial -amylases: a biotechnological perspective. *Process Biochem,* 2003. **38**: p. 1599-1616.

[126] van der Maarel, M. J., et al., Properties and applications of starch-converting enzymes of the alpha-amylase family. *J Biotechnol,* 2002. **94**: p. 137-155.

[127] Viikari, L., et al., *FEMS Microbiol. Rev.* 1994. **13**: p. 335.

[128] Singh, S., et al., Biotechnological applications of industrially important amylase enzyme. *International Journal of Pharma and Bio Sciences. India,* 2011. **2**(1): p. 486-496.

[129] Bradbury, E. J., et al., Chondroitinase ABC promotes functional recovery after spinal cord injury. *Nature,* 2002. **416**(6881): p. 636-640.

[130] Ensor, C. M., et al., Pegylated arginine deiminase (ADI-SS PEG20, 000 mw) inhibits human melanomas and hepatocellular carcinomas in vitro and in vivo. *Cancer research,* 2002. **62**(19): p. 5443-5450.

[131] Blain, F., et al., Expression system for high levels of GAG lyase gene expression and study of the hepA upstream region in Flavobacterium heparinum. *Journal of bacteriology,* 2002. **184**(12): p. 3242-3252.

[132] *Genencor International website,* http:// www.genencor.com/ wt/ gcor/ adv therapeutics.

[133] Lee-Huang, S., et al., Lysozyme and RNases as anti-HIV components in β-core preparations of human chorionic gonadotropin. *Proceedings of the National Academy of Sciences,* 1999. **96**(6): p. 2678-2681.

[134] Fusetti, F., et al., Structure of human chitotriosidase implications for specific inhibitor design and function of mammalian chitinase-like lectins. *Journal of Biological Chemistry,* 2002. **277**(28): p. 25537-25544.

[135] Zimmer, M., et al., The murein hydrolase of the bacteriophage φ3626 dual lysis system is active against all tested Clostridium perfringens strains. *Applied and environmental microbiology,* 2002. **68**(11): p. 5311-5317.

In: Microbial Catalysts. Vol. 2
Editors: Shadia M. Abdel-Aziz et al.

ISBN: 978-1-53616-088-8
© 2019 Nova Science Publishers, Inc.

Chapter 7

INDUSTRIAL APPLICATIONS OF MICROBIAL ENZYMES

Kandasamy Selvam[1], Balakrishnan Senthilkumar[2], Arumugam Sengottaiyan[1], Thangasamy Selvankumar[1,] and M. Govarthanan[3,†]*

[1]PG & Research Department of Biotechnology, Mahendra Arts and Science College (Autonomous), Namakkal, Tamil Nadu, India
[2]Department of Medical Microbiology, College of Health and Medical Sciences Haramaya University, Harar, Ethiopia
[3]Department of Environmental System Engineering, University of Seoul, Seoul, Republic of Korea

ABSTRACT

Enzymes play a crucial role as metabolic catalysts, leading to their great importance in the development of industrial bioprocesses. Current applications focuses on many different markets including cleaning (detergents), textiles, starch processing, brewing, leather, baking, paper, food products, animal feeds, cosmetics and personal care. The use of enzyme technology is attractive because enzymes are highly efficient to specific, and work under mild conditions. The use of enzymes results in reduce the process time, energy, water, and also improved product quality and potential process integration. The chapter covers the most important applications of microbial enzymes in various industrial processes.

Keywords: biocatalysis, microbial enzymes, applications, industry

---

* Corresponding Author's E-mail: selvankumar75@gmail.com;.
† Author's E-mail: gova.muthu@gmail.com.

1. INTRODUCTION

Enzymes are the bio-catalysts playing an important role in all stages of metabolism and biochemical reactions. Certain enzymes are of special interest and utilized as organic catalysts in numerous processes an industrial scale. Microbial enzymes are superior enzymes obtained from different microorganisms, especially for the industrial usage on commercial scales. The enzymes were discovered from microorganisms in the 20th century. The studies on their isolation, characterization of properties, production on bench-scale to pilot-scale and their application in bio-industry have continuously progressed, and the knowledge has regularly been updated. Many enzymes from microbial sources are already used in various commercial processes. Selected microorganisms including bacteria, fungi and yeasts have been globally studied for the bio-synthesis of economically viable preparations of various enzymes for commercial applications.

Microbial enzymes are widely used in industrial processes based on potential productivity, chemical stability, environmental protection, plasticity and vast availability [1, 2]. *Bacillus* species like *Bacillus subtilis*, *Bacillus amyloliquefaciens* and *Bacillus licheniformis* are used as bacterial workhorses in industrial microbial cultivations for the production of a variety of enzymes and fine biochemicals. A large quantity (20 - 25g/L) of extracellular enzymes has been produced by the various *Bacillus* strains which has been placed among the most significant industrial enzyme producers. The fungal origins like *Aspergillus, Conidiobolus, Mucor, Paecilomyces, Penicillium, Rhizopus* are used in industrial microbial cultivations for the production of a variety of enzymes which are advantageous due to the ease of cell removal during downstream processing. In recent years, the potential use of microorganisms as biotechnological sources of industrially relevant enzymes has stimulated interest in the exploration of extracellular enzymatic activity in several microorganisms [3].

Industrially important enzymes have traditionally been obtained from submerged fermentation (SmF) because of the ease to handling and great control of environmental factors such as temperature and pH. Solid-state fermentation (SSF) constitutes an interesting alternative since the metabolites produced are concentrated and the purification procedures are less costly [4]. SSF is preferred to SmF because of the simple technique, low capital investment, lower levels of catabolite repression end product inhibition, low waste water output, better product recovery, and high quality production [5, 6, 7]. On dry basis, agricultural substrates like corn, wheat, sorghum, coffee pulp waste and other cereals and grains contain around 60 – 75% (by mass) starch, hydrolysable to glucose, with a significant mass increase, which offers a good resource in many fermentation processes. With the advent of biotechnological innovations, mainly in the area of enzyme and fermentation technology, many new areas have been opened for their utilization as raw materials for the production of value added fine products [6].

2. HISTORY OF ENZYME TECHNOLOGY

The history of enzyme technology began in 1874 when the Danish chemist Christain Hansen produced the first specimen of rennet by extracting dried claves with saline solution, which was the first enzyme used for industrial purposes. The significant event had been preceded by a lengthy evolution. The digestion of meat by the secretions and conversion of starch to sugar by plant extracts and saliva were known. The fermentative activity of microorganisms was discovered in 18th century by the French scientist Louis Pasteur. In 1878 German physiologist Wilhelm Kuhne (1837-1900) coined the term enzyme from Latin word, which literally mean "in yeast". In 1897 Eduard Buchner began to study the ability of yeast extracts that lacked any living yeast cells to ferment sugar. In a series of experiments at the University of Berlin, he found that the sugar was fermented even when there is no living yeast cells in the mixture. He named the enzyme that brought about the fermentation of sucrose "zymase".

The first application of the cell free enzymes used the rennin aspartic protease isolated from calf or lamb stomach in cheese making. The first commercial enzyme (trypsin) was prepared by Rohm in Germany in 1914; isolated from animals and used in the detergents to degrade proteins. Introduction of microbial proteases in washing powder have made a real breakthrough in the detergent industries. The first commercial bacterial *Bacillus* protease was marked in1959 and was a big commercial success when Novozymes in Denmark started to manufacture it in 1965.

In 1930 enzymes were used in fruit juice manufacturing for clarification of juices. The major usage started in 1960s in starch industries. The traditional acid hydrolysis of starch was completely replaced by α-amylases and glucoamylases, which could convert starch with over 95%, yield to glucose. Starch industries became the second largest user of enzymes next to detergent industry [8, 9, 10, 11].

The use of enzymes is beneficent in higher product quality, low manufacturing cost, less waste and reduced energy consumption. More traditional chemical treatments produced undesirable side effects and more waste disposal problems. The degree to which a desired technical effect is achieved by an enzyme can be controlled through various means, dose, temperature, and time. As the enzymes are catalysts, the amount added to accomplish a reaction is relatively small. Enzymes used in food processing are generally destroyed during subsequent processing steps and not present in the final food product.

3. ENZYMES IN GLOBAL MARKET

The global market of industrial enzymes is estimated at \$3.3 billion in 2010. This market reached \$4.4 billion in 2015, a compound annual growth rate (CAGR) of 6% over the 5-year forecast period. Technical enzymes valued in 2010 was $1 billion. This sector increased to 6.6% compound annual growth rate (CAGR) and reached $1.5 billion in 2015. The highest sales of technical enzymes occurred in the leather market, followed by the bio-ethanol market. The food and beverage enzymes segment is reached about $1.3 billion in 2015, from a value of $975 million in 2010, rising at a compound annual growth rate (CAGR) of 5.1%. Within the food and beverage enzymes segment, the milk and dairy market had the highest sales, with $401.8 million in 2009.

The global industrial enzymes market is very competitive with Novozymes being the largest role in the industry, followed by DSM, and DuPont (after it acquired a majority stake in Danisco and its Genencor division), among others. The companies mainly compete on the basis of product quality, performance, use of intellectual property rights, and the ability to innovate, among other such factors. North America and Europe are the largest consumers of industrial enzymes although the Asia Pacific region will undergo a rapid increase in enzyme demand in China, Japan and India, reflecting the size and strength of these country's economies.

4. INDUSTRIAL ENZYMES APPLICATIONS

Enzymes are applied in various fields, including technology, food manufacturing, animal nutrition, cosmetics, medicine, and as tools for research and development. Almost 4000 enzymes are known, and out of these, approximately 200 microbial original types are used commercially. However, only 20 enzymes are produced on truly industrial scale. With the improved understanding of the enzyme production biochemistry, fermentation processes, and recovery methods, an increasing number of industrial enzymes can be foreseeable. The world enzyme demand is satisfied by about 12 major producers and 400 minor suppliers. Nearly 75% of the total enzymes are produced by three top enzyme companies, i.e., Denmark-based Novozymes, US-based DuPont (through the May 2011 acquisition of Denmark-based Danisco) and Switzerland-based Roche. The market is highly competitive, and has small profit margins and is technologically intensive.

The use of enzymes to produce goods for human consumption date back at least 2000 years, when microorganisms were used in processes such as leavening bread and saccharification of rice in koji production [12]. The mechanism of the enzymes was unknown until 1877, when Moritz Traube proposed that "protein-like materials catalyze fermentation and other chemical reactions." Later, the historic demonstration by Buchner in 1897 showed that alcoholic fermentation could be carried out using cell-free yeast extract, and appeared to be the first application of biocatalysis. The word 'zymase' was coined to describe this cellfree extract [13, 14], which was the initial recognition of what is now called an 'enzyme'. There are currently around 5500 known enzymes [15], classified based on the type of reaction they catalyze (oxidoreductases, transferases, hydrolases, lyases, isomerases, and ligases). Specific enzyme names refer to the substance on which they act. An enzyme that acts on cellulose, for example, is known as cellulase, and an enzyme that acts on protein is named protease, etc. [16].

Enzymes for industrial use are produced by growing bacteria and fungi in submerged orSSF. When submerged being the primary fermentation mode, the unit operations in enzyme production involve fermentation followed by cell disruption and filtration. The crude enzyme is further purified by precipitation followed by centrifugation and vacuum drying or lyophilization, collectively known as "downstream processing" [14].

4.1. Amylase

Amylases are a group of hydrolases that can specifically cleave the α-1,4-glycosidic bonds in starch. Two important groups of amylases are glucoamylase and α-amylase. Glucoamylase (exo-1,4-α-D-glucan glucanohydrolase, E.C. 3.2.1.3) hydrolyzes single glucose units from the nonreducing ends of amylose and amylopectin in a stepwise manner [17]. Whereas α-amylases (endo-1,4-α-D-glucan glucohydrolase, E.C. 3.2.1.1) are extracellular enzymes that randomly cleave the 1,4-α-D-glucosidic linkages between adjacent glucose units inside the linear amylose chain [17, 18, 19].

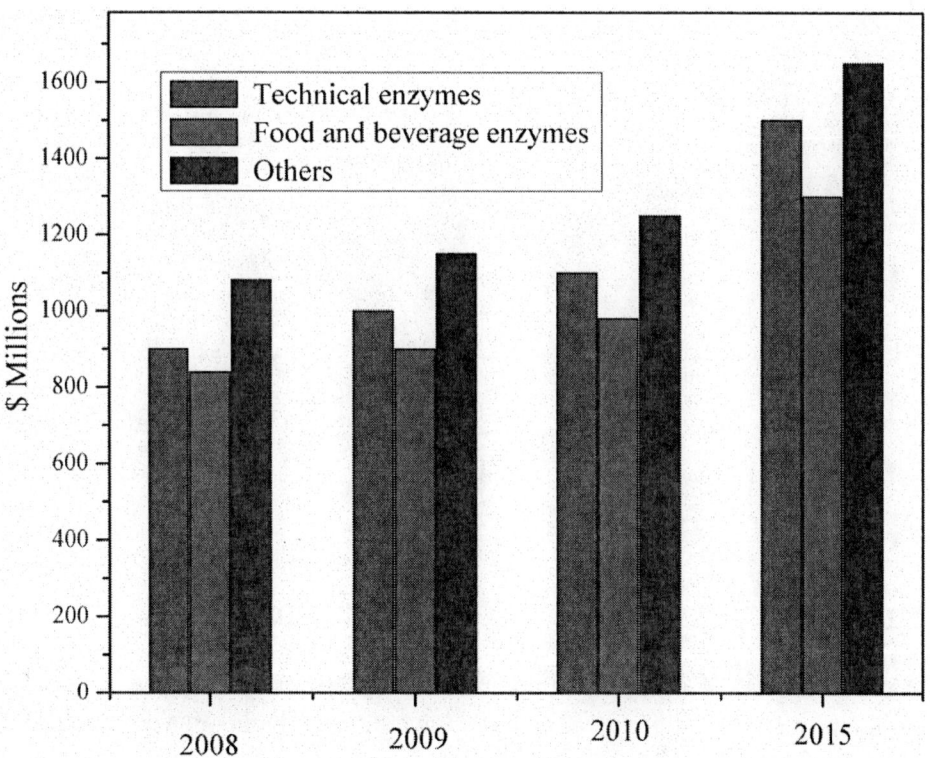

Figure 1. Global industrial enzyme market 2008-2015 ($millions).

Table 1. gives the representative examples of enzyme applications in various industries

Industry	Enzyme	Applications
Cleaning (Detergent)	Amylase	Starch stain removal
	Protease	Protein stain removal
	Lipase	Lipid stain removal
	Cellulase	Cleaning, color clarification, anti-redeposition (cotton)
	Mannanase	Mannanan stain removal (reappearing stains)
Textile	Cellulase	Denim finishing, cotton softening
	Amylase	De-sizing
	Pectate lyase	Scouring
	Catalase	Bleach termination
	Laccase	Bleaching
	Peroxidase	Excess dye removal
Starch and fuel	Amylase	Starch liquefaction and saccharification
	Amyloglucosidase	Saccharification
	Glucose-isomerase	Cyclodextrin production
	Xylanase	Viscosity reduction (fuel and starch)
	Protease	Protease (yeast nutrition – fuel)
Leather	Protease	Unhearing, bating
	Lipase	De-pickling
Baking	Amylase	Bread softness and volume, flour adjustment
	Xylanase	Dough conditioning
	Lipase	Dough stability and conditioning (*in situ* emulsifier)
	Phospholipase	Dough stability and conditioning (*in situ* emulsifier)
	Glucose oxidase	Dough strengthening
	Lipoxygenase	Dough strengthening, bread whitening
	Protease	Biscuits, cookies
	Transglutaminase	Laminated dough strengths
Pulp and paper	Lipase	Pitch control, contaminant control
	Protease	Biofilm removal
	Amylase	Starch-coating, de-inking, drainage improvement
	Xylanase	Bleach boosting
	Cellulase	De-inking, drainage improvement, fiber modification
Food (including dairy)	Protease	Milk clotting, infant formulas (low allergenic), flavor
	Lipase	Cheese flavor
	Lactase	Lactose removal (milk)
	Pectin methyl esterase	Firming fruit-based products
	Pectinase	Fruit-based products
	Transglutaminase	Modify visco-elastic properties
Animal feed	Phytase	Phytate digestibility – phosphorus release
	Xylanase	Digestibility
	Glucanase	Digestibility
Beverage	Pectinase	De-pectinization, mashing
	Amylase	Juice treatment, low calorie beer
	Glucanase	Acetolactate decarboxylase Maturation (beer) Mashing
	Laccase	Clarification (juice), flavor (beer), cork stopper treatment
Personal care	Amyloglucosidase	Antimicrobial (combined with glucose oxidase)
	Glucose oxidase	Bleaching, antimicrobial
	Peroxidase	Antimicrobial
pharmaceutical	Lipase, amylase and protease	Treatment of pancreatic insufficiency
Fine chemicals	Lipase	Amino-butanoic acid production

Amylases constitute a class of industrial enzymes representing approximately 30% of the world enzyme production [20, 21]. They have diverse applications in a wide variety of industries such as food, fermentation, textile, paper, detergent and sugar industries. It can be used in the fields related with biotechnology such as: removing environmental pollutant, conversion of starch to desired substrate by many microorganisms, infiltration of waste containing starch and production of biochemical material with helping starch substrate. With the advent of new frontiers in biotechnology, the spectrum of amylase application has expanded into many other fields, clinically, medicinally and analytically. Interestingly, the first enzyme produced industrially was an amylase from a fungal source in 1894, which was used as a pharmaceutical aid for the treatment of digestive disorders. Although amylases can be derived from several sources, including plants, animals and microorganisms. The microbial amylases have almost completely replaced chemical hydrolysis of starch in starch processing industry [22]. Microbial enzymes generally meet industrial demands [23, 24].

4.2. Proteases

Proteases are enzymes which catalyze the hydrolysis of peptide bonds. Proteases are essential constituents of all life forms on earth including prokaryotes, fungi, plants, and animals. Proteases are highly exploited enzymes in food, leather, detergent, pharmaceutical, diagnostics, waste management, and silver recovery. Proteases (serine protease, cysteine protease, aspartic proteases and metalloprotease) constitute one of the most important groups of industrial enzymes, accounting for about 60% of the total enzyme market. Among the various protease, bacterial proteases are the most significant, compared with animal and fungal proteases and among bacteria, *Bacillus* sp. are specific producers of extracellular proteases [25]. These enzymes have wide industrial application, including the pharmaceutical industry, leather industry, and manufacture of protein hydrolysates, food industry and waste processing industry [26].

4.3. Cellulase

Bioconversion of cellulose containing raw materials is an important problem of current biotechnology due to the increasing demand for energy, food, and chemicals. Cellulases are enzymes which hydrolyze the β-1,4- glycosidic linkage of cellulose and synthesized by microorganisms during their growth on cellulosic materials [27]. The complete enzymatic hydrolysis of cellulosic materials needs different types of cellulase; namely endoglucanase, (1,4-D-glucan-4-glucanohydrolase; EC 3.2.1.4), exocellobiohydrolase (1, 4-D-glucan glucohydrolase; EC 3.2.1.74) and glucosidase (D-glucoside glucohydrolase; EC 3.2.1.21).

Enzymatic process to hydrolyze cellulosic materials could be accomplished through a complex reaction of these various enzymes. Two significant attributes of these enzyme-based bioconversion technologies are reaction conditions and the production cost of the related enzyme system. There has been many research works focused on obtaining new microorganisms producing celluloytic enzymes with higher 105 specific activities and greater efficiency worldwide [28]. Enzymes produced by marine microorganisms can provide numerous advantages over traditional enzymes due to the wide range of environments [29].

4.4. Lipases

Lipases have emerged as one of the leading biocatalysts with proven potential for contributing to the multibillion dollar underexploited lipid technology bio-industry and have been used in *in-situ* lipid metabolism and *ex-situ* multifaceted industrial applications [30]. Lipases are triacylglycerol acyl hydrolases (EC 3.1.1.3) that catalyze the hydrolysis of triacylglycerol to glycerol and fatty acids. They often express other activities such as phospholipase, lysophospholipase, cholesterol esterase, cutinase, amidase and other esterase type of activities [31]. Microbial lipases have gained special industrial attention due to their ability towards extremes of temperature, pH, and organic solvents, and chemo-, region-, and enantioselectivity. Lipases are ubiquitous in nature and are produced by several plants, animals, and microorganisms [32]. Some important lipase-producing bacterial genera are *Bacillus*, *Pseudomonas* and *Burkholderia* [33] and fungal genera include [34] *Aspergillus*, *Penicillium*, *Rhizopus*, *Candida*. Although considerable progress has been made over the recent years towards the developing cost-effective systems for lipases, the high cost of production of this enzyme remains the major challenge associated with large-scale industrial applications.

4.5. Pectinase

Pectinase are a group of enzymes which contribute the breakdown of pectin. It is a structural polysaccharide found in primary cell wall and middle lamina of fruits and vegetables. Pectolysis is one of the most important processes for plant, as it plays a role in cell elongation and growth as well as fruit ripening. Microbial pectolysis is important in plant pathogenesis, symbiosis and decomposition of plant deposits [35]. The main source of the microorganisms that produce pectinolytic enzymes are yeast, bacteria and large varieties of fungi and particularly *Aspergillus* sp. endopolygalacturonase production was first reported in 1951 using *Saccharomyces fragilis* [36].

4.6. Laccase

Laccase is an enzyme that has potential ability of oxidation. It belongs to those enzymes, which have innate properties of reactive radical production, and its utilization in many fields has been ignored because of its unavailability in the commercial field. There are diverse sources of laccase producing organisms like bacteria, fungi, and plants. Laccases use oxygen and produce water as by-product. They can degrade a range of compounds including phenolic and non-phenolic compounds. They also have ability to detoxify a range of environmental pollutants. Their property to act on a range of substrates and also to detoxify a range of pollutants have made them to be usable for several purposes in many industries including paper, pulp, textile and petrochemical industries.

4.7. Mannanase

Mannanase enzyme is important in paper industry including bioleaching pulp [37] waste bioconversion of biomass to fermentable sugars [38], increasing the quality of feed quality [39] and reduces viscosity of coffee extracts [40]. Endo-β-1,4- mannanases (EC.3.2.1.78) randomly hydrolyze the main chain of hetero mannans, the major softwood hemicellulose [41]. Mannanases have been tested in several industrial processes, such as extraction of vegetable oils from leguminous seeds, viscosity reduction of extracts during the manufacture of instant coffee and manufacture of oligosaccharide [42, 43] as well as applications in the textile industry [41]. In paper industry, mannanases have synergistic action in the bio-bleaching of the wood pulp, significantly reducing the amount of chemicals used [44].

5. ENZYMES IN DETERGENT INDUSTRIES

Detergent industries are the primary consumers of enzymes, in terms of both volume and value. The use of enzymes in detergents formulations enhances the detergents ability to remove tough stains and making the detergent environmentally safe [45]. Amylases are the second type of enzymes used in the formulation of enzymatic detergent, and 90% of all liquid detergents contain these enzymes. As colored stains, their removal is of interest in both detergent and dishwashing contexts. Removal of starch from surfaces is also important in providing a whiteness benefit since it is known that starch can be an attractant for many types of particulate soils. The suitability of any hydrolytic enzymes for inclusion in detergent formulation is dependent on its stability and compatibility with detergent components.

Alkaline proteases have contributed greatly to the development and improvement of modern household and industrial detergents. They are effective at the moderate temperature and pH values that characterize modern laundering conditions in industrial and institutional cleaning [46]. Of these, alkaline protease finds a major application as detergent additives because of their ability to hydrolyze and remove proteinacious stains like blood, egg, gravy, milk, etc. in high pH conditions [47].

Cellulases, in particular EG III and CBH I, are commonly used in detergents for cleaning textiles several reports [48, 49] disclose that EG III variants, in particular from *Trichoderma reesei* are suitable for the use in detergents. *Trichoderma viride* and *Trichoderma harzianum* are also industrially utilized natural sources of cellulases, like *Aspergillus niger*. Cellulase preparations, mainly from species of *Humicola* (*Humicola insolens* and *Humicola griseathermoidea*) that are active under mild alkaline conditions and at elevated temperatures, are commonly added in washing powders, and in detergents.

A tremendous increase in the significance of the biotechnological applications of lipases since the last two decades display amazing versatility in catalytic behavior. The latest trend in detergent industry is towards lower wash temperatures which not only save energy, but also help to maintain the texture and quality of fabrics [50].

6. ENZYMES IN TEXTILE INDUSTRIES

In textile industry, strength of the textile is improved by warping the starch paste to textile weaving. It also prevents the loss of string by friction, cutting and generation of static electricity on the string by giving softness to the surface of string due to laid down warp. After weaving the cloth, the starch is removed and the cloth goes to scouring and dyeing. The starch on cloth is usually removed by application of α-amylase. The α-amylases remove selectively the size [51]. Amylase from *Bacillus* strain was employed in textile industries for quite a long time [52].

Cellulases has become the third largest group of enzymes used in the industry since their introduction a decade ago [53]. They are used in the bio-stoning of denim garments for producing softness and the faded look of denim garments replacing the use of pumice stones which were traditionally employed in the industry [54, 55]. They act on the cellulose fiber to release the indigo dye used for coloring the fabric producing the faded look of denim. *Humicola insolens* cellulase is most commonly employed in the equally good cellulases are utilized for digesting off the small fiber ends protruding from the fabric resulting in a better finish cellulases, used in softening defibrillation, and in processes for providing localized variation in the color density of fibers.

Textile processing has benefited greatly in both environmental and product quality aspects through the use of enzymes. Prior to weaving of yarn in to fabric, the warps yarns are coated with a sizing agent to lubricate and protect the yarn from abrasion during weaving.

The main sizing agent used for cotton fabrics has been starch because of its excellent film-forming capacity, availability, and reality low cast. Before the fabric is dyed, the sizing agent is applied and the natural non-cellulosic materials present in the cotton is removed. Before the discovery of amylase enzymes, the only way to remove the starch-based sizing was extended treatment with casting soda at high temperature. The chemical treatment was not effective in removing the starch and also resulted in a degradation of the cotton fiber resulting in distraction of the natural soft feel or 'hand' of the cotton the use enzyme such as pectinase in conjugation with amylases, lipases, cellulases, and other hemicellulolytic enzymes. The removal sizing agents has decreased the use of harsh chemicals in textile industry, resulting in a lower discharge of waste chemicals to the environment, improving both the safety of working conditions for textile workers and the quality of the fabric. Laccases-mediator system finds potential application in enzymatic modification of dye bleaching in the textile and dyes industries [56].

Threads of raw silk must be degummed to remove sericin, a proteinaceous substance that covers the silk fiber. Traditionally, degumming is performed in an alkaline solution containing soap. It also has other disadvantages of high energy consumption, time consumption and also loss in luster of silk due to the amounts of water used in this process [57].

7. ENZYMES IN STARCH AND FUEL INDUSTRIES

The major market for α-amylases is the starch industry, which is used for starch hydrolysis in the starch liquefaction process that converts starch into fructose and glucose syrups. The enzymatic conversion of all starch includes: gelatinization, which involves the dissolution of starch granules, thereby forming a viscous suspension; liquefaction, which involves partial hydrolysis and loss in viscosity; and saccharification, involving the production of glucose and maltose via further hydrolysis. This process requires the use of a highly thermostable α-amylase for starch liquefaction, which acts at temperature between 70 - 100°C depending upon the temperature [58]. Initially, the α-amylase of *Bacillus amyloliquefaciens* was used but it has been replaced by the α-amylase of *Bacillus stearothermophilus* or *Bacillus licheniformis* [59]. The enzymes from the *Bacillus* species are of special interest for large-scale biotechnological processes due to their remarkable thermostability and because efficient expression systems are available for these enzymes [60].

Ethanol is the most utilized liquid biofuel for the bioethanol production and starch is the most used substrate due to its low price and easy availability of raw material in most of the. In this production, starch has to be solubilized and then submitted to two enzymatic steps in order to obtain fermentable sugars [61]. The conventional process for the bioconversion of starch into ethanol involves saccharification, where starch is converted into sugar using an amylolytic microorganism or enzymes such as glucoamylase and α-amylase, followed by fermentation, where sugar is converted into ethanol using an ethanol fermenting microorganism such as yeast *Saccharomyces cerevisiae* [62]. A potential application of cellulase is the conversion of cellulosic materials to glucose and other fermentable sugars, which in turn can be, used as microbial substrates for the production of single cell proteins or a variety of fermentation products like ethanol [63].

8. ENZYMES IN LEATHER INDUSTRY

Soaking, dehairing of hides and skins and bating have been traditionally being carried out by using different chemicals which poses a high tannery waste pollution threat. Hence, proteases with a pH optimum around 9-10 are widely used in soaking to facilitate the water uptake of the hides or skin. Alkaline proteases with elastolytic and keratinolytic activity are used for dehairing and bating process to obtain a desired grain, softness and tightness of leather in a short time. Alkaline proteases with keratinolytic activity have been reported for remarkable dehairing properties [64, 65, 66]. A novel protease showing keratinolytic activity from *Bacillus subtilis* have been studied as a potential for replacing sodium sulfide in the dehairing process of leather industry [667]. Verma et al. [68] showed the use of protease from *Thermoactinomyces* sp. RM4 for dehairing goat hides.

9. ENZYME IN BACKING INDUSTRIES

With the development of modern biotechnology, the food industry has undergone great changes. There are many reports about the genetic engineering enzymes that have been used safely in the food industry. Amylases are extensively employed in the processed-food industry such as baking, brewing, preparation of digestive aids, production of cakes, fruit juices and starch syrups [69].

In the bread-making process laccases affix bread and enhance dough-enhancement additives to the bread dough, which results in improved freshness of the bread texture, flavour and the improved machinability [70].

Lipases have immense application in food industry such as in cheese ripening, flavor development and EMC technology [71]. Lipases are used *ex-situ* to produce flavors, and to modify the structure by inter- or transesterification, in order to obtain products of increased nutritional value, or suitable for parental feeding [72]. Lipases have also been used for addition in food to modify flavor by synthesis of esters of short chain fatty acids and alcohols, which are known flavor and fragrance compounds [73].

10. ENZYME IN PULP AND PAPER

The use of α-amylases in the pulp and paper industry is the modification of starch for coated paper, i.e., for the production of low-viscosity, high molecular weight starch. The coating treatment serves to make the surface of paper sufficiently smooth and strong, to improve the writing quality of the paper. In this application, the viscosity of the natural starch is too high for paper sizing and this can be altered by partially degrading the polymer with α-amylases in a batch or continuous processes [74].

Cellulases and hemicellulases have been employed for biomechanical pulping for modification of the coarse mechanical pulp and hand sheet strength properties de-inking of recycled fibers improved drainage and run ability of paper mills. Cellulases are employed in the removing of inks coating and toners from paper bio characterization of pulp fibers is another application where microbial cellulases are employed. Cellulases are also used in preparation of easily biodegradable cardboard. The enzyme is employed in the manufacture of soft paper including paper towels and sanitary paper and preparations containing cellulases are used to remove adhered paper [75].

The pulp and paper industry processes huge quantities of lignocellulosic biomass every year. The technology for pulp manufacture is highly diverse, and numerous opportunities exist for the application of microbial enzymes. Historically, enzymes have found some uses in the paper industry, but these have been mainly confined to areas such as modifications of raw starch. The enzymatic pitch control method using lipases have been in use in a large-scale paper-making process as a routine operation since early 1990s. 'Pitch' or the hydrophobic components of wood (mainly triglycerides and waxes), causes severe problems in pulp and paper manufacture. Lipases are used to remove the pitch from the pulp produced for paper making [76]. Nippon Paper Industries, Japan, have developed a pitch control method that uses the *Candida rugosa* fungal lipase to hydrolyze up to 90% of the wood triglycerides.

The advancement of biotechnology and increased reliance of paper and pulp industries on the use of microorganisms and their enzyme for biobleaching and paper making, and the use of enzyme other than xylanases and ligninases, such as mannanase, pectinases is increasing in the paper and pulp industries in many countries [77, 78]. During paper making pectinase can depolymerize polymers of galacturonic acids, and subsequently lower the cationic demand of pectin solutions and the filtrate from peroxide bleaching [79, 80]. An overall bleach-boosting of eucalyptus Kraft pulp was obtained when alkaline pectinase from *Streptomyces* sp. QG- 11-3 was used in combination with xylanase from the same organism for biobleaching [81]. The ability of polygalacturonic acid to complex cationic polymers depends strongly on the degree of polymerization. Pectinases depolymerise polygalacturonic acids and consequently decrease the cationic demand in the filtrate from peroxide bleaching of thermo mechanical pulp [80].

11. ENZYMES IN ANIMAL FEED

Alkaline proteases are widely used for production of protein hydrolysates for more than 40 years. Hydrolysates can be used as additives to food and mixed feed to improve their nutritional value. In medicine, they are administered to patients with digestive disorders and food allergies [82].

In food industry, cellulases are used in extraction and clarification of fruit and vegetable juices. Production of fruit nectars and purees and in the extraction of olive oil. Glucanases are added to improve the malting of barley in beer manufacturing and in wine industry. Better maceration and color extraction is achieved by use of exogenous hemicellulases and glucanases. Cellulases are also used in carotenoid extraction in the production of food coloring agents. Enzyme preparations containing hemicellulase and pectinase in addition to cellulases are used to improve the nutritive quality of forages [83]. Improvements in feed digestibility and animal performance are reported with the use of cellulases in feed processing describes the feed additive use of *Trichoderma* cellulases in improving the feed conversion ratio and increasing the digestibility of a cereal-based feed.

In the field of biotechnology there are many industrial applications that result in biotech products that are used every-day at home. Some of these food science applications utilize enzymes to produce or make improvements in the quality of different foods. Lipases have immense application in food industry such as in cheese ripening, flavor development and EMC technology [71]. Lipases are used *ex situ* to produce flavors, and to modify the structure by inter- or transesterification, in order to obtain products of increased nutritional value, or suitable for parental feeding [72]. Lipases is used for food processing to modify flavor by synthesis of esters of short chain fatty acids and alcohols, which are known as flavor and fragrance compounds [65].

Intensive research use varies enzymes in animal and poultry feeds in early 1980s. The first commercial success was addition of α-glucanase in barley-based feed diets. Usually a feed enzyme preparation is a multi enzyme cocktail containing glutanases, xylanases, proteinases, pectinases and amylases. Enzyme addition reduces viscosity which increases absorption of nutrients, librates nutrients either by hydrolysis of non degradable fibers, or by librating nutrients blocking by these fibers, and reduces the amount of faces.

12. ENZYMES IN MEDICINAL USES

Alpha-amylases is potentially useful in the pharmaceutical and fine chemicals industries if enzymes with suitable properties could be prepared [84]. Interestingly, the first enzyme produced industrially was an amylase from a fungal source in 1894, which was used as a pharmaceutical aid for the treatment of digestive disorders [85].

The use of immobilized alkaline protease from *Bacillus subtilis* possessing therapeutic properties has been studied for development of soft gel-based medicinal formulas, ointment compositions, gauze, non-woven tissues, and new bandage materials [86]. Oral administration of proteases from *Aspergillus oryzae* has been used as a diagnostic aid to correct certain lytic enzyme deficiency syndromes [87].

Lipases isolated from the wax moth (*Galleria mellonella*) were found to have a bacteriocidal action on *Mycobacterium tuberculosis* H37Rv. This preliminary study may be regarded as part of global unselected screening of biological and other materials for detecting new promising sources of drugs [88]. Lipase from *Candida rugosa* has been used to synthesize Iovastatin, a drug that lower serum cholesterol level. The asymmetric hydrolysis of 3-phenylglycidic acid ester which is a key intermediate in the synthesis of diltiazem hydrochloride, a widely used coronary vasodilator, was carried out with *S. mmcescens* lipase [89].

Laccases have been used for the synthesis of several products of pharmaceutical industry [90]. The first chemical of the pharmaceutical importance has been prepared using laccase enzyme is actinocin which was prepared from 4-methyl-3-hydroxyanthranilic acid. This compound has anticancer capability and works by blocking the transcription of DNA from the tumor cell [91].

13. ENZYMES IN CHEMICAL INDUSTRY

A high stability in the presence of organic solvents is a feature which is highly desired in applications involving biocatalysis in non-aqueous medium for peptide synthesis. Alkaline proteases from *Aspergillus flavus*, *Bacillus pseudofirmus* SVB1, *Pseudomonas aeruginosa* PseA has shown promising results for potential of peptide synthesis due to their organic solvent stability [92, 93, 94].

14. ENZYMES IN WASTE MANAGEMENT

Wastes from poultry processing industry and leather industry are recalcitrant commonly known proteolytic enzymes due to presence of keratin-rich wastes whose polypeptide is densely packed and strongly stabilized by several hydrogen bonds and hydrophobic interactions in addition to several disulfide bonds [95]. Chemical and mechanical hydrolysis of keratin wastes is successful but they have several disadvantages of being energy intensive, polluting and leading to loss of essential amino acids. Hence, enzymatic degradation using alkaline proteases with keratinolytic activity is an attractive method [95, 96]. *Bacillus* species is the most widely reported bacterial source of keratinases for feather degradation [97, 98, 99, 100, 101]. Other reported bacterial sources of keratinases are *Pseudomonas* sp. MS21, *Microbacterium* sp., *Chryseobacterium* sp. and *Streptomyces* sp. [102, 103, 104, 105].

Environmentally, the treatment of wastewater from citrus processing industries containing pectic substances is carried out in multiple steps, including physical dewatering, chemical coagulation, direct activated sludge treatment and chemical hydrolysis, which lead to formation of methane. These have several disadvantages, such as the high cost of treatment and longer treatment times in addition to environmental pollution from the use of chemicals. Thus, an alternative, cost-effective, and environmentally friendly method is the use of pectinases from bacteria, which selectively remove pectic substances from the wastewater. The pretreatment of pectic wastewater from vegetable food processing industries with alkaline pectinase and alkalophilic pectinolytic microbes facilitates removal of pertinacious material and renders it suitable for decomposition by activated sludge treatment [106, 107]. An extracellular endo-pectate lyase from an alkalophilic soil isolate, *Bacillus* sp. GIR 621, is used effectively to remove pectic substances from industrial waste water [107].

Employment of lipases in bioremediation processes is a new aspect in lipase biotechnology. The wastes of lipid- processing factories and restaurants can be cleaned by the help of lipases from different origins. In this sector, lipases could be used by either *ex-situ* or *in-situ* [108]. Due to the rapid development observed in industries, environmental pollution became more and more critical. Lipase producing strains play a key role in the enzymological remediation of polluted soils [109].

15. ENZYMES AS DIAGNOSTIC TOOLS

Lipases are also important drug targets or marker enzymes in the medical sector. They can be used as diagnostic tools and their presence or increasing levels can indicate certain infection or disease. Lipases are used in the enzymatic determination of serum triglycerides to generate glycerol which is subsequently determined by enzyme linked colorimetric reactions. The level of lipases in blood serum can be used as a diagnostic tool for detecting conditions such as acute pancreatitis and pancreatic injury [110]. Acute pancreatitis usually occurs as a result of alcohol abuse or bile duct obstruction. Serum trypsin level, ultrasonography, computed tomography, and endoscopic retrograde cholangiopancreatography are the most accurate laboratory indicators for pancreatitis.

Serum amylase and lipase levels are still used to confirm the diagnosis of acute pancreatitis [111]. Diagnosis of chronic pancreatitis and revealing the presence of exocrine pancreatic insufficiency has also been determined by measuring serum amylase, pancreatic isoamylase, lipase, trypsinogen and elastase [112]. Lipase of pathogenic bacteria such as *Propionibacterium acnes* [113], *Corynebacterium acnes* and *Staphylococcus aureus* [114] has also been found to have an influence on skin rashes in acne patients.

CONCLUSION

This chapter describes the historical perception of enzymes and their applications in the various sector was explained. Further, its express the various enzymes and their applications in cleaning (detergents), textiles, starch processing, brewing, leather, baking, paper, food products, animal feeds, cosmetics and personal care have been emphasized. This chapter also gives more valuable information about enzymes its bi-products in the global market. This will provide a valuable message for researchers and industrialist to understand the significance of specific enzyme scale-up process and the value of enzymes utilization in sectors of healthcare and environment.

REFERENCES

[1]   Burhan, Arikan., Nisa, Unaldi., Gokhan, Coral., Omer, Colak., Ashabil, Aygan. & Osman, Gulnaz. (2003). "Enzymatic properties of a novel thermophilic, alkaline and chelator resistant amylase from an alkalophilic *Bacillus* sp. Isolate ANT-6." *Process Biochemistry*, *38*, 1397–403.

[2]   Mishra, S. & Behera, N. (2008). "Amylase activity of starch degrading bacteria is isolated from soil receiving kitchen wastes." *African Journal of Biotechnology*, *7*, 3326–331.

[3]   Charles, P., Devanathan, V., Anbu, P., Ponnuswamy, M. N., Kalaichelvan, P. T. & Hur, B. (2008). "Purification, characterization and crystallization of an extracellular alkaline protease from *Aspergillus nidulans* HA-10." *Journal of Basic Microbiology*, *48*, 347–52.

[4]   Ashok, Pandey., Poonam, Nigam., Carlos, R. Soccol., Vanete, T. Soccol., Dalel, Singh. & Radjiskumar, Mohan. (2000). "Advances in microbial amylases." *Biotechnology and Applied Biochemistry*, *31*, 135–52.

[5]   Selvankumar, T., Govarthanan, M. & Govindaraju, M. (2011). "Endoglucanase production by *Bacillus amyloliquefaciens* using coffee pulp as substrate in solid

state fermentation." *International Journal of Pharma and Bio Sciences, 2,* B-355-62.

[6]   Govarthanan, M., Park, S. H., Kim, J. W., Lee, K. J., Cho, M., Kamala-Kannan, S. & Oh, B. T. (2014). "Statistical optimization of alkaline protease production from brackish environment *Bacillus* sp. SKK11 by SSF using horse gram husk." *Preparative Biochemistry and Biotechnology, 44,* 119.

[7]   Selvam, K., Govarthanan, M., Kamala-Kannan, S., Govindharaju, M., Senthilkumar, B., Selvankumar, T. & Sengottaiyan, A. (2014). "Process optimization of cellulase production from alkali-treated coffee pulp and pineapple waste using *Acinetobacter* sp. TSK-MASC." *RSC Advances, 4,* 13045.

[8]   Pandey, A. (1995). "Glucoamylase research: an overview." *Starch/Strack, 47,* 439-45.

[9]   Pandey, A., Nigam, P., Soccol, C. R., Singh, D., Soccol, V. T. & Mohan, R. (2000). "Advances in microbial amylases." *Biotechnology and applied biochemistry, 31,* 135-52.

[10]   Pandey, A., Soccol, C. R. & Mitchel, D. A. (2000). "New developments in Solid state fermentation: I bioprocess and products." *Process Biochemistry, 35,* 173-79.

[11]   Sivaramakrisnan, S., Gangadhran, D., Nampoothiri, K. M., Soccol, C. R. & Pandey, A. (2006). "Alpha amylase from microbial source- an overview on recent developments" *Food technology and biotechnology, 44,* 173-84.

[12]   Demain, A. L. & Fang, A. (2000). "The natural functions of secondary metabolites." In: Scheper, T., Fiechter, A. (Eds.), History of Modern Biotechnology. Springer-Verlag, Berlin.

[13]   Bornscheuer, U. T. & Buchholz, K. (2005). "Highlights in biocatalysis e historical landmarks and current trends." *Engineering in Life Sciences, 4,* 309-23.

[14]   Soetaert, W. & Vandamme, E. J. (2010). "The scope and impact of industrial biotechnology. In: Soetaert, W., Vandamme, E.J. (Eds.), *Industrial Biotechnology.* WILEYVCH Verlag GmbH and Co, Weinheim, 1-16.

[15]   BRENDA. (2012). Institute of Biochemistry and Bioinformatics at the Technical, University of Braunschweig. Germany. www.brenda-enzymes.org.

[16]   IUB. (1961). International Union of Biochemistry e Report of the Commission on Enzymes. Pergamon Press, Oxford.

[17]   Anto, Hema., Trivedi, U. B. & Patel, K. C. (2006). "Glucoamylase production by solid-state fermentation using rice flake manufacturing waste products as substrate." *Bioresource Technology, 97*(10), 1161-66.

[18]   Castro, Aline Machado de., Carvalho, Daniele Fernandes., Freire, Denise Maria Guimaraes. & Castilho, Leda dos Reis. (2010). Economic analysis of the production of amylases and other hydrolases by *Aspergillus awamori* in solid-state fermentation of babassu cake. *Enzyme Research.* doi:10.4061/2010/576872.

[19] Pandey, A., Webb, C., Soccol, C. R. & Larroche, C. (2005). Enzyme Technology, New Delhi, Asiatech publishers, Inc., *197*.

[20] Calik, P. & Ozdamar, T. H. (2001). "Carbon sources affect metabolic capacities of *Bacillus* species for the production of industrial enzymes theoretical analyses for serine and neutral proteases and alpha-amylase." *Biochemical Engineering Journal*, *8*, 61-81.

[21] Selvam, K., Govarthanan, M., Senthilkumar, B., Senbagam, D., Selvankumar, T., KamalaKannan, S. & Sudhakar, C. (2016a). "Optimization of protease production from surface modified coffee pulp waste and corncobs using *Bacillus* sp. by SSF". *3 Biotech*, *6*(2), 167.

[22] Sumitani, J., Nagae, H., Kawaguchi, T. & Arai, M. (1998). "*Bacillus* animal type alpha-amylase: Cloning and sequencing of the gene, and comparison of the deduced amino acid sequence with that of other amylases." *Journal of Fermentation and Bioengineering*, *85*, 428-32.

[23] Ibrahim, C. D. (2008). "Development of applications of industrial enzymes from Malaysian indigenous microbial sources." *Bioresource Technology*, *99*, 4572-582.

[24] Veille, C. & Zeikus, G. J. (2001). "Hyperthermophilic enzymes: sources, use, and molecular mechanisms for thermostability." *Microbiology and Molecular Biology Reviews*, *65*, 1–43.

[25] Ramakrishna, D. P. N., Gopi, R. N. & Raja Gopal, S. V. (2012). "Solid state fermentation of alkaline protease by *Bacillus subtilis* KHS-1 (MTCC No-10110) using different agro-industrial residues." *International Journal of Pharmacy and Pharmaceutical Sciences*, *4*(1), 512-17.

[26] Selvam, K., Selvankumar, T., Radhika, R., Srinivasan, P., Sudhakar, C., Senthilkumar, B. & Govarthanan, M. (2016b). "Enhanced production of amylase from *Bacillus* sp. using groundnut shell and cassava waste as a substrate under process optimization: Waste to wealth approach". *Biocatalysis and Agricultural Biotechnology*, *7*, 250-56.

[27] Lee, S. M. & Koo, Y. M. (2001). "Pilot-scale production of cellulose using *Trichoderma reesei* Rut C-30 in fed-batch mode." *Journal of Microbiology and Biotechnology*, *11*, 229-33.

[28] Johnvesly, B., Virupakshi, S., Patil, G. N., Ramalingam, N. & Naik, G. R. (2002). "Cellulase-free thermostable alkaline xylanase from thermophillic and alkalophillic *Bacillus* sp. JB-99." *Journal of Microbiology and Biotechnology*, *12*, 153-56.

[29] Rasmussnen, R. S. & Morrissey, M. T. (2007). "Marine biotechnology for production of food ingredients." *Advances in Food and Nutrition Research*, *52*, 237-92.

[30] Sharma, C. K. & Kanwar, S. S. (2012). "Purification of a Novel Thermophilic Lipase from *Bacillus licheniformis* MTCC-10498." *Research Journal of Biological Sciences Introduction, 1*(3), 43-48.

[31] Svendsen, A. (2000). "Review: Lipase protein engineering." *Biochemia et Biophysica Acta, 1543*, 223-38.

[32] Thakur, S. (2012). "Lipases, its sources, properties and applications: A review." *International Journal of Scientific and Engineering Research, 3*(7), 1-29.

[33] Gupta, R., Gupta, N. & Rathi, P. (2004). "Bacterial lipases: An overview of production, purification and biochemical properties." *Applied Microbiology and Biotechnology, 64*, 763–81.

[34] Singh, A. K. & Mukhopadhyay, M. (2012). "Overview of fungal lipase: A review." *Applied Biochemistry and Biotechnology, 166*, 486-20.

[35] Lang, C. & Dornenburg, H. (2000). "Perspectives in the biological function and the technological application of polygalacturonases." *Applid Microbialogy and Biotechnology, 53*, 366-75.

[36] Luh, B. S. & Pha, H. J. (1951). "Studies on polygalacturonase of certain yeast." *Archives of Biochemistry and Biophysics, 33*, 213-27.

[37] Jahwarhar, I. A., Noraini, S. & Wan, M. W. Y. (2012). "Screening and optimization of medium composition for mannanase production by *Aspergillus terreus* SUK-1 in solid state fermentation using statistical experimental methods." *Research Journal of Microbiology, 7*(5), 242-55.

[38] Chandrakant, Priya. & Bisaria, V. S. (1998). "Simultaneous bioconversion of cellulose and hemicellulose to ethanol." *Critical Reviews in Biotechnology, 18*, 295-31.

[39] Marini, A. M., Yatim, A. M., Babji, A. S., Annuar, B. O. & Noraini, S. (2006). "Evaluation of nutrient contents and amino acid profiling of various types of PKC." *Journal of Science and Technology, 2*, 135-41.

[40] Hagglund, P., Eriksson, T., Collen, A., Nerinckx, W., Claeyssens, M. & Stalbrand, H. A. (2003). "Cellulose binding module of the *Trichoderma reesei* β-mannanase Man5A increases the mannan-hydrolysis of complex substrates." *Journal of Biotechnology, 101*, 37-48.

[41] Mabrouk, M. E. M. & El Ahwany, A. M. D. (2008). "Production of β- mannanase by *Bacillus amylolequifaciens* 10A1 cultured on potato peels." *African Journal of Biotechnology, 7*(8), 1123-128.

[42] Phothichitto, K., Nitisinprasert, S. & Keawsompong, S. (2006). "Isolation, screening and identification of mannanase producing microorganisms." *Kasetsart Journal (Natural Science), 40*, 26-38.

[43] Sanchez, O. J. & Cardona, C. A. (2008). "Trends in biotechnological production of fuel ethanol from different feedstocks." *Bioresource Technology, 99*, 5270-295.

[44] Ninawe, S. & Kuhad, R. C. (2006). "Bleaching of wheat straw-rich soda pulp with xylanase from a thermoalkalophilic *Streptomyces cyaneus* SN32." *Bioresource Technology, 94,* 2291- 295.

[45] Hmidet, N., Bayoudh, A., Berrin, J. G., Kanoun, S., Juge, N. & Nasri, M. (2008). "Purification and biochemical characterization of a novel alpha-amylase from *Bacillus licheniformis.* NH1: cloning, nucleotide sequence and expression of amyN gene in *Escherichia coli." Process Biochemistry, 43,* 499-10.

[46] Ito, S., Kobayashi, T., Ara, K., Ozaki, K., Kawai, S. & Hatada, Y. (1998). "Alkaline detergent enzymes from alkaliphiles: enzymatic properties, genetics, and structures." *Extremophiles, 2*(3), 185-90.

[47] Saeki, K., Ozaki, K., Kobayashi, T. & Ito, S. (2007). "Detergent alkaline proteases: enzymatic properties, genes, and crystal structures." *Journal of Bioscience and Bioengineering, 103*(6), 501-08.

[48] Flower, T. EGIII-like Cellulase composition. DNA encoding such EGIII Compositions and methods for obtaining same, *Int Pat WO 2000/014208 A1,* 16 March.

[49] Clarkson, K. A., Weiss, G. L. & Larenas, E. A. (2000). "Detergent compositions containing substantially pure EGIII Cellulase" *Eur Pat EP 0 586 375 B1* (to Genencore International Inc USA), 15 March.

[50] Divya, Bajpai. &Tyagi, V. K. (2007). "Laundry detergents: An overview." *Journal of Oleo Science, 56,* 327-40.

[51] Feitkenhauer, Heiko. (2003). "Anaerobic digestion of desizing wastewater: influence of pretreatment and anionic surfactant on degradation and intermediate accumulation." *Enzyme and Microbial Technology, 33,* 250–58.

[52] Haq, I., Ali, S., Javed, M. M., Hameed, U., Saleem, A. & Adnanf-Qadeer, M. A. (2010). "Production of alpha amylase from a randomly induced mutant strain of *Bacillus amyloliqefaciens* and its application as a desizer in textile industry." *Pakistan Journal of Botany, 42*(1), 473- 84.

[53] Xia, L. & Cen, P. (1999). "Cellulaes production by solid state fermentation on liginocellulosic waste from the xylose industry." *Process Biochemistry, 34,* 909-12.

[54] Bhat, M. K. (2000). "Cellulase and related enzymes in biotechnology." *Biotechnology Advances, 18,* 355-83.

[55] Olson, L. A. (1990). "Treatment of denim with cellulase to produce a stone washed appearance." *US Pat. 4912056* (to Ecolab Inc, USA), 27 March.

[56] Kunamneni, A., Ghazi, I., Camarero, S., Ballesteros, A., Plou, F. J. & Alcalde, M. (2008). "Decolorization of synthetic dyes by laccase immobilized on epoxy-activated carriers." *Process Biochemistry, 43*(2), 169-78.

[57] Gulrajani, M. L., Agarwal, R. & Chand, S. (2000). "Degumming of silk with fungal protease." *Indian Journal of Fibre and Textile Research, 25*(2), 138–142.

[58] Regulapati, R., Malav, P. N. & Gummadi, S. N. (2007). "Production of thermostable alpha-amylase by solid state fermentation- A review." *American Journal of Food Technology*, *2*(1), 1-11.

[59] Ogasahara, K., Imanishi, A. & Isemura, T. (1970). "Studies on thermophilic aamylase from *Bacillus stearothermophilus*." I. Some general and physico-chemical properties of thermophilic a-amylase. *Journal of Biochemistry*, *67*, 65-75.

[60] Prakash, O. & Jaiswal, N. (2010). "alpha-Amylase: An Ideal Representative of Thermostable Enzymes." *Applied Biochemistry and Biotechnology*, *162*(7), 2123-4.

[61] Moraes, L. M. P., Filho, S. A. & Ulhoa, C. J. (1999). Purification and some properties of an α- amylase glucoamylase fusion protein from *Saccharomyces cerevisiae. World Journal of Microbiology and Biotechnology*, *15*, 561-64.

[62] Oner, E. T. (2006). "Optimization of ethanol production from starch by an amylolytic nuclear petite *Saccharomyces cerevisiae* strain." *Yeast*, *23*, 849-56.

[63] Martin, J. W. (1978). Method of removing paper adhered to a surface, *US Pat 4092175* (William Zinnsser & Co, Inc., USA), 30 May.

[64] Giongo, J., Lucas, F., Casarin, F., Heeb, P. & Brandelli, A. (2007). "Keratinolytic proteases of *Bacillus* species isolated from the Amazon basin showing remarkable dehairing activity." *World Journal of Microbiology and Biotechnology*, *23*(3), 375-82.

[65] Macedo, A. J., da Silva, W. O. B., Gava, R., Driemerier, D., Henriques, J. A. P. & Termignoni, C. (2005). "Novel keratinase from *Bacillus subtilis* S14 exhibiting remarkable dehairing capabilites." *Applied and Environmental Microbiology*, *71*(1), 594–96.

[66] Riffel, A., Ortolan, S. & Brandelli, A. (2003). "De-hairing activity of extracellular proteases produced by keratinolytic bacteria." *Journal of Chemical Technology and Biotechnology*, *78*(8), 855-59.

[67] Arunachalam, C. & Sarita, K. (2009). "Protease enzyme: an eco-friendly alternative for leather industry." *Indian Journal of Science and Technology*, *2*(12), 29–32.

[68] Verma, A., Pal, H. S., Singh, R. & Agarwal, S. (2011). "Potential of alkaline protease isolated from *Thermoactinomyces* sp. RM4 as an alternative to conventional chemicals in leather industry dehairing process." *International Journal of Agriculture, Environment and Biotechnology*, *2*(4), 173-78

[69] Zeman, N. & McCrea, J. (1985). "Alpha-amylase Production Using a Recombinant DNA Organism." *Cereal food world*, *30*, 777-80.

[70] Minussi, R. C., Pastore, G. M. & Duran, N. (2002). "Potential applications of laccase in the food industry." *Trends in Food Science and Technology*, *13*, 205–16.

[71] Saxena, R. K., Ghosh, P. K., Gupta, R., Davidson, W. S., Bradoo, S. & Gulati, R. (1999). "Microbial lipases: Potential biocatalysts for the future industry." *Current Science*, *77*(1), 101–15.

[72] Reetz, M. T. (2002). "Lipases as practical biocatalysts" *Current Opinion in Chemical Biology*, *6*(2), 145-50.

[73] Macedo, G. A., Lozano, M. M. S. & Pastore, G. M. (2003). "Enzymatic synthesis of short chain citronellyl esters by a new lipase from *Rhizopus* sp." *Electronic Journal of Biotechnology*, *6*(1), 72-75.

[74] Fryer, P. J. & Asteriadou, K. A. (2009). "Prototype cleaning map: a classification of industrial cleaning processes." *Trends in Food Science and Technology*, *20*, 255-62.

[75] Pere, J., Paavilainen, L., Siika-Aho, M., Cheng, Z. & Viikari, L. (1996). "Potential use of enzymes in drainage control of nonwood pulps." in *Proc 3rd Int Non-wood Fibre Pulping and Paper Making Conf*, vol 2, (Beijing, China), 421-30.

[76] Jaeger, K. E. & Reetz, T. M. (1998). "Microbial lipases from versatile tools for biotechnology. *Trends in Biotechnology*, *16*, 396–403.

[77] Bajpai, Pratima. (1999). "Application of enzymes in the pulp and paper industry." *Biotechnolpgy Progress*, *15*, 147- 57.

[78] Kirk, T. K. & Jefferies, T. W. (1996). "Role of microbial enzymes in pulp and paper processing." *American chemical society*, Washington D.C., pp. 1-4.

[79] Reid, J. & Ricard, M. (2000). "Pectinase in paper making: Solving retention problems in mechanical pulps bleached with hydrogen peroxide." *Enzyme and Microbial Technology*, *26*, 115-23.

[80] Viikari, L., Tenkanen, M. & Suranakki, A. (2001). "Biotechnology in the pulp and paper industry." *Journal of Biotechnology*, *50*, 523-40.

[81] Beg, Q. K., Kapoor, M., Tiwari, R. P. & Hoondal, G. S. (2001). "Bleach-boosting of eucalyptus kraft pulp using compination of xylanase and pectinase from *Streptomyces* sp.QG-11-3." *Research bulletin of the Panjab University Science*, *51*, 71-78.

[82] Neklyudov, A., Ivankin, A. & Berdutina, A. (2000). "Properties and uses of protein hydrolysates (Review)." *Applied Biochemistry and Microbiology*, *36*(5), 452-59.

[83] Cinar, Inci. (2005). "Effects of cellulase and pectinase concentrations on the colour yield of enzyme extracted plant carotenoids." *Process Biochemistry*, *40*, 945-49.

[84] Fogarty, W. M. & Kelly, C. T. (1980). "In Economic Microbiology, Microbial Enzymes and Bioconversions. Academic Press, 115–70.

[85] Pandey, A., Nigam, P. R., Sccol, C. T., Soccol, V., Singh, D. & Mohan, R. (2000). Advances in microbial amylases. *Journal of Biotechnology*, *31*, 135-52.

[86] Davidenko, T. I. (1999). "Immobilization of alkaline protease on polysaccharides of microbial origin." *Pharmaceutical Chemistry Journal*, *33*(9), 487-89.

[87]   Rao, M. B., Tanksale, A. M., Ghatge, M. S. & Deshpande, V. V. (1998).
       "Molecular and biotechnological aspects of microbial proteases." *Microbiology
       and Molecular Biology Reviews*, *62*(3), 597-35.

[88]   Annenkov, G. A., Klepikov, N. N., Martynova, L. P. & Puzanov, V. A. (2004).
       "Wide range of the use of natural lipases and esterases to inhibit *Mycobacterium
       tuberculosis.*" *Problemy tuberkuleza i boleznei legkikh*, *6*, 52-56.

[89]   Matsumae, H., Furui, M. & Shibatani, T. (1993). "Lipase-catalyzed asymmetric
       hydrolysis of 3-phenylglycidic acid ester, the key intermediate in the synthesis of
       diltiazem hydrochloride." *Journal of Fermentation and Bioengineering*, *75*, 93-98.

[90]   Arora, Daljit Singh. & Sharma, Rakesh Kumar. (2010). "Ligninolytic Fungal
       Laccases and Their Biotechnological Applications." *Applied Biochemistry and
       Biotechnology*, *160*, 1760–88.

[91]   Burton, Stephanie G. (2003). "Laccases and phenol oxidases in organic synthesis."
       Current Organic Chemistry, *7*, 1317-331.

[92]   Yadav, S. K., Bisht, D. Shikha. & Darmwal, N. S. (2011). "Oxidant and solvent
       stable alkaline protease from *Aspergillus flavus* and its characterization." *African
       Journal of Biotechnology*, *10*(43), 8630-640

[93]   Anush, Gupta. & Khare, Sunil. (2007). "Enhanced production and characterization
       of a solvent stable protease from solvent tolerant *Pseudomonas aeruginosa* PseA."
       Enzyme and Microbial Technology, *42*(1), 11-16.

[94]   Sen, S., Dasu, V. V., Dutt, K. & Mandal, B. (2011). "Characterization of a novel
       surfactant and organic solvent stable high-alkaline protease from new *Bacillus
       pseudofirmus* SVB1." *Research Journal of Microbiology*, *6*(11), 769-83.

[95]   Ramnani, P., Singh, R. & Gupta, R. (2005). "Keratinolytic potential of *Bacillus
       licheniformis* RG1: structural and biochemical mechanism of feather degradation."
       Canadian Journal of Microbiology, *51*(3), 191-96.

[96]   Selvam, K., Senbagam, D., Sudhakar, C., Senthilkumar, B., Selvankumar, T.,
       Govarthanan, M., Sengottaiyan, A. & Palanisamy, S. (2018). "Molecular modeling
       and docking of protease from *Bacillus* sp. for the keratin degradation".
       Biocatalysis and Agricultural Biotechnology, *13*, 95-104.

[97]   Ni, H., Chen, Q., Chen, F., Fu, M., Dong, Y. & Cai1, H. (2011). "Improved
       keratinase production for feather degradation by *Bacillus licheniformis*
       ZJUEL31410 in submerged cultivation." *African Journal of Biotechnology*,
       10(37), 7236-244.

[98]   Kojima, M., Kanai, M., Tominaga, M., Kitazume, S., Inoue, A. & Horikoshi, K.
       (2006). "Isolation and characterization of a feather-degrading enzyme from
       Bacillus pseudofirmus FA30-01." *Extremophiles*, *10*(3), 229-35.

[99]   Cortezi, Mariana., Contiero, Jonas., Lima, Cristian J. B. de., Lovaglio Roberta, B.
       & Monti, Rubens. (2008). "Characterization of a feather degrading by *Bacillus

amyloliquefaciens protease: A new strain." *World Journal of Agricultural Sciences, 4* (5), 648-56.

[100] Govarthanan, M., Selvankumar, T. & Arunprakash, S. (2011). Production of keratinolytic enzyme by a newly isolated feather degrading *Bacillus* sp. from chick feather waste. *International Journal of Pharma and Bio Sciences, 2*(3), 259-265.

[101] Govarthanan, M., Selvankumar, T., Selvam, K., Sudhakar, C. & Vincent Aroulmoji, Kamala-Kannan, S. (2015). "Response surface methodology based optimization of keratinase production from alkali-treated feather waste and horn waste using *Bacillus* sp. MG-MASC-BT". *Journal of Industrial and Engineering Chemistry, 27*, 25-30.

[102] Tork, S., Aly, M. M. & Nawar, L. (2010). "Biochemical and molecular characterization of a new local keratinase producing *Pseudomomanas* sp., MS21." *Asian Journal of Biotechnology, 2*(1), 1-13

[103] Thys, R. C. S. & Brandelli, A. (2006). "Purification and properties of a keratinolytic metalloprotease from *Microbacterium* sp." *Journal of Applied Microbiology, 101*(6), 1259-268.

[104] Brandelli, Adriano. & Riffel, Alessandro. (2005). "Production of an extracellular keratinase from *Chryseobacterium* sp. growing on raw feathers." *Electronic Journal of Biotechnology, 8*(1), 35-42.

[105] Tapia, D. M. T. & Simoes, M. L. G. (2008). "Production and partial characterization of keratinase produced by a microorganism isolated from poultry processing plant wastewater." *African Journal of Biotechnology, 7*(3), 296-300.

[106] Horikoshi, K. (1990). Enzymes from alkalophiles. In: Fogarty WM, Kelly CT (eds) Microbial enzymes and biotechnology, 2nd edn. Elsevier, Ireland., 275-95.

[107] Tanabe, H., Kobayashi, Y. & Akamatsu, T. (1988). "Pretreatment of pectic waste water in pectate lyase from an alkalophilic *Bacillus* species." *Agricultural and Biological Chemistry, 52*, 1853-856.

[108] Pandey, A., Benjamin, S., Soccol, C. R., Nigam, P., Krieger, N. & Soccol, V. T. (1999). "The realm of microbial lipases in biotechnology" *Applied Biochemistry and Biotechnology, 29*(2), 119– 31.

[109] Lin, J. F., Lin, Q., Li, J., Fei, Z. A., Li, X. R., Xu, H., Qiao, D. R. & Cao, Y. (2012). "Bacterial diversity of lipase-producing strains in different soils in southwest of China and characteristics of lipase," *African journal of microbiology research, 6*(16), 3797-806.

[110] Lott, J. A. & Lu, C. J. (1991). "Lipase isoforms and amylase isoenzymes-assays and application in the diagnosis of acute pancreatitis." *Clinical Chemistry, 37*, 361–68.

[111] Munoz, A. & Katerndahl, D. A. (2000). "Diagnosis and management of acute pancreatitis." *American Family Physician, 62*, 164–74.

[112] Pezzilli, R., Talamini, G. & Gullo, L. (2000). "Behaviour of serum pancreatic enzymes in chronic pancreatitis." *Digestive and Liver Disease, 32*(3), 233–37.

[113] Higaki, S., Kitagawa, T., Kagoura, M., Morohashi, M. & Yamagishi, T. (2000). "Correlation between *Propionibacterium acnes* biotypes, lipase activity and rash degree in acne patients." *The Journal of Dermatology, 27*, 519–22.

[114] Simons, J. W. F. A., Adams, H., Cox, R. C., Dekker, N., Gotz, F., Slotboom, A. J. & Verheij, H. M. (1996). "The lipase from *Staphylococcus aureus*: Expression in *Escherichia coli*, large-scale purification and comparison of substrate specificity to *Staphylococcus hyicus* lipase." *European Journal of Biochemistry, 242*, 760–69.

In: Microbial Catalysts. Vol. 2
Editors: Shadia M. Abdel-Aziz et al.

ISBN: 978-1-53616-088-8
© 2019 Nova Science Publishers, Inc.

Chapter 8

PHYTASES: POTENTIAL BIOTECHNOLOGICAL APPLICATIONS AND FUTURE CHALLENGES

Shadia M. Abdel-Aziz[1], Khalid T. Biobaku[2,], El-Sayed E. Mostafa[1], Ismail A. Odetokun[3,*] and Abhinav Aeron[4]*

[1]Microbial Chemistry Department, National Research Centre, Dokki, Giza, Egypt
[2]Department of Veterinary Pharmacology and Toxicology,
University of Ilorin, Nigeria
[3]Department of Veterinary Public Health and Preventive Medicine,
University of Ilorin, Nigeria
[4]Department of Biotechnology, School of Engineering and Technology,
Sharda University, Greater Noida, Uttar Pradesh, India

ABSTRACT

Phytases, a group of enzymes are capable of releasing phosphate from phytates. These enzymes are wide-spread in nature, occurring in microorganisms, plants as well as in some animal tissues. To date, many phytases have been reported from bacteria belong to enterobactericeae such as *Bacillus* spp., *E. coli*, *Enterobacter*, *Klebsiella* spp., *Erwinia caratovore*, and *Yesinia intermedia*. Phytases have been widely used in animal feed to improve phosphorus nutrition and reduce phosphorus pollution in animal waste. Unfortunately, phytates are not hydrolyzed in the monogastric gut, and the phytate-associated phosphates remain unabsorbed, requiring the exogenous addition of phosphate to avoid phosphorus deficiency. Furthermore, phytates are a strong chelator of cations and bind minerals such as Ca^{2+}, Zn^{2+} and Fe^{2+} making them unavailable for absorption in the intestine of the monogastric animals. In addition, phytates are known to form complexes with proteins under both acidic and alkaline pH conditions. These interactions are found to affect the protein structure, thus decreasing enzymatic activity, protein

---

* Corresponding Authors' Emails: odetokun.ia@unilorin.edu.ng; ismail23us@gmail.com.

solubility, and proteolytic digestibility. So far, phytases have been mainly, if not solely, used as a feed supplement in diets for animals and poultry, and to some extent for fish. The inclusion of phytases in animal feed is attractive and attracts more attention.

Keywords: phytate, phytase, animal feed, human nutrition, development strategies

1. Introduction

Phosphorus (P) is a macronutrient for plant nutrition and plays important roles in the biosynthesis of nucleic acids, cell membranes, and regulation of many enzymes. Plants store organic P in the form of phytate (salts of phytic acid), carrying 6 phosphate groups. However, this bound P, present in seed grain as phytate, is generally unavailable to human and mono-gastric animals (chicken, swine, fish, etc.) due to either absence or insufficient secretion of intrinsic phytase activity. Phosphorus is also a vital element for animal nutrition; considering growth of bones and tissues (80% P is found in bones and teeth). Phytate, being chelates metal ions, reduces energy uptake and behaves as an antinutrient. Phytates, existing in a metal–complex form, are insoluble, difficult for human to be hydrolyzed during digestion, and thus are nutritionally less available for absorption [3]. The antinutrient behavior of phytate complex has been considered as a threat in human diet due to the presence of phytic acid which is known as a strong chelator of divalent minerals such as Ca^{2+}, Mg^{2+}, Zn^{2+} and Fe^{2+}.

Phytic acid, known as inositol hexakisphosphate [IP6], inositol polyphosphate, or phytate, has a great affinity for binding with proteins, amino acids, and/or multivalent cations or minerals in foods. Reduction of phytic acid in phytates can be achieved through enzymatic reaction by phytases, which are widely produced in nature by bacteria, fungi, yeast, plants, and animals. These groups of enzymes are capable of releasing phosphate from phytate. Phytases have been intensively studied due to the great interest in such enzymes for reducing phytate content and increasing the available P for animal feed and human foods. A number of studies have already shown that addition of phytate-degrading enzymes enhances phosphate utilization from phytate [4, 5, 6]. The phytate P is generally unavailable to monogastric animals (chicken, swine, fish, and human) due to either absence or insufficient secretion of enzymes essential for phytate hydrolysis in digestive tract [2]. Consequently, P remains unabsorbed in the digestive tract and gets excreted with faeces, leading to the environmental pollution. Microbial phytases have been applied mainly for animal feed, and human foodstuffs in order to improve food processing, phosphorus nutrition and bioavailability. Although commercial production of phytases has focused on the fungus *Aspergillus*, studies have suggested bacterial phytases to be an alternative to the fungal enzymes because of their higher substrate specificity, greater resistance to proteolysis, and better catalytic efficiency [7].

Grain products, a daily-bread products, are the basis for nutrition and the main source of fibres and minerals such as iron, calcium, magnesium, selenium, zinc, as well as folic acid, iodine, and B vitamins group. As indicated, phytic acid form phytate complex with minerals that are insoluble at physiological pH, therefore, minerals and phosphate become unavailable for absorption in the human intestine. This especially creates a problem in production of wheat baked goods. The human body suffers specific metabolic disorders that cause various diseases due to the lack of minerals in the daily diet and insufficient intake from grain products [8]. When assessing the potential nutritional problems, the search for new biotechnological tools to improve daily intake of minerals from grain products is relevant. This chapter summarizes the importance and applications of phytases in food and feed, potential biotechnological applications of phytases, and strategies used for the development of new and effective phytases with improved properties.

2. PHYTATES

Phytate is the principal storage form of P, and a variety of minerals in plants is representing approximately 75–80% of the total P in plant seeds [9]. Phytate can exist in a metal-free form or in metal–phytate complex at acidic and neutral pH values [10]. Phytate content of some plant-derived foods are represented in Table 1.

Table 1. Phytate content in some plant or plant-derived foods [11]

Food	wPhytate (mg/g)	Food	wPhytate (mg/g)
Cereal-based		*Legume-based*	
Whole wheat bread	3.2-7.3	Chickpea (cooked)	2.9-11.7
Whole rye bread	1.9-4.3	Cowpea (cooked)	3.9-13.2
Unleavened wheat bread	10.6-3.2	Black beans (cooked)	8.5-17.3
Maize bread	4.3-8.3	White beans (cooked)	9.6-13.9
unleavened wheat bread	12.2-19.3	Faba beans (cooked)	8.2-14.2
Oat bran	7.3-2.1	Soybeans	9.2-16.7
Oat flakes	8.4-12.1	Tofu	8.9-17.8
Oat porridge	6.9-10.2	Green peas (cooked)	1.8-11.5
Maize	9.8-21.3	Peanuts	9.2-19.7
Cornflakes	0.4-1.5	*Others*	
Wild rice (cooked)	12.7-21.6	Sesame seeds (toasted)	39.3-57.2
Sorghum	5.9-11.8	Soy protein concentrate	11.2-23.4
Wild rice (cooked)	12.7-21.6	Amaranth grain	10.6-15.1

2.1. Phytate as an Antinutrient

Phytate behaves in a broad pH range as a highly negatively charged ion and has therefore a tremendous affinity for food components with positive charge(s), such as minerals, trace elements and proteins. These interactions have nutritional consequences and affect the yield and quality of food ingredients such as starch, corn steep liquor or plant protein isolates [11]. The major concern about presence of phytate in human diet is its negative effect on mineral uptake such as zinc, iron, calcium, magnesium, manganese and copper. Formation of insoluble mineral-phytate complexes at physiological pH values is a major reason for the poor mineral bioavailability, because these complexes are essentially non-absorbable from the human gastrointestinal tract. Furthermore, the small intestine in human has only a very limited capability to hydrolyze phytate due to the lack of endogenous phytate-degrading enzymes and the limited microbial population in the upper part of the digestive tract [11]. An overview of the negative interactions of phytate with other nutrients is represented in Table 2 [10].

2.2. Enzymatic Hydrolysis of Phytate

Enzymatic hydrolysis of phytate occurs by the sequential release of orthophosphate groups from the inositol ring of phytic acid to produce free inorganic P along with a series of intermediate myo-inositol phosphates. Phytase not only releases P from plant-based diets but also makes calcium, magnesium, protein and lipid available. Thus, by releasing the bound P in feed ingredients of vegetable origin, phytase makes more P available for bone growth, and protects the environment against P pollution [11].

Table 2. Negative interaction of phytate and nutrients in food [10]

Nutrients	Mode of action
Protein	Formation of nonspecific phytate–protein complex, not readily hydrolyzed by proteolytic enzymes [12].
Carbohydrate	Formation of phytate-carbohydrate complex making the carbohydrate less degradable. Inhibition of amylase activity by complexing with Ca^{2+} ion [13].
Lipid	Formation of lipophytin complexes may lead to metallic soaps in gut lumen, resulting in lower lipid availability [14].
Mineral ions (zinc, iron, calcium, magnesium, manganese and copper)	Formation of insoluble phytate–mineral complexes decreases mineral availability [15].

3. PHYTASES

3.1. Definition

A phytase (*myo*-inositol hexakisphosphate phosphohydrolase) is a type of phosphatase enzyme that catalyzes the hydrolysis of phytic acid (*myo*-inositol hexakisphosphate) and releases the usable form of phosphorus (Figure 1). While phytases have been found to occur in animals, plants, fungi, and bacteria, they are most commonly detected and characterized from fungi [16]. Phytate-degrading enzymes have a great interest for reducing the phytate content in animal feed and foods for human consumption.

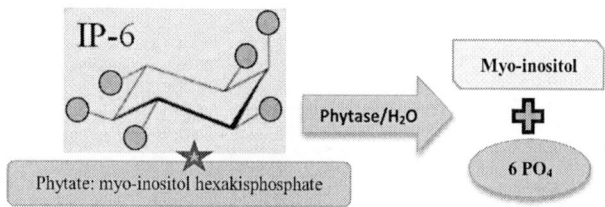

Figure 1. Releasing of available phosphorus (6PO₄) from Phytate (IP-6) by phytase.

3.2. Classification

Phytases have been classified as: *a*) 3-phytases (EC 3.1.3.8) and 6-phytases (EC 3.1.3.26) based on the position of first phosphate hydrolyzed. The 3-phytases initiates dephosphorylation of phytic acid at the 3 position of phytic acid, and 6-phytases at position 6. The 3-phytases are the largest group of phytases which are generally found in bacteria and fungi. Plant phytases acts preferentially at the C6 carbon and are 6-phytase [5]; *b*) classification as acid phytases and alkaline phytases on the basis of pH optimum; c) classification as histidine acid phosphatase (HAP), Beta-Propeller phytase (BPP), cysteine phosphatase (CPH) and purple acid phosphatase (PAP), on the basis of catalytic property [16]. Most bacterial, fungal, and plant phytases belong to the HAPs which can initiate hydrolysis of phytic acid on either the C3 or the C6 position of the inositol ring and produce *myo*-inositol monophosphate as the final product [5].

3.3. Occurrence and Environmental Protection

Phytates are occurring in some animal tissues and plants such as barley, maize, rice, wheat, soybean as well as spelt, rape seed, and lily [17]. Concern about the impact of P in the environment has greatly increased over the past decade. Over application of P-containing materials (whether inorganic P-based fertilizers or organic residues such as crop residues, manures, composts, and municipal bio-solids) can result in high levels of extractable soil inorganic phosphate [18]. If these soils are close to surface water such as streams, lakes, and rivers, P can enter the water (either as soluble P in runoff water or as P absorbed by eroded soil particles), thereby increasing P availability in these water bodies [19]. Excess of P can potentially lead to eutrophication in water columns. Excessive richness of nutrients in a lake or other body of water, frequently due to runoff from the land, causes a dense growth of plant life and death of animal life from lack of oxygen. Animal management has an important role in reducing P losses in the environment. This could be achieved by using technologies to increase the bioavailability of P in manures and excreta to produce an organic fertilizer with large amounts of available P [18].

3.4. Milestones in Phytase Commercialization

History and milestones of phytase commercialization was reported [4]. It was not until 1962, that first efforts were done to make phytase as a commercial product. In 1968, Sheih and Ware isolated *Aspergillus ficuum*, which produced the highest yield of phytase with two pH optima, 5.5 and 2.5, and deposited it as NRRL 3135 [20]. Through 1968 to 1991, gene cloning and overexpression of *phyA* from *A. niger* NRRL 3135, along with amyloglucosidase promoter, resulted in 52-fold improvement of phytase yield. In 1996, after receiving approval from several countries, the Food and Drug Administration has approved the use of phytase in foods as GRAS (generally recognized as safe). Phytase is being marketed as food additive in the United States since January 1996 as Natuphos [4]. Overexpression of phytases from *Trichoderma reesei* (marketed as Finase-F), and *Aspergillus oryzae* IFO 4177 (marketed as Bio-Feed phytase by DSM) was achieved in 1997 [4]. Up to now, several organisms have been screened for production of phytase with higher catalytic properties.

4. Sources of Phytases

Phytases occur widely among plants, animals, and microorganisms. In the last 20 years, research has indicated that several strains of bacteria, yeast, and fungi can produce high yield of phytases with applications at the industrial scale. Phytase was purified and expressed in a wide range of hosts using various biochemical methods. Depending on the source and/or expression host, phytases can exhibit different biophysical and biochemical properties [2]. Some microbial sources for phytase production are represented in Table, 3 [2].

4.1. Fungal Phytases

One of the first systematic studies on fungal phytase was reported by Shieh and Ware, where various microorganisms where tested for extracellular phytase production and a strain of *A. niger* known as *Aspergillus ficuum* NRRL 3135 was identified as the most efficient [20]. This strain exhibited highest phytase activity without sporulation which is a prerequisite in large scale production. An acid phosphatase, referred to as *phy-B* phytase, has been noted for the pH optimum 2.5 [21]. Two pH optima, at 2.5 and 5.0–5.5, was observed for the *A. niger* NRRL 3135 phytase, *phy-A* [22]. The genus *Aspergillus* (*A. niger* in particular) continues to be preferred for production of phytase [2]. Fungal phytase are reported from different strains; over 200 fungal isolates belonging to genera *Aspergillus*, *Mucor*, *Penicillium* and *Rhizopus* have been tested for phytase production. The most active fungi for phytase production were identified as *A. niger*, *A. oryzae*, *A. amstelodami*, *A. candidus*, *A. flavus* and *A. repen* [5].

4.2. Bacterial Phytases

Phytases have been detected in several types of bacteria, such as *bacilli*, *enterobacteria*, anaerobic ruminal bacteria, and *Pseudomonas* [23, 24]. After 1980s, several bacterial strains (wild or genetically modified) were applied for phytase synthesis and were found to hydrolyze phytate such as *Lactobacillus amylovorus*, *Escherichia coli*, *B. subtilis*, *Bacillus amyloliquefaciens* and *Klebsiella* spp. [25]. Phytases belonging to the *Enterobactericeae* have been reported for bacteria such as *E. coli*, *Enterobacter*, *Klebsiella* spp., *Erwinia caratovore*, and *Yesinia intermedia* [11]. In general, phytases from bacteria have a pH optimum between neutral and alkaline, and have temperature optima from 40 up to 70°C [26, 27, 28]. Within bacterial phytases, an enzyme with high thermal stability (*Bacillus* phytase) or high proteolytic stability (*E. coli* phytase) does exist, and such bacterial phytases will be developed for their favorable properties as feed additives [2, 29].

4.3. Yeast Phytases

Yeasts are ideal candidates for phytase and phosphatase research due to their nonpathogenic and GRAS status; however, they have not been utilized to their full potential. To date, only few studies have been published on yeast phytase, such as *S. cerevisiae*, *Saccharomyces castellii* and *Arxula adeninivorans* [5, 30]. Among numerous yeast species, *Pichia spartinae* and *Pichia rhodanensis* exhibited the highest levels of extracellular phytase with optimal temperatures at 75–80°C and 70–75°C, and optimum pH at 3.6–5.5 and 4.5–5.0, respectively [31].

Phytase can be produced by microorganisms using submerged fermentation (SmF) or solid-state fermentation (SSF). Submerged fermentation has largely been employed as a production technology for commercial phytases. In recent years, however, the SSF has gained significant interest for production of phytase, especially from fungi. The type of strain, culture conditions, nature of the substrate, and availability of the nutrients are critical factors affecting the yield and should be considered when selecting a particular production technique [10]. For example, some fungi in SmF are exposed to hydrodynamic forces but in SSF the surface of the solid particles acts as the matrix for the culture. Varied substrates, such as wheat bran, full-fat soybean flour, canola meal, cane molasses, and oil cakes when used, the higher titers of enzyme production, extracellular nature of enzyme, and low protease production are observed [32]. Several studies have compared phytase productivity in different fermentation systems. In such comparisons, important aspects such as medium composition, the strain morphology, phytase production, growth patterns, as well as the enzymatic productivity, culture conditions, and type of strain were studied to explain how the fermentation system can affect the strain physiology and productivity [10, 26]. There is a complex relationship between the morphology of some fungi, transport phenomena, viscosity of the cultivation broth, and related productivity. The morphological characteristics vary between the freely dispersed mycelia and the distinct pellets of aggregated biomass with every growth form having a distinct influence on broth rheology [10]. Hence, the advantages and disadvantages for mycelial or pellet cultivation need to be balanced out carefully. Fungal morphology is often a bottleneck of productivity in industrial production due to the inadequate understanding of the morphogenesis of filamentous microorganisms [33]. Better understanding of the molecular and cell biology of microorganisms as well as the relevant approaches in biochemical engineering are of significant importance to obtain an optimized production process.

Table 3. Sources of microorganisms reported for phytase production

Source	Strains	Reference
Fungi	*Aspergillus terreus, Aspergillus oryzae* AK9, *Aspergillus niger, Aspergillus fumigatus, Aspergillus ficuum* NTG-23, *Penicillium oxalicum* PJ3, *Penicillium minioluteum, Mucor racemosus, Mucor hiemalis, Rhizopus oryzae Saccharomyces cerevisiae, Pichia anomala,*	[34], [35], [36], [37] [38], [39], [40], [41], [42], [43]
Yeast	*Pichia spartinae, Pichia rhodanensis, Debaryomyces castellii, Kodamaea ohmeri, Hansenula fabianii*	[30], [44], [45], [31], [46], [47], [48]
Bacteria	*B. amyloliquefaciens,* US573, *Bacillus licheniformis, Bacillus subtilis, Bacillus subtilis* B.S.46, *Bacillus cereus, Escherichia coli, Serratia plymothica*	[49], [50], [24], [51], [52], [53], [54], [55], [56], [57]

4.4. Plant Phytases

Many plant seeds contain significant amounts of phytic acid that is degraded during germination by one or more phytases. Seeds contain both constitutive phytase activity and phytases that are synthesized again during germination; however, the last mechanism is not well understood [2]. Activity of phytase has been well reported from *Arabidopsis thaliana* AtPAP15 [58], *Glycine max* GmPhy [59], and *Medicago truncatula* MtPHY1 [60]. The optimum temperature and pH measured for most plant phytases ranges from 45 to 60°C and from 4.0 to 7.2 pH, respectively. Alkaline phytases with unique catalytic properties have been identified in plants. Purified alkaline phytase from pollen grains of *Lilium longiflorum* is reported [61]. Unique properties of such alkaline phytase render it to be potentially useful as a feed and food supplement [2].

4.5. Animal Phytases

Investigations of animal phytases are limited. Existence of the first animal phytase was demonstrated in the blood and liver of calves in 1908 by McCollum and Hart. Since then, argument has persisted regarding the existence of phytases in the digestive tract of animals (especially monogastric animals). Other investigators failed to find phytase in the extracts of intestine, pancreas, kidney, bone, liver and blood of several species of animals [62]. Preliminary studies on activity of phytase produced by rumen microorganisms were initiated by Raun et al. [63], and undertaken again by Yanke et al. [64]. They further examined presence of phytase activity in species of obligatory anaerobic ruminal bacteria and concluded that the most highly active strain was *Selenomonas ruminantium*. Schlemmer et al., have outlining the complete system of phytate degradation in the gut of human and the enzymes involved [65]. In this study, pigs were used as a model for human, and it was concluded that negligent amounts of endogenous phytase activity were found in stomach chyme and small intestine. Intestinal bacteria with endogenous phytase activity were discovered in several species of fish. Huang et al., have screened the intestinal contents of grass carp and found that phytate-degrading isolates, are belonged to *Pseudomonas*, *Bacillus*, and *Shewanella* species [66].

5. STRATEGIES APPLIED FOR IMPROVING PHYTASE PRODUCTION

The levels of phytases produced by naturally occurring microorganisms are too low to be considered as economically viable. Improvement in phytase production could be achieved through biotechnological development of strains to attain high and economical production. Strategies employed for improvement and economic production of phytases by microorganisms are discussed below [10].

5.1. Classical Mutagenesis

A phytase that has the desirable characteristics for application in animal feed industry should be: *a*) cheap; *b*) active in the stomach and efficient as a catalyst; *c*) thermostable during animal feed processing and storage; *d*) resistant to proteases; *e*) easily processed by the feed manufacturer for its suitability as an animal feed additive [67, 68]. Genetic manipulation techniques such as site directed mutagenesis could be employed for further improvement of these properties [69]. Strain improvement by mutagenesis and selection is a highly developed technique, and plays a central role in the commercial development of microbial fermentation processes. Mutagenic procedures can be carried out in terms of type of mutagen and dose to obtain mutant types, which may be screened for improved phytase, such as in *A. niger*, using physical and chemical mutagenesis [70]. The structure of phytase and its genetic manipulation, before designing an ideal enzyme is, however, very important. The specific activity of the heat-stable *A. fumigatus* phytase, pH stabilility of *A. niger* PhyA phytase, and thermostability of *E. coli* AppA phytase have been improved using mutagenesis approach [69, 71, 72, 73].

5.2. Genetic Improvement

Genetic modification technique can be used efficiently to reduce phytic acid content in cereals by cloning the genes of phytase, and by creating transgenic plant with modified genome encoding for phytase [5]. Although phytases are widely distributed in nature, the economic production in wild-type microorganisms is so far not viable. Hence, cloning and expression of phytase genes in suitable host organisms are necessary to reach higher productivities. Because cost-effective production of phytase is a major limiting factor for its application, different heterologous expression systems have been evaluated in plants, bacteria, and fungi, including yeast [10]. Phytase genes from *A. niger*, *A. terreus*, *A. fumigatus*, *E. nidulans*, and *M. thermophila* have all been expressed and secreted as active enzymes. However, some fungi systems secrete active phytases but along with high level of undesired proteases. This requires further purification or inhibition of proteolysis that adds to the production cost [10].

5.3. Protoplast Fusion

Protoplast fusion has significant potential for strain improvement, enhancing phytase production, and has been applied for various industrially important microorganisms.

Protoplast fusion may be used to produce intergeneric hybrids, and is an important tool because it can overcome the limitations of conventional mating systems in gene manipulation. However, it is still an emerging area in phytase research with only few reports of interspecific protoplast fusion between two auxotrophic mutants, *A. niger* CFR 335 ala, and *A. ficuum* SGA01 [10].

5.4. Directed Evolution

Engineering of enzymes using directed evolution is a successful strategy, especially in improving their thermostability and catalytic properties. This involves the construction of a mutant library through random mutagenesis or *in vitro* recombination techniques, followed by the selection of mutants with a desired characteristic using a high through-put screening technique. The desirable mutants are selected and identified by directional selection methods with excluding the mutants of non-interest [10].

5.5. Response Surface Methodology

Response surface methodology is an advantaged strategy because it involves the systematic, efficient and simultaneous interaction of variables. Optimization of culture conditions is important for maximizing production and yield, with minimizing the costs. Effect of culture conditions, particularly inoculum age, media composition (wheat bran and full-fat soybean flour) and duration of solid-state fermentation (SSF) on phytase production has extensively been studied with fungi [10]. However, low productivity and difficulties associated with operating and up-scaling of SSF conditions were observed. Phytase production using yeast cultures has generally been carried out in SmF systems. The strains used include *Schwanniomyces castellii*, *Pichia*, *Arxula adeninivorans* and *Candida kruzei*. Galactose and glucose were the preferred carbon sources. Production of extracellular phytase by lactic acid bacteria of the genera Lactobacillus and Streptococcus was evaluated by 19 strains which exhibited enzyme activity in submerged fermentation medium using glucose and inorganic phosphate [74]. By optimization of recombinant phytase expression in shake flask culture, *P. pastoris* was found to be a suitable host for high-level expression of phytase, and it can possess high potential for industrial applications [75]. Response surface methodology was applied to study the interaction effects of assay conditions to obtain optimum value for maximizing phytase activity by *Bacillus subtilis* B.S.46. The optimization resulted in 137% increase in phytase activity under optimum condition [76].

6. PHYTASE MANUFACTURE AND MARKET TREND

Many enzymes have emerged as a big-feed supplements. Feed enzymes such as protease, xylanase, phytase, amylase, cellulase, lipase, and β-glucanase are growing fast in the animal-nutrition market. However, only about 6% of manufactured animal feed contains enzymes, against 80 - 90% of vitamins, which is considered the largest animal nutrition category [10]. Phytase with desirable characteristics for application in the animal feed industry can be called an "ideal phytase" which should satisfy the following points: *a*) be effective in releasing phytate P in the digestive tract; *b*) be stable to proteases (trypsin and pepsin); *c*) resist inactivation by heat during feed pelleting and storage; and *d*) achieved by low-cost production, high yield, and purity by a relatively inexpensive system [10]. It is well known that a single phytase may never be ideal for all feeds and foods, i.e., phytase with different desired features are not present within a single phytase [10]. For example, the stomach pH in finishing pigs is much more acidic than that of the weaning pigs [77]. Thus, phytase with optimum pH close to 3.0 will perform better in the finishing pigs. For poultry, an enzyme would be beneficial if it is active over a broad pH range from acidic (stomach) to neutral (crop) [78]. Phytases used for aquaculture applications require a lower temperature than those used for swine or poultry. Therefore, the choice of an organism for production and development of phytases is dependent upon the target application using directed protein engineering [10].

Table 4. Examples for commercial-phytase products [79, 80]

Company	Country	Trademark	Donor microorganism[†]	Production microorganism
Roal	Finland	Fianase	*Aspergillus awamori*	*Trichoderma reesei*
Finnfeeds International	Finland	Avezyme	*Aspergillus awamori*	*Trichoderma reesei*
Novozymes	Denmark	Ronozyme	*Aspergillus oryzae*	*Aspergillus oryzae*
AB enzymes	Germany	Fianase	*Aspergillus awamori*	*Trichoderma reesei*
BASF	Germany	Natuphos	*Aspergillus niger*	*Aspergillus niger*
AB Enzymes	Germany	Quantum	*Escherichia coli*	*Pichia pastoris*[*]
AB Vista	Germany	Quantum Blue	*Escherichia coli*	*Trichoderma reesei*
DSM	USA	BioFeed	*Peniophora lycii*	*Aspergillus oryzae*
Genencor International	USA	Rovabio	*Penicillium simplicissimum*	*Penicillium funiculosum*
Alltech	USA	Allzyme phytase	*Aspergillus niger*	*Aspergillus oryzae*
Danisco Animal Nutrition	Europe, Africa USA, and Asia	Phyzyme XP	*Escherichia coli*	*Schizosaccharomyces pombe*
Huvepharma	Europe, Africa and Asia	OptiPhos	*Escherichia coli*	*Pichia pastoris*[*]
Fermic	Mexico	Phyzyme	*Aspergillus oryzae*	*Aspergillus oryzae*

[†] The donor microorganism refers to the microoorganism from which the gene encoding the phytase was derived.

[*] Now renamed as *Komagaetella pastoris* [80].

Several companies produce accessible P to animals, and release other associated nutrients, making animals generally more efficient at digesting their feed. This leads to less excretion of undigested phosphate, which as a result, helps to reduce water pollution and to be friendly for environment. Additionally, phytase-production markets ensure a very fast and efficient release of phytate P and other valuable nutrients, considering cost savings through a more nutritive diet [79, 80, 81]. Examples of commercial-phytase products are represented in Table 4.

7. BIOTECHNOLOGICAL APPLICATIONS OF PHYTASES

Due to the anti-nutritional effect of phytate, phytases are routinely added to the livestock feed to release phosphate from the plant molecule, phytate. Use of phytase reduces the addition of inorganic phosphate to diets and decreases the anti-nutritional effects of phytate [80]. The importance of phytases as potential biotechnological tools has been recognized in various fields. High dietary P bioavailability reduces the need for supplemental inorganic P such as mono- and di- calcium-phosphate. Furthermore, inorganic phosphate is a non-renewable resource, and it has been estimated that the easily-accessible phosphate on earth will be depleted in the next 50 years. Thus, employment of microbial phytase is an effective tool for natural resource management of P on a global scale. Organic P hydrolysis by microbial phytases has been extensively considered in a variety of biotechnological applications, including environmental protection, animal, and human nutrition [80, 82]. Because of the potential value of phytases in improving the efficiency of P use, rapid development in biotechnology of heterologous gene expression was observed, with large amounts of phytases at relatively low cost.

7.1. Phytases as Animal Feed

In animal nutrition, P plays a key metabolic role with more physiological functions than any other mineral. These functions include P as a major constituent of nucleic acids and cell membranes, major constituent of the structural components of skeletal tissues (80% P found in the bones and teeth). It is directly involved in cell biology such as: *a*) all energy-producing cellular reactions; *b*) maintenance of osmotic pressure; *c*) protein synthesis and transport of fatty acids; *d*) amino acid exchange; *e*) efficiency of feed utilization, growth, and cell differentiation; *f*) appetite control, and efficiency of feed utilization and fertility [1, 2, 83].

Diets with low P content can be considerably improved by the use of P feed supplement in the form of compound feed or as separate mineral supplements. P supplements are manufactured in many chemical and physical forms to suit different feeding and handling practices [2].

During the past two decades, there has been significant increase in use of phytases as a feed additive in poultry, fish, lambs, and pig diets. In numerous studies, the efficacy of microbial phytases to release phytate-bound P has been demonstrated in various animals [82, 84, 85]. Phytases are also found to enhance the utilization of different minerals. Phytases from different sources have been evaluated individually and in combination for their efficacy as feed additives in poultry. Use of both bacterial and fungal phytases together as feed additive would be another promising alternative in improving the P utilization and alleviation of mineral deficiency due to their synergistic activities throughout the gastrointestinal tract of the animals. Ruminants have the ability to utilize phytate P, partially as a result of the phytase activity of the ruminal microflora. However, monogastric animals, such as swine, fish, and poultry, show negligible or no phytase activity in the digestive tract, and thus, large amounts of P are discharged into the environment where they pose serious P pollution problems [86].

More than 25 years ago, several companies were established to market a phytase for standard feed and help animal (poultry, turkey, and pigs) for better utilization of P and other important nutrients such as proteins and calcium (Raga 2016). Phytases have been marketed as an animal feed enzyme in the US since 1996 [10]. In recent years, considerable efforts have been achieved to improve the nutritive value of animal feedstuff through supplementation with exogenous enzymes. The global market for feed enzymes is a promising segment in the enzyme industry. The main characteristics that define the best phytase supplements should include: *a*) high stability; *b*) high catalytic activity in the gut; *c*) effective in releasing phytate phosphate in the digestive tract; and *d*) high industrial production levels. Further, the enzyme should display high levels of activity at the gut pH (with the pH changing throughout the digestive system). The enzyme should be stable during manufacture processes and formulation, where the activity can be lost through thermal denaturation during feed pellet production. In addition, the enzyme may be exposed to high temperatures, during storage and transportation and through degradation by proteases and the acidic conditions encountered in the gut [80]. Requirement for stability at high temperatures during pelleting coupled with optimal activity at the lower temperatures in the gut presents is a major challenge. This phenomenon is often observed in enzymes from thermophilic microorganisms that are stable at very high temperatures but have relatively low or non-existent activities at gut temperature [80].

7.2. Phytases in Aquaculture

Utilization of dietary phosphates, which critically affects fish growth, is a major concern in aquaculture as well as in the aquatic environment. An efficient utilization of feed, leading to optimum fish growth, serves as a benchmark of successful aquaculture worldwide [87]. Studies using phytase as a feed additive in aquaculture have established that phytase supplementation could enhance the bioavailability of P, nitrogen, and other minerals, thereby decreasing P load in aquatic environment [88, 89]. Bacterial phytase could potentially be used as feed supplement to boost cellular immune response of fish and could be employed as a health management strategy in culture systems [90]. Thus, these may have significant impact on the development of feed supplements and health management in aquaculture systems.

7.3. Phytases in Plant Growth Promotion

Phosphorus is one of the most abundant elements on earth. It exists in soil either in dissolved or solid form, which is dominant. In solid form, P is classified as inorganic P (P bound to Al, Fe, Ca, Mg etc. as complex salts) and organic P (P bound to organic material such as dead and living plant material and microorganisms, soil organic matter, etc.). Mineral soil contains 33–90% of total P in inorganic form. The typical range for total P content of agricultural soils is estimated between 0.2 to 2.0 g/kg [2]. Dissolved P is typically less than 0.1% of the total soil P and usually exists as orthophosphate ions, inorganic polyphosphates and organic P. Extracellular phytase activities have been reported under phosphate stress conditions in diverse plant species, namely, tobacco, barley, tomato, alfalfa [10]. The ability of plants to use phosphorus from soil is improved by inoculating the soil with microorganisms possessing the ability to release phytase.

The novel *Enterobacter cancerogenus* MSA2, a plant growth promoting gamma-proteobacterium, was isolated from the rhizosphere of *Jatropha curcas* (a potentially important biofuel feed stock plant). This strain is the first plant growth promoting bacterium produces ACC deaminase enzyme and promotes plant growth with *Jatropha curcas* [91]. On other side, the effect of fungal phytase on plant growth at pot and tray level was studied, compared with commercial fertilizers pertaining to chemical and physiological parameter and as soil amendment. It was found that, phytase was efficient in reducing the phytic acid content of soil by about 30% while simultaneously increasing the phytate phosphate availability by 1.18-fold [92].

7.4. Phytases for Human Nutrition

Phytase is already used as a supplement in diets for monogastric animals to improve phosphate utilization from phytate, the major storage form of phosphate in plant seeds. In recent years, this class of enzymes has also found interesting for use in processing and manufacturing of food for human consumption [95]. Different strategies could be applied to optimize phytate degradation during food processing and digestion, in the human alimentary tract, such as: *a*) adjustment of more favorable conditions during food processing for the phytases naturally occurring in the raw material; *b*) addition of isolated phytases to the production process; *c*) use of raw material with a high intrinsic phytate-degrading activity either naturally present or introduced by genetic engineering; and *d*) use of recombinant food-grade microorganisms as carriers for phytate-degrading activity in the human gastrointestinal tract [95]. Furthermore, phytases may find application in production of functional foods or food supplements with health benefits. Technological improvements are expected to occur due to phytate degradation during processing as shown for bread making, production of plant protein isolates, corn wet milling, and the fractionation of cereal bran [95]. Phytases are added in foods to improve the public health and decrease malnutrition. The limited bioavailability of cereals mineral content, due to the relatively low mineral levels and presence of phytic acid, offers challenges in nutrition point of view [96]. Antinutritional factors reduce the bioavailability of cereals mineral content to 5–15%. The FDA (The Food and Drug Administration) has approved the use of phytase in food as GRAS "generally recognized as safe."

8. CURRENT CHALLENGES AND FUTURE PERSPECTIVES

Recent trends in the market have clearly shown phytase as an important enzyme and feed supplement. Due to the serious concerns about environmental pollution, 22 countries have adopted the use of *phyA*, produced from *A. niger* NRRL 3135, as a feed additive. At present, phytase constitutes about 20% of the total enzyme use in the livestock or allied sector, which is expected to increase many fold in the future due to increasing productive research leads. Industrial production of phytase currently utilizes the soil fungus *Aspergillus*, on which considerable research has been conducted [2]. Studies focused on microbial production of phytase, downstream processing, and application-oriented research will help in developing an integrated technological solution to phytase production [10]. Such studies present new insights in the biological and engineering facets of phytase-producing microbes, and reveal a new era in phytase biotechnology. The new era in phytase biotechnology may target the following purposes: *a*) retaining the activity of phytases during storage and use which is the core aim of a viable process; *b*) technique such as immobilization that will help in solving the problem of stabilizing phytase formulations; *c*) Downstream processing as an essential aspect of phytase bioprocessing; *d*) Rapid and economic methods such as liquid–liquid extraction are forthcoming promising alternatives; and *e*) Development of efficient, scalable and economical process for phytase bio-separation to overcome the techno-economic limitations of conventional downstream processes [10]. Possible roles of microbes present in soil in P management is important. Beneficial P solubilizing bacteria in the rhizosphere converted P to the plant in an accessible form mainly by means of organic acid production. In this regard, a few studies on addition of phytase or phytase-producing microbes to the soil revealed increased P uptake by plants. Further studies with specified target for maintaining soil P level and decreasing application of fertilizer might be highly promising considering environmental P pollution and increasing cost of fertilizers [10, 97, 98].

Despite of the so many beneficial effects reported so far, the actual usefulness of such enzymes are limited due to: *a*) high variation in activity of phytase production; *b*) lack of farmer awareness regarding the use of phytase; *c*) cost factor and low enzyme production; *d*) nonexistence of a single phytase with applicability in all kind of feeds or applications; *e*) low storage stability and narrow pH and substrate specificity; and *f*) dose response variation and high processing temperature susceptibility [2]. Therefore, development of novel ideal phytase with required industrial properties is desirable. Production of canola seeds with improved phytase activity has paved the way for future research to develop transgenic soybean, cotton, sunflower and other grain and cereal plants which have potential uses in fish feed [2]. These plants with improved phytase production from roots might have better P utilization efficiency.

The worldwide interest on phytases has led to the development of recombinant microbial phytase producers and plant bioreactors. The *in vitro* manipulations in microbial phytases and recombinant DNA technology have generated phytases with improved properties. Phytases have been explored for their utility as a food and feed additive, however, other applications, such as in soil amendment and environmental protection are yet to be adequately explored. A phytase with all desirable characteristics, such as, thermostability, acid stability, protease insensitivity, adequate activity at animal body temperature, and high specific activity in a single phytase is also still a challenge. To achieve better P management with reduced environmental pollution, the following recommendations are reported: *a*) regulation for optimum P concentration is required for determining actual P concentration and the dose of fertilizer to be applied; *b*) farmers should be familiar, through awareness campaigns, with the use of manure products, its proper use, disposal and related P content, the methodology and advantages in the use of phytases as a new approach; *c*) cost monitoring of phytase- supplemented animal feed is required for its maximum utilization and reach to the farmers; and *d*) development of biofertilizers that aid in better P solubility and its availability to plants will certainly be useful in the long-term control of P pollution [2].

CONCLUSION

Controlled degradation of phytate, by phytase during food processing and/or food digestion in the human stomach and upper small intestine, results in an improvement of the bioavailability of essential minerals such as iron and zinc. This is seen as a way to reduce the risk of mineral deficiency, especially in vulnerable groups such as child-bearing women, strictly vegetarians and inhabitants of developing countries. Phytases not only supply numerous amounts of P in diet, but its use is also ecofriendly. Given the importance of P in bone formation and consequently growth, addition of phytase can be expected to provide additional benefits. Phytases are produced and marketed by many countries worldwide, there is however, no one ideal phytase which can fulfill all requirements. Thus, genetic engineering of microorganisms with improved thermo-stability, pH optimum, and substrate specificity should be performed. Furthermore, research aiming at improving food texture and digestibility at a cost-effective level should be continued. The use of phytase in human food and animal feed will seize the antinutritional effects of phytate and phytic acid, decrease environmental pollution, and increase availability of P, starch, protein, amino acids, and calcium. Moreover, use of phytase in animal feed abolish the surplus addition of inorganic phosphate in animal feed.

CONFLICT OF INTEREST

The authors have declared no conflicts of interest.

REFERENCES

[1]   National Research Council, 1993. *Nutrient requirements of fish*. National Academy Press, Washington, DC, 114 pp.

[2]   Kumar V, Singh D, Punesh S, et al. 2015. Management of environmental phosphorus pollution using phytases: current challenges and future prospects. In: *Applied Environmental Biotechnology*: *Present Scenario and Future Trends*: 97-114. DOI 10.1007/978-81-322-2123-4_7.

[3]   Afinah S, Yazid A, Shobirin A, et al. 2010) Phytase: application in food industry. *Intr Food Res J.* 17:13-21.

[4]   Joshi J. 2014. Phytase - A key to unlock phytate complex. *Intr J Pure App Biosci.* 2:304-313.

[5]   Gupta R, Shivraj S, and Singh N. 2015. Reduction of phytic acid and enhancement of bioavailable micronutrients in food grains. *J Food Sci Technol.* 52:676-684.

[6]   von Sperber C, Tamburini F, Brunner B, et al. 2015. The oxygen isotope composition of phosphate released from phytic acid by the activity of wheat and *Aspergillus niger* phytase. *Biogeosci.* 12:4175-4184.

[7]   Singh N, Joshi D. and Gupta R. 2013. Isolation of phytase producing bacteria and optimization of phytase production parameters. *Jundishapur J. Microbiol.* 6: e6419-e6424.

[8]   Cizeikiene D. 2015. *Bioproducts of Bacteriocins producing lactic acid bacteria, their antimicrobial and phytase activities, and applications*. Kauno Technologijos University. Technological Sciences, Chemical Engineering (05T).

[9]   Tyagi P, Verma S. 1998. Phytate phosphorus content of some common poultry feed stuffs. *Indian J Poult Sci*. 33: 86-88.

[10] Bhavsar K, Khire J. 2014. Current research and future perspectives of phytase bioprocessing. RSC *Adv*. 4:26677-26691.

[11] Greiner R, Konietzny U. 2006. Phytase for Food Application. *Food Technol Biotechnol*. 44:125-140.

[12] Kemme P, Age W, Mroz Z, et al. 1999. Digestibility of nutrients in growing-finishing pigs is affected by *Aspergillus niger* phytase, phytate and lactic acid levels 1. Apparent ileal digestibility of amino acids. *Livest Prod Sci*. 58:107-117.

[13] Sharon E, Rickard S. and Thompson L. 1997. Interactions and Biological Effects of Phytic Acid. In: *Antinutrients and phytochemicals in food*. Shahidi F. (ed.) ACS, Washington, DC, p. 294-312.

[14] Leeson S. 1993. *Recent advances* in fat utilization by *poultry*. In: *Recent Advances in Animal Nutrition in Australia*. (ed. Farrell DJ) University of New England, Armidale, NSW, Australia. p. 170.-181.

[15] Angel R, Tamim N, Applegate T, et al. 2002. Phytic acid chemistry: influence on phytin-phosphorus availability and phytase efficacy. *J Appl Poult Res*. 11:471-480.

[16] Mullaney E, Ullah A. 2003. The term phytase comprises several different classes of enzymes. *Biochem Biophys Res Commun*. 312:179-184.

[17] Konietzny U, Greiner R. 2002. Molecular and catalytic properties of phytate-degrading enzymes (phytases). *Intr J Food Sci Technol*. 37:791-812.

[18] Fuentes B, Bolan N, Naidu R, et al. 2006. Phosphorus in organic waste-soil systems. *J Soil Sci Plant Nutr*. 6:64–83.

[19] Mullen M. 2005. Phosphorus in soils: biological interactions, p. 210–215. In D. Hillel (ed.), *Encyclopedia of Soils in the Environment*. Elsevier Ltd., Oxford.

[20] Shieh T, Ware J. 1968. Survey of microorganisms for the production of extracellular phytase. *Appl Microbiol*. 16:1348-1351.

[21] Ehrlich K, Montalbano B, Mullaney E, et al. 1993. Identification and cloning of a second phytase gene (phyB) from *Aspergillus niger (ficuum)*. *Biochem Biophys Res Commun*. 195:53–57.

[22] Wodzinski R, Ullah A. 1996. Phytase. *Adv Appl Microbiol*. 42:263–302.

[23] Jorquera M, Martinez O, Maruyama F, et al. 2008. Current and future biotechnological applications of bacterial phytases and phytase producing bacteria. *Microbes Env*. 23:182–191.

[24] Kumar V, Singh P, Jorquera M, et al. 2013. Isolation of phytase-producing bacteria from Himalayan soils and their effect on growth and phosphorus uptake of Indian mustard (*Brassica juncea*). *World J Microbiol Biotechnol*. 29:1361–1369.

[25]   Pandey A, Szakacs G, Soccol C, et al. 2001. Production, purification and properties of microbial phytases. *Biores Technol.* 77:203–214.

[26]   Vats P, Banerjee U. 2004. Production studies and catalytic properties of phytases (myo-inositol-hexakisphosphate phosphohydrolases): An overview. *Enz Microb Technol.* 35:3-14.

[27]   Kim Y, Kim H, Bae K, et al. 1998. Purification and properties of a thermostable phytase from *Bacillus sp.* DS11. *Enz Microb Technol.* 22:2–7.

[28]   Cho J, Lee C, Kang S, et al. 2003. Purification and characterization of a phytase from *Pseudomonas syringae* MOK1. *Curr Microbiol.* 47:290–294.

[29]   Igbasan F, Manner K, Miksch G, et al. 2000. Comparative studies on the *in vitro* properties of phytases from various microbial origins. *Arch Tierernahr.* 53:353–373.

[30]   Nuobariene L, Arneborg N. and Hansen A. 2014. Phytase active yeasts isolated from bakery sourdoughs. In *9th Baltic Conference on Food Science and Technology. Food for Consumer Well-Being: FOODBALT- conference proceedings.* (pp. 223-227). Latvia University of Agriculture.

[31]   Nakamura Y, Fukuhara H. and Sano K. 2000. Secreted phytase activities of yeasts. *Biosci Biotechnol Biochem.* 64: 841–844.

[32]   Bhavsar K, Ravi K. and Khire J. 2011. High level phytase production by *Aspergillus niger* NCIM 563 in solid state culture. *J Ind Microbiol Biotechnol.* 38:1407-1417.

[33]   Thomas W, Hestler T. and Krull R. 2011. Morphology engineering osmolality and its effect on *Aspergillus niger* morphology and productivity. *Microb Cell Fact.*10: 58-72.

[34]   Gontia-Mishra I, Deshmukh D, Niraj T, et al. 2013. Isolation, morphological and molecular characterization of phytate-hydrolysing fungi by 18S rDNA sequence analysis. *Braz J Microbiol.* 44: 317-323.

[35]   Kwanjira C, Ayuthaya D, Suthum I, et al. 2005. Phytase activity from *Aspergillus oryzae* AK9 cultivated on solid state soybean meal medium. *Process Biochem.* 40: 2285–2289.

[36]   Vats P, Banerjee U. 2005. Biochemical characterization of extracellular phytase (*myo*-inositol hexakisphosphate phosphohydrolase) from a hyper-producing strain of *Aspergillus niger* van Teighem. *J Ind Microbiol Biotechnol.* 32:141–147.

[37]   Mullaney E, Daly C. and Ullah A. 2000. Advances in phytase research. *Adv Appl Microbiol.* 47:157-199.

[38]   Zhang G, Dong X, Wang Z, et al. 2010. Purification, characterization, and cloning of a novel phytase with low pH optimum and strong proteolysis resistance from *Aspergillus ficuum* NTG-23. *Biores Technol.* 101: 4125-4131.

[39] Lee S, Cho J, Bok J, et al. 2015. Characterization, gene cloning, and sequencing of a fungal phytase, phyA, from *Penicillium oxalicum* PJ3. *Preparative Biochem Biotechn*. 45: 336-347.

[40] Alves N, Luis H, Roberta H, et al. 2016. Production and partial characterization of an extracellular phytase produced by *Muscodor* sp. under submerged fermentation. *Adv Microbiol*. 6: 23-32.

[41] Bogar B, Szakacs G, Pandey A, et al. 2003. Production of phytase by *Mucor racemosus* in solid-state fermentation. *Biotechnol Prog*. 19:312–319.

[42] Boyce A, Walsh, G. 2007. Purification and characterization of an acid phosphatase with phytase activity from *Mucor hiemalis*. *Wehmer J Biotechnol*. 132:82–87.

[43] Rani R, Ghosh S. 2011. Production of Phytase under Solid-State Fermentation Using *Rhizopus oryzae*: Novel Strain Improvement Approach and Studies on Purification and Characterization. *Biores Technol*. 102: 10641-10649.

[44] Haraldsson A, Veide J, Andlid T, et al. 2005. Degradation of phytate by high-phytase *Saccharomyces cerevisiae* strains during simulated gastrointestinal digestion. *J Agric Food Chem* 53:5438–5444.

[45] Vohra A, Satyanarayana T. 2001. Phytase production by the yeast *Pichia anomala*. *Biotechnol Lett*. 23:551–554.

[46] Ragon M, Neugnot-Roux V, Chemardin P, et al. 2008. Molecular gene cloning and overexpression of the phytase from *Debaryomyces castellii* CBS 2923. *Protein Expr Purif*. 58:275–283.

[47] Li X, Liu Z, Chi Z, et al. 2009. Molecular cloning, characterization and expression of the phytase gene from marine yeast *Kodamaea ohmeri* BG3. *Mycol Res*. 113:24–36.

[48] Watanabe T, Ozaki N, Iwashita K, et al. 2008. Breeding of wastewater treatment yeasts that accumulate high concentrations of phosphorus. *Appl Microbiol Biotechnol*. 80:331–338.

[49] Ines B, Ameny F, Blibech M, et al. 2015. Characterization of an extremely salt-tolerant and thermostablephytase from *Bacillus amyloliquefaciens* US573. *Intr J Biol Machrom*. 80:581-587.

[50] Chouayekh I. 2016. Characterization of an extremely salt-tolerant and thermostable phytase from *Bacillus amyloliquefaciens* US573. *Intr J Biol Macromol*. 80:581–587.

[51] Kumar V, Sangwan P, Verma A, et al. 2014. Molecular and biochemical characteristics of recombinant β-propeller Phytase from *Bacillus licheniformis* strain PB-13 with potential application in aquafeed. *Appl Biochem Biotechnol*. 173:646–659.

[52] Shamna K, Rajamanikandan K, Mukesh K, et al. 2012. Extracellular production of Phytases by a Native Bacillus subtilis Strain. *Ann Biol Res*. 3: 979-987.

[53]   Karim R, Hashemib M, Safaria M, et al. 2016. A novel phytase characterized by thermostability and high pH tolerance from rice phyllosphere isolated *Bacillus subtilis* B.S.46. *J Adv Res*. 7:381-390.

[54]   Dahiya S, Singh N. 2016. Effect of varying doses of phytase enzyme from a novel strain of Bacillus cereus MTCC 10072 in animal feed. *Eur J Biotechnol Biosci*. 4:11-14.

[55]   Yan L, Zhou T, Jang H, et al. 2009. Comparative Effects of Phytase Derived from Escherichia coli and *Aspergillus niger* in Sixty-Eight-week old Laying Hens Fed Corn-soy Diet. *Asian-Aust J Anim Sci*. 22:1391 – 1399.

[56]   Shedova E, Lipasova V, Velikodvorskaya G, et al. 2008. Phytase activity and its regulation in a rhizospheric strain of *Serratia plymuthica*. *Folia Microbiol*. 53:110-114.

[57]   Zhang R, Yang P, Huang H, et al. 2011. Two types of phytases (histidine acid phytase and β-propeller phytase) in *Serratia sp*. TN49 from the gut of *Batocera horsfieldi* (Coleoptera) larvae. *Curr Microbiol*. 63: 408–415.

[58]   Li R, Lu W, Guo C, et al. 2012. Molecular characterization and functional analysis of OsPHY1, a purple acid phosphatase (PAP)-type phytase gene in rice (*Oryza sativa* L.). *J Integr Agric*. 11:1217–1226.

[59]   Hegeman C, Grabau E. 2001. A novel phytase with sequence similarity to purple acid phosphatases is expressed in cotyledons of germinating soybean seedlings. *Plant Physiol*. 126:1598–1608.

[60]   Xiao K, Harrison M. and Wang Z. 2005. Transgenic expression of a novel *M. truncatula* phytase gene results in improved acquisition of organic phosphorus by Arabidopsis. *Planta*, 222:27–36.

[61]   Garchow B, Jog S, Mehta B, et al. 2006. Alkaline phytase from *Lilium longiflorum*: purification and structural characterization. *Protein Expr Purif*. 46:221–232.

[62]   Rapoport S, Leva E. and Guest G. 1941. Phytase in plasma and erythrocytes of various species of vertebrates. *J Biol Chem*. 139:621–632

[63]   Raun A, Cheng E, Burroughs W. 1956. Ruminant nutrition, phytate phosphorus hydrolysis and availability to rumen microorganisms. *J Agric Food Chem*. 4:869–871.

[64]   Yanke L, Bae H, Selinger L, et al. 1998. Phytase activity of anaerobic ruminal bacteria. *Microbiol*. 144:1565–1573.

[65]   Schlemmer U, Jany K, Berk A, et al. 2001. Degradation of phytate in the gut of pigs: pathway of gastro-intestinal inositol phosphate hydrolysis and enzymes involved. *Arch Anim Nutr*. 55:255–280.

[66]   Huang H, Shi P, Wang Y, et al. 2009. Diversity of beta-propeller phytase genes in the intestinal contents of grass carp provides insight into the release of major phosphorus from phytate in nature. *Appl Env Microbiol*. 75:1508–1516.

[67] Lei X, Stahl C. 2001. Biotechnological development of effective phytases for mineral nutrition and environmental protection. *Appl Microbiol Biotechnol.* 57:474–481.

[68] Singh B, Kunze G, Satyanarayana T. 2011. Developments in biochemical aspects and biotechnological applications of microbial phytases. *Biotechnol Mol Bio Rev.* 6:69–87.

[69] Mullaney E, Daly C, Kim T, et al. 2002. Site-directed mutagenesis of *Aspergillus niger* NRRL 3135 phytase at residue 300 to enhance catalysis at pH 4.0. *Biochem Biophys Res Commun.* 297:1016–1020.

[70] Bhavsar K, Gujar P, Shah P, et al. 2012. Combinatorial approach of statistical optimization and mutagenesis for improved production of acidic phytase by *Aspergillus niger* NCIM 563 under submerged fermentation condition. *Appl Microbiol Biotechnol.* 97:673–679.

[71] Tomschy A, Tessier M, Wyss M, et al. 2000. Optimization of the catalytic properties of *Aspergillus fumigatus* phytase based on the three-dimensional structure. *Protein Sci.* 9:1304–1311.

[72] Rodriguez E, Wood Z, Karplus P, et al. 2000. Site-directed mutagenesis improves catalytic efficiency and thermostability of *Escherichia coli* pH 2.5 acid phosphatase/phytase expressed in *Pichia pastoris. Arch Biochem Biophys.* 382:105–112.

[73] Lie X, Porres J. 2003. Phytase enzymology, applications and biotechnology. *Biotechnol Lett.* 25:1787–1794.

[74] Sreeramulu G, Srinivasa D, Nand K, et al. 1996. *Lactobacillus amylovorus* as a phytase producer in submerged culture. *Lett Appl Microbiol.* 23:385–388.

[75] Akbarzadeh A, Ehsan D, Mojtaba A, et al. 2015. Optimization of Recombinant Expression of Synthetic Bacterial Phytase in *Pichia pastoris* Using Response Surface Methodology. *Jund J Microbiol.* 8: e27553: doi: 10.5812/jjm.27553.

[76] Rocky-Salimi K, Hashemi M, Safari M, et al. 2016. A novel phytase characterized by thermostability and high pH tolerance from rice phyllosphere isolated Bacillus subtilis B.S.46. *J Adv Res.* 7:381–390.

[77] Jongbloed A, Mroz Z. and Kemme P. 1992. The effect of supplementary *Aspergillus niger* phytase in diets for pigs on concentration and apparent digestibility of dry matter, total phosphorus, and phytic acid in different sections of the alimentary tract. *J Animal Sci.* 70:1159-1168.

[78] Quan C, Tian W, Fan S, et al. 2004. Purification and properties of a low-molecular-weight phytase from *Cladosporium* sp. FP-1. *J Biosci Bioeng.* 97:260-266.

[79] Mittal A, Gupta V, Singh G, et al. 2013. Phytase: A Boom in Food Industry. *Octa. J. Biosci.* 1:158-169.

[80] Robert S. 2016. Biotechnology in the development of improved phytases. In *27th Annual Australian Poultry Science Symposium*, 14 -17 February 2016, Sydney, N.S.W. Qeensland Univ. Technology, Australia.

[81] Raga M. 2016. *The phytase pioneer, BASF, sets a new benchmark in animal nutrition with Natuphos*. http://www.basf.com.

[82] Buendia G, Mendoza G, Juan M, et al. 2016. Influence of supplemental phytase on growth performance, digestion and phosphorus balance of lambs fed sorghum-based diets. *Italian J Anim Sci*. 13:187-189.

[83] Dobrota C. 2004. The biology of phosphorous. In: *Phosphorus in environmental technologies*. Valsami-Jones, E. (ed). IWA, London, pp 51–74.

[84] Daniel L, Albert G. 2015. Use of phytases in fish and shrimp feeds: a review. *Rev Aquacul*. 9:1-7.

[85] Wealleans A, Barnard L, Romero L, et al. 2016. A value based approach to determine optimal phytase dose: A case study in turkey poults. *Anim Feed Sci Technol*. 216:288–295.

[86] Yueming D, Ajay A, Hagen S, et al. 2014. Phytase in non-ruminant animal nutrition: a critical review on phytase activities in the gastrointestinal tract and influencing factors. *J Sci Food Agric*. 95: 878–896.

[87] Kumar V, Sinha A, Makkar H, et al. 2012. Phytate and phytase in fish nutrition. *J Anim Physiol Anim Nutr*. 96:335-364.

[88] Nwanna L, Schwarz F. 2007. Effect of supplemental phytase on growth, phosphorus digestibility and bone mineralization of common carp (*Cyprinus carpio* L). *Aquacult Res*. 38:1037-1044.

[89] Akpoilih B, Omitoyin B. and Ajani E. 2016. Dietary microbial phytase improves growth, phosphorus digestibility and serum phosphorus, but not bone mineralization, in *Juvenile Clarias gariepinus* Fed Roasted and Oil-pressed Groundnut-Based Diet. *Fish. Aquat. Sci*. 11: 108-130.

[90] Lazado C, Marlowe C, Caipang A, et al. 2010. Responses of Atlantic cod Gadus morhuahead kidney leukocytes to phytase produced by gastrointestinal-derived bacteria. *Fish Physiol Biochem*. 36:883.

[91] Jha C, Patel B. and Saraf M. 2012. Stimulation of the growth of Jatropha curcas by the plant growth promoting bacterium Enterobacter cancerogenus MSA2. *World J Microbiol Biotechnol*. 28: 891-899.

[92] Gujar P, Bhavsar K. and Khire J. 2013. Effect of phytase from *Aspergillus niger* on plant growth and mineral assimilation in wheat (*Triticum aestivum* Linn.) and its potential for use as a soil amendment. *J Sci Food Agric*. 93:2242–2247.

[93] Juan M, Frontela C, Ros G, et al. 2012. Application of Bifidobacterial Phytases in Infant Cereals: Effect on Phytate Contents and Mineral Dialyzability. *J Agric Food Chem*. 60: 11787–11792.

[94] Troesch B, Egli I, Zeder C, et al. 2009. Optimization of a phytase-containing micronutrient powder with low amounts of highly bioavailable iron for in-home fortification of complementary foods. *Amr J Clin Nutr*. 89: 539–544.

[95] Troesch B. 2015. How can phytase improve public health nutrition? *Sight & Life* 29:108-110.

[96] Das A, Raychaudhuri U. and Chakraborty R. 2011. Cereal based functional food of Indian subcontinent: a review. *J Food Sci Tech*. 49: 665-672.

[97] Ushasree M, Vidya J. and Pandey A. 2017. Other enzymes: phytases, p309–333. In: *Current developments in biotechnology and bioengineering*. Pandey A, Negi S, Soccol C (eds). Elsevier, Amsterdam, Netherlands.

[98] Abbasi F, Tahmina F. and Liu J. 2019. Low digestibility of phytate phosphorus, their impacts on the environment, and phytase opportunity in the poultry industry. *Env Sci Poll Res*. 26: 9469–9479.

In: Microbial Catalysts. Vol. 2　　　　　ISBN: 978-1-53616-088-8
Editors: Shadia M. Abdel-Aziz et al.　　　© 2019 Nova Science Publishers, Inc.

Chapter 9

SKIN PIGMENTATION DISORDERS: CAUSAL ENZYME AND SAFE TREATMENT

Shadia M. Abdel-Aziz[1,], Hoda A. Hamed[1], D. Sukmawati[2] and Neelam Garg[3]*

[1]Department of Microbial Chemistry, National Research Centre, Dokki, Giza, Egypt
[2]Department of Biology, Universitas, Negeri Jakarta, Indonesia
[3]Department of Microbiology, Faculty of Life Sciences, University of Kurukshetra, Kurukshetra, Haryana, India

ABSTRACT

Melanocytes are the cells responsible for skin color. These cells are located at the bottom of the epidermis and are able to inject pigment into the surrounding skin cells in packets known as melanosomes. The melanocytes do this when signaled by an enzyme called Tyrosinase. This enzyme is switched "on" through a variety of stimulatory factors. Melanin is one of the most important pigments which exist ubiquitously from microorganisms, plants and animals. It is secreted by melanocyte cells and determines the color of skin and hair in mammalian. It protects the skin from photo-carcinogenesis by absorbing UV sunlight and removing reactive oxygen species. Excessive level of melanin pigmentation causes various dermatological disorders including hyperpigmentation such as aged lentigo, melasma, freckles, age spots and sites of actinic damage which can give rise to aesthetic problems. Inhibitors of the enzyme tyrosinase can be used to prevent or treat melanin hyperpigmentation disorders. Therefore, tyrosinase has increasingly become important in cosmetic and medical products. Great advances have been made in understanding the cellular and biochemical mechanisms in pigment biology and the processes underlying skin pigmentation for developing various skin lightening agents to reduce skin hyper-pigmentation. Natural product companies are formulating creams and serums that contain mild natural lighteners.

---

* Corresponding Author's E-mail: abdelaziz.sm@gmail.com.

Keywords: skin pigmentation, hyper-pigmentation disorders, tyrosinase, melanin, microbial tyrosinase

1. INTRODUCTION

The skin has many different types of skin cells. Cells responsible for skin color are melanocytes (skin cells that produce the dark pigment). These cells are located at the bottom of the epidermis and are able to inject pigment into the surrounding skin cells in packets known as melanosomes. Melanocytes do this when signaled by an enzyme called "Tyrosinase" (EC 1.14.18.1). This enzyme is switched "on" through a variety of stimulatory factors [1]. These stimulating factors are; UV radiation for the sun, burns or injury, irritation from acids, oral and topical medications such as birth control pills or antibiotics [2]. A tan is a normal reaction to the sun, and over two to three days a tan develops. The sun stimulates the skin to produce tyrosinase for protecting the deeper layers of skin, the dermis. As a result, the enzyme signals the melanocytes to inject melanosomes into the surrounding cells [3]. This is the body's way of protecting the deeper layers of skin. For people with hyper-pigmentation though, the skin has become dysfunctional and does not dissipate the pigment normally. For these people, any slight stimulation can lead to big pigmentation problems. The basal layer sometimes can't stop producing melanin in certain areas and keeps overproducing melanin. Sometimes melanin can be deposited into the dermis, through trauma or excessive inflammation, and is not possible to improve with skin care regimes [3, 4].

Skin pigmentation is more common in women than men, this might be because of hormones or because many women spend their holidays sunbathing. As one gets older, chances of having skin pigmentation tend to increase. However, it is also possible for the younger in the form of melisma, and this means that there may not be any age limit to those who can end up with this skin condition. Skin pigmentation is originally caused by the UV rays from the sun. Normally, the skin's natural protection is able to resist the UV rays, but with aging, immunity of the body will reduce. This will make it possible for the years of accumulated sun exposure to take its toll on the skin [5]. When this happens, the sun will cause some genetic mutation of the genes that are responsible for formation of the skin pigment; melanin. This genetic mutation can also occur on skins of individuals that are naturally having light skin due to reduction in melanin content on such skins. The genetic mutation experienced in such individuals can be passed across to their offspring and the offspring too will end up with skin pigmentation. In individuals with light skin, the genetic situation responsible for reduced melanin will prevail in their children and this will make it possible for the offspring to have the skin condition during their lifetime. Thus, in some cases, skin pigmentation is hereditary [5]. Skin whitening, in all application forms, is a significant trend. Many studies for treatment of pigmentation disorders and tyrosinase inhibition have been reported [1, 2, 6, 7, 8]. Hyper-pigmentation disorders are commonly encountered in dermatology clinics. Botanical and natural ingredients have gained popularity as alternative depigmenting products. Clinical studies evaluating the use of different natural products in treating hyper-pigmentation [9]. Various processes, in the complex mechanism of skin pigmentation, can be regulated individually or concomitantly to alter complexion coloration and thus relieve skin complexion diseases. This chapter will discuss skin pigmentation disorders, tyrosinase inhibitors, and regulation of processes that control skin complexion coloration.

2. PIGMENTATION

Pigmentation is the coloring of a person's skin, which will appear normal in color with health. In the case of illness or injury, skin may change color, becoming darker (hyper-pigmentation) or lighter (hypo-pigmentation). Melanin, the natural color in our skin, is synthesized by melanocytes. Melanin is formed through a series of oxidative reactions involving the amino acid tyrosine in presence of the enzyme tyrosinase [6].

2.1. Melanin

Melanin is a heterogeneous, insoluble non-protein polymer composed of subunits that are products of enzymatically modified tyrosine. Melanin plays a crucial role in the absorption of free radicals generated within the cytoplasm and in shielding the host from various types of ionizing radiations, including UV light [12]. Melanin is secreted by melanocyte cells, and determines the color of skin and hair in mammalians [11]. Melanogenesis is a process to synthesize melanin. Upon exposure of the skin to UV irradiation, melanogenesis is initiated with the first step of tyrosine oxidation through tyrosinase [6, 9, 10]. Up to 10% of skin cells in the basal layer of the epidermis produce melanin. Skin whitening agents often inhibit the activity of tyrosinase by competitive or non-competitive inhibition of its catalytic activity, by inhibiting its maturation, or by accelerating its degradation [6].

Melanin protects the skin from photo-carcinogenesis by absorbing UV sunlight and removing reactive oxygen species [6, 13, 14]. The modifications in melanin biosynthesis occur in many disease states. The excessive level of melanin pigmentation causes various dermatological disorders including hyper-pigmentations such as senile lentigo, melasma, post-inflammatory melanoderma, freckles, ephelide, age spots, and sites of actinic damage which can give rise to esthetic problems [15, 16]. Hyper-pigmentation usually becomes a big problem as people age because darker spots will start to be seen on the face, arms and body. In addition, hormonal changes such as pregnancy and drugs manipulating hormone levels may cause hyper-pigmentation.

2.2. Pigmentation Disorders

Pigmentation disorders is a general term includes hyper-pigmentary disorders (darkening of the skin) and hypo-pigmentary disorders (decrease in the normal skin color). The most common disorders being melasma (hyper-pigmentary) and vitiligo (hypo-pigmentary). All forms of skin hyper-pigmentation essentially involve the same mechanism. However, various factors can provoke melanocytes to go into overdrive, and these different root causes are what distinguish the different types of brown spots. Melasma, freckles, and age spots are just different names for the different characteristics of skin hyper-pigmentation [17]. Common causes of skin hyper-pigmentation include: *a*) Sun exposure which is the main cause of chronic hyper-pigmentation due to naturally production of pigments; *b*) Biological effects and hormonal changes, e.g., pregnancy and aging, can alter the body and lead to skin dysfunction. Melasma (pregnancy mask) is a common development for women after pregnancy; *c*) Botched skin treatments: certain medications like birth control pills and oral antibiotics can stimulate discoloration of the skin. Sensitivity of skins to medications varies with aging, where the skin becomes more sensitive; and 4) Heredity: Dark skinned individuals simply have more melanocytes in the basal layer of the skin [5, 18]. Melanocyte cells are also very reactive to any stimulation source like sun exposure, sensitizing topical preparations, or skin injury.

Many underlying conditions are responsible for skin changes and pigmentation disorders, where the most recognized cause is sun. Numerous medications, especially hormones, have been implicated in skin discoloration disorders [19]. These drugs can sensitize the skin to the sun. Enhancement of melanocyte stimulating hormone by some medications like birth control pills or even pregnancy can be another cause. Finally, hormonal therapy by itself can have a direct impact on skin coloration. In these situations, the medication either stimulates more melanin to be produced, or suppresses the process that decreases melanin production. The result is increased discoloration regardless of the skin type or color [20]. When this occurs in patients that already have some natural increased color or freckles, the risk of skin discoloration increases that much more. Freckles (and some skin spots) are genetically programmed into the skin and many times are not even visible to the eye until some process makes them darken. Examples for skin discoloration are discussed below.

2.2.1. Melasma

An example of hyper-pigmentation is melasma (also known as chloasma). This condition is characterized by tan or brown patches, most commonly on the face (Figure

1A). Melasma can occur in pregnant women and is often called the "mask of pregnancy". Melasma frequently goes away after pregnancy. It is thought to affect 5-6 million women in the USA, 50 to 75% of pregnant women in the USA and 80% of pregnant women in Mexico [21]. It is considered a chronic disease that can sometimes appear before pregnancy, disappear a few months after giving birth, reappear in subsequent pregnancies or last for several years. Pigmentation usually affects the face and mainly the cheeks, forehead and upper lips. Melasma may become permanent if not treated early. The disease is generally treated with a combination of active ingredients for topical application; the most efficient being a triple combination drug, which must be used every day, while applying a broad-spectrum sunscreen several times a day [21].

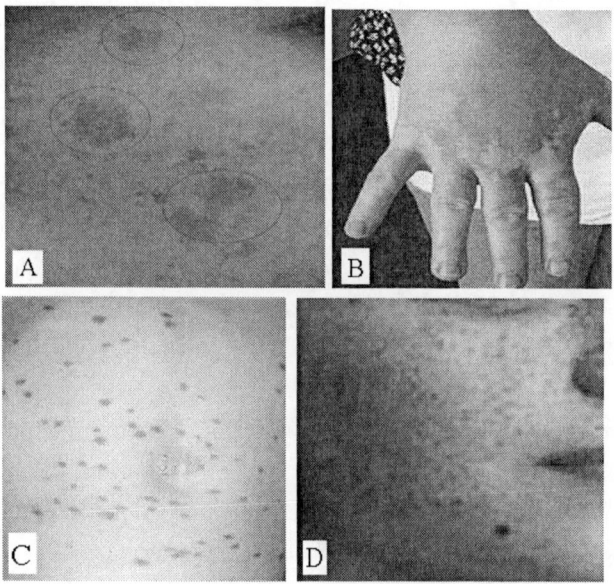

Figure 1. Examples for skin discoloration: (A), melasma; (B), vitiligo; (C), brown spots; and (D), freckles (Wikipedia).

2.2.2. Vitiligo

Vitiligo, an example of hypo-pigmentation, is a chronic skin disease characterized by portions of the skin losing their pigment (Figure 1B). It occurs when skin pigment cells die or are unable to function. It may appear anytime from shortly after birth to old age and affects 1.3% of the population [22, 23]. The onset can be precipitated by specific life events (physical injury, sunburn, emotional injury, illness, pregnancy). Both sexes are affected equally. Lesions are uniformly milky-white, round, oval or linear in shape, vary in size and appear anywhere on the body, but mainly in areas of repeated trauma, pressure or friction (elbows, knees, fingers and toes). Genetic, immunologic, environmental and stress factors are thought to be involved [22]. Vitiligo is a difficult disease to treat. When it is of limited extent, topical treatments can be used; but when a large surface of the body is affected, ultraviolet light (UV) therapy is recommended, alone or in combination with topical treatments [21].

2.2.3. Brown Spots

Brown spots on the face are essentially sun damage. It may appear as freckles in young ages, but they get larger and often darker as one got older (Figure 1C). The first thing to do about brown spots and age spots on the face or hands is to prevent more of them by applying sunscreen daily [20]. Bear in mind that continued sun exposure will stimulate the growth of brown spots, so that the bleaching creams will not protect against the regrowth of brown spots or the development of new ones.

2.2.4. Freckles

Freckles are small, brown flat dots (Figure 1D), caused by both genes and much more sun exposure. On the upside, these are pretty easy to treat because they're closer to the surface of the dermis. Freckles can be easily treated with lasers, peels, and some skin care products [19]. As a person ages, they generally fade away. "Sunburn freckles" are another variety often seen in some people, and these appear larger and darker than standard freckles, with a medium-brown color. The edges are irregular, but the pigment is even throughout the area.

2.2.5. Age Spots

Age spots (also known as liver spots) are small, flat pigmented spots which look similar to freckles. However, unlike freckles, age spots generally don't fade without treatment. Liver spots are more common in older people, looks like large freckles (Figure 2). These are most often seen on sun-exposed skin after age 40, usually on the face, shoulders, neck, ear, or back of the hands [20]. Like freckles, age spots are primarily caused by sun exposure, but poor nutrition and abnormal liver function are said to worsen the problem.

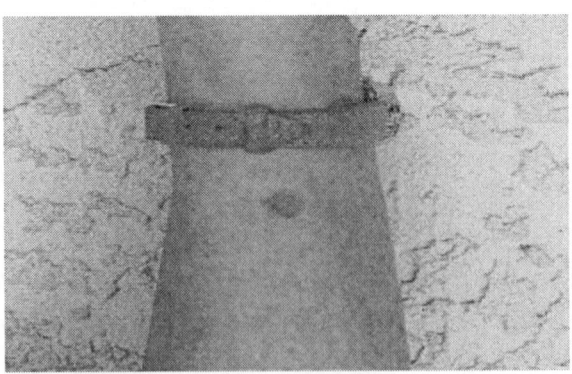

Figure 2. A 10 mm liver spot on the forearm of a 66-year old woman (Wikipedia). Liver spots form after years of exposure to the sun.

2.2.6. Albinism

Albinism is a congenital disorder characterized by the complete or partial absence of pigment in the skin, hair and eyes due to absence or defect of tyrosinase. Albinism is a genetic deficiency of melanin pigment production, i.e., Albinos have an abnormal gene that restricts the body from producing melanin. Albinism is a rare inherited disorder caused by the absence of the main enzyme that produces melanin; tyrosinase [24]. Production is rarely totally absent but perhaps 1-10% of normal, which results in a complete lack of pigmentation in skin, hair, or eyes. Approximately one in 17,000 people have one of the types of albinism [24]. There is no cure for albinism. People with albinism should use a sunscreen at all times because they are much more likely to get sun damage and skin cancer. This disorder is most common among whites. Symptoms of albinism include visual problems that are an important feature of albinism [25]. Melanin is reduced or absent where it is normally present in the eye, skin, hair and brain, and this causes maldevelopment of neural pathways related to vision [26]. Severe nystagmus, photophobia, strabismus, and reduced visual acuity are common features [26]. Albinos have red eyes but the color of the iris varies from a dull grey to blue and even brown. A brown iris is common in ethnic groups with darker pigmentation. Under certain lighting conditions, there is a reddish or violet hue reflected through the iris from the retina and the eyes appear red (similar to the red eye in flash photography).

3. TYROSINASE INHIBITORS

Tyrosinase is a bifunctional copper-containing enzyme that catalyzes both the hydroxylation of tyrosine into o-diphenols (monophenolase activity) and the oxidation of o-diphenols into o-quinones (diphenolase activity) [27]. This *o*-quinone can then be transformed into melanin pigments through a series of enzymatic and non-enzymatic reaction [28, 29, 30]. There are two types of melanin pigments that can be produced by the melanocytes (Figure 3), namely, "eumelanin" (black or brown) and "pheomelanin" (yellow or red). In melanosomes, tyrosinase catalyzes the transformation of substrates like tyrosine into activated melanin precursors such as indole-5,6-quinone (red spheres, Figure 3). The precursors are oriented by amyloid fibrils for polymerization into melanin [28]. In mammals, mixtures of both types are typically found [31].

Tyrosinase is known to be a key enzyme for melanin biosynthesis. Copper ions in the active site of the enzyme is essential for its inhibition and the ability of diminishing the human melanin synthesis. Tyrosinase inhibitors, therefore, can be clinically useful for the treatment of some dermatological disorders associated with melanin hyper-pigmentation [6, 32]. They also find uses in cosmetics for whitening and depigmentation after sunburn [27, 33]. An effective skin lightening treatment combines several of tyrosinase inhibitor ingredients as they tend to work well together to produce the desired results [34, 35]. The

U.S. Food, Drugs and Cosmetic Act, defines cosmetics by their intended use to be rubbed, poured, sprinkled, or sprayed on, introduced into, or otherwise applied to the human body for cleansing, beautifying, promoting attractiveness, or altering the appearance [36]. It is important to understand that cosmetics do not alter the structure or function of the skin [37]. Cosmetics can be divided into two broad groups: *Make-up* and *Skin care products*. Cosmeceuticals are topical cosmetic-pharmaceutical hybrids that enhance the beauty through constituents provide additional health-related benefit. They are applied topically as cosmetics, but contain ingredients that influence the skin's biological function [38]. Cosmeceuticals are, in fact, a bridge between personal-care products and pharmaceuticals and are commonly used for hyper- pigmentation [38]. Pigmentary disorders are the third most common dermatologic disorder and cause significant psychosocial impairment [39, 40]. These disorders are generally difficult to treat, hence, the need for skin lightening agents includes cosmeceuticals. These agents selectively target hyperplastic melanocytes and inhibit key regulatory steps in melanin synthesis. With the recent safety concern regarding use of hydroquinone, the need for alternative natural, safe and efficacious skin lightening agents is becoming all the more necessary. Cosmeceuticals include tyrosinase inhibitors such as phenolic compounds (hydroquanine and its derivatives) or natural skin-treatment agents. Tyrosinase inhibitors (Phenolic and Non-phenolic compounds) are discussed below.

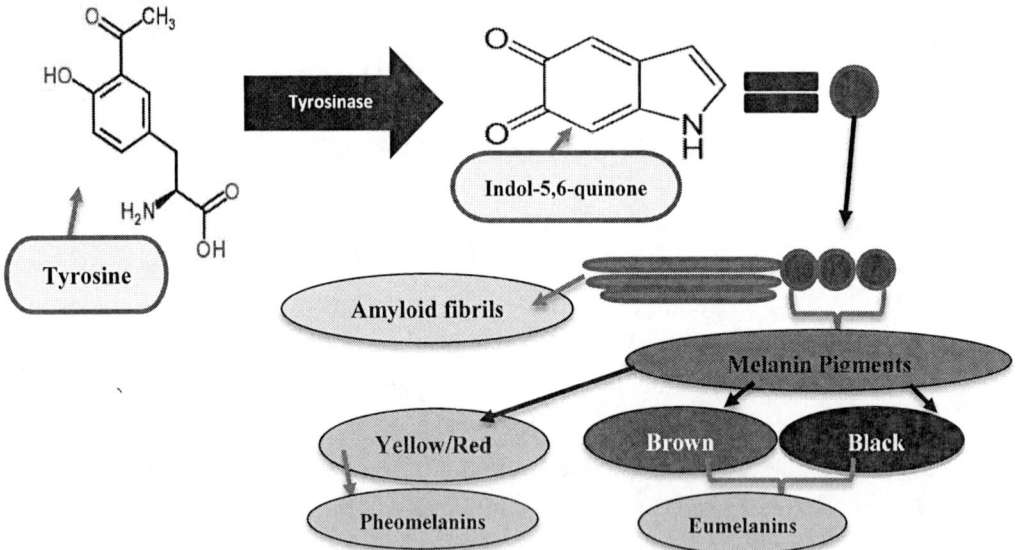

Figure 3. Production of different pigments by melanosomes. In melanosomes, tyrosinase catalyzes the transformation of substrates like tyrosine into activated melanin precursors such as indole-5,6-quinone (red spheres). The precursors are oriented by amyloid fibrils for polymerization into melanin [29, Zaidi et al. 2014].

3.1. Phenolic Compounds

3.1.1. Hydroquinone

Hydroquinone is a dihydric phenol with two important derivatives *viz.*, monobenzyl and monomethyl ether of hydroquinone. Hydroquinone competitively inhibits melanin synthesis by inhibiting sulfhydryl groups and acting as a substrate for tyrosinase [41]. Melanosomes and melanocytes are damaged by the semi-quinone free radicals released during the above reaction [41, 42]. Hydroquinone is considered the gold standard for the treatment of hyper-pigmentation. It is commonly used at concentrations of 2-4%. Clinical studies report that well to excellent responses induced by 2% hydroquinone. Higher concentrations are effective but can cause irritation; hydroquinone can cause DNA damage. This carcinogenic effect has raised concerns regarding its use. Hence, the International Agency for Research on Cancer has placed hydroquinone as not classifiable as to its carcinogenicity in humans [40, 43]. Due to the side-effect and safety profile, hydroquinone is not used as a component of cosmeceuticals available in the market for the treatment of hyper-pigmentation [44].

3.1.2. Mequinol

Mequinol (4-hydroxyanisole, hydroquinone monomethyl ether), is a derivative of hydroquinone. Its mechanism of action is unclear. It acts as a substrate for tyrosinase, thereby inhibiting the formation of melanin precursors [45]. Combination of mequinol at a concentration of 2% with 0.01% tretinoin can cause erythema, burning, pruritus, desquamation, skin irritation, halo hypo-pigmentation [46]. Combination with sunscreens can reduce the incidence of adverse effects [47].

3.1.3. N-Acetyl-4-S-Cysteaminylphenol

N-acetyl-4-S-cysteaminylphenol is a phenolic agent that inhibits tyrosinase activity by acting as an alternative substrate. It is more stable and causes less irritation than hydroquinone. Clinical response is evident after 2-4 weeks. Various studies using 4% of N-acetyl-4-S-cysteaminylphenol have found marked improvement in patients with melisma [41].

3.2. Non-Phenolic Compounds

3.2.1. Kojic Acid

Kojic acid is derived from Koji (a Japanese mushroom). Kojic acid (5-hydroxy-2 hydroxymethyl-4-pyrone) lightens the skin by inhibiting tyrosinase which in turn reduces the amount of melanin produced. Kojic acid is one of the most popular natural ingredients found in skin lightening products. However, Kojic acid has the potential to induce skin sensitization according to the Scientific Committee on Consumer Products and Safety (one of the independent scientific committees managed by the Directorate-General for Health and Consumer Protection of the European Commission, European Union) [48].

3.2.2. Mulberry

Morus, a genus of flowering plants in the family *Moraceae*, comprises 10–16 species of deciduoustrees commonly known as mulberries. The closely related genus *Broussonetia* is also commonly known as mulberry, notably the Paper Mulberry, *Broussonetia papyrifera* [49]. This skin lightening ingredient is extracted from the roots of the paper mulberry plant. Mulberry is said to be more effective than hydroquinone and kojic acid in the sense, that significantly lower concentrations are needed to have the same effect as higher concentrations of kojic acid and hydroquinone.

3.2.3. Arbutin

Arbutin is a glycosylated hydroquinone extracted from the bearberry plant in the genus *Arctostaphylos*. It inhibits tyrosinase and thus prevents the formation of melanin. Arbutin is therefore used as a skin-lightening agent. Arbutin is found in wheat, and is concentrated in pear skins. It is also found in *Bergenia crassifolia* [50]. The *Alpha-arbutin* has a stronger effect than *Beta-arbutin* and is also commonly found in skin lightening products as a safer alternative to hydroquinone [51]. *Beta* arbutin (Bearberry Extract) is derived from the leaves of bearberry, cranberry and blueberry plants. It works in a similar way to kojic acid, in that it inhibits the production of tyrosinase to restrict the amount of pigment produced. Although naturally derived, it can cause skin irritation in some people with sensitive skin.

3.2.4. Glutathione

Glutathione is a powerful antioxidant that has several health benefits including boosting the immune system and cleansing the liver. It is naturally occurring in the body and is a part of the natural defense system. Glutathione is supporting skin whitening by the direction of melanin production towards the light and easy soluble pheomelanin (Melanin, the end-product of complex multistep transformations of l-tyrosine, is polymorphous and multifunctional biopolymer, represented by eumelanin, pheomelanin, neuromelanin, and mixed melanin pigment) [52]. It is commonly found in skin lightening pills but can also be used topically.

3.2.5. Licorice Root

Licorice, or Liquorice is the root of *Glycyrrhiza glabra* from which a sweet flavor can be extracted [53]. The licorice plant is a legume native to southern Europe, India, and parts of Asia. It is not botanically related to anise, star anise, or fennel, which are sources of similar flavoring compounds. Licorice is commonly used in the skin lightening industry. Licorice also works to inhibit the enzyme tyrosinase to limit the amount of pigment produced. It has additionally anti-inflammatory properties and is particularly effective at fading sun tan.

3.2.6. Papaya

Papaya is a tropical fruit shaped like an elongated melon, with edible orange flesh and small black seeds. The papain enzyme found in papaya works to gently exfoliate dead skin cells and reveal new, brighter skin cells beneath. Orange and green papaya are both effective but green papaya contains more of the papain enzyme [54]. This ingredient is most often found in soaps and sometimes in skin lightening creams too.

3.2.7. Vitamin A

Vitamin A is a group of unsaturated nutritional organic compounds, that include retinol, retinal, retinoic acid, and several provitamin A carotenoids, among which beta-carotene is the most important [55]. Vitamin A, also known as retinol, increases the rate at which skin cells are renewed.

3.2.8. Vitamin B3

Vitamin B3, also known as nicotinamide, has antioxidant and anti-inflammatory properties. A study has done on pigmented reconstructed epidermis showed that niacinamide interferes with the interaction between keratinocytes and melanocytes, thereby inhibiting melanogenesis [56]. Niacinamide is effective topical skin lightener that works by disrupting the transfer of melanosome from the melanocyte to the keratinocyte. It also modulates the protease-activated receptor (PAR-2) that is involved in the transfer of melanosomes from melanocytes to surrounding keratinocytes. Clinical trials using 2% niacinamide have shown that it significantly reduces the total area of hyper-pigmentation and increases skin lightness after 4 weeks of treatment. Daily use of niacinamide with sunscreen was effective in reducing hyper-pigmentation and in increasing lightness of basal skin color compared with sunscreen alone [56]. Vitamin B3 is found in some skin whitening products and sunscreens.

3.2.9. Vitamin C

Vitamin C or L-ascorbic acid, or simply ascorbate (the anion of ascorbic acid), is an essential nutrient for humans and certain other animal species. Vitamin C refers to a number of vitamers that have vitamin C activity in animals, including ascorbic acid and its salts, and some oxidized forms of the molecule like dehydroascorbic acid [57]. Vitamin C suppresses the production of pigment in the skin. Magnesium Ascorbyl Phosphate is a derivative of Vitamin C and this is the form of the ingredient that is commonly found in skin whitening products. Vitamin C can also protect the skin from ultraviolet rays.

3.3. Mechanisms of Melanogenesis Inhibition

Controlling pigmentation via the regulation of tyrosinase activity is an important approach. The enzyme could be regulated by: a) transcription of its messenger RNA; b) its maturation via glycosylation; c) its trafficking to melanosomes; d) modulation of its catalytic activity and/or stability. [58, 59]. Mechanisms of melanogenesis inhibition are discussed below.

3.3.1. Inhibition of Tyrosinase mRNA Transcription

Melanin synthesis is directly regulated by the enzymatic function of tyrosinase and thus by transcription of its encoding gene [59]. Microphthalmia associated transcription plays a role in the development, survival, and function of certain cell types. To carry out this role, the protein attaches to specific areas of DNA and helps control the activity of particular genes [60]. Microphthalmia-associated transcription factor helps for controlling the development and function of melanocytes. Within these cells, this protein also controls production of the pigment melanin, which contributes to hair, eye, and skin color. This protein is the master regulator of melanogenesis-related gene expression [61, 62].

3.3.2. Aberrant Tyrosinase Maturation

Aberration of tyrosinase glycosylation in the endoplasmic reticulum or Golgi inhibits its folding and maturation [63, 64].

3.3.3. Inhibition of Tyrosinase Catalytic Activity

Numerous reports have described the inhibition of tyrosinase activity, many of them used mushroom tyrosinase as a model. Tyrosinase is a multi-copper enzyme which is widely distributed in different organisms and plays an important role in the melanogenesis and enzymatic browning. Therefore, its inhibitors can be attractive in cosmetics and medicinal industries as depigmentation agents and also in food and agriculture industries as anti-browning compounds [65, 66]. Inhibitors of tyrosinase function can be divided into two groups, competitive inhibitors and non-competitive inhibitors. Early studies revealed that hydroquinone and azelaic acid are competitive inhibitors of tyrosinase activity as well having cytotoxic effects on melanocytes [67], both being potent therapies for hyper-pigmentary disorders [68]. Tyrosinase activity depends on the binding and function of two copper atoms at the active site which is facilitated by a copper transporter [69]. Therefore, chelating copper inhibits the activity of purified or recombinant tyrosinase. Another non-competitive method of inhibiting tyrosinase activity is the reduction of its phosphorylation. Therefore, inhibiting tyrosinase phosphorylation reduces tyrosinase activity in cultured melanocytes as well as lightening pigmentation in the skin and hair.

3.3.4. Acceleration of Tyrosinase Degradation

The synthesis and the degradation of tyrosinase are tightly coupled to its function, and are influential parameters that regulate melanin synthesis. Degradation of tyrosinase could be altered by environmental factors surrounding melanocytic cells. Studies of tyrosinase degradation have revealed that a variety of intrinsic factors in the epidermis have a potency to regulate tyrosinase degradation [59]. Linoleic acid is an unsaturated fatty acid, and is a major component of biological cell membranes. Topical application of linoleic acid has been shown to decrease UV-induced hyper-pigmentation of the skin. Fatty acids can regulate tyrosinase degradation in contrasting manners [59].

4. MICROBIAL TYROSINASES

Tyrosinase is essential for many living organisms to carry out various functions, including melanin biosynthesis as a defense against the harmful effects of UV light. In plants, it is required for the biosynthesis of phenolic polymers such as lignin, flavinoids, and tannins [70]. Tyrosinases also play an important role in regulation of the oxidation-reduction potential of cell respiration and in wound healing in plants [71]. Due to the ability of tyrosinases to react with phenols, these enzymes have been proposed for uses in a variety of biotechnological, biosensor and biocatalysis applications [72, 73]. For example, tyrosinases can be applied in: *a*) detoxification of phenol-containing waste-water and contaminant soils [74]; *b*) synthesis of L-3,4-dihydroxyphenylalanin which is known as one of the preferred drugs for the treatment of Parkinsons disease [75, 76, 29]; or *c*) as additives in food processes due to their cross-linking abilities [77]. Commercial production of bacterial tyrosinases is extensively reported. Production of tyrosinase from the common mushroom *Agaricus bisporus* is widely reported because of its commercial availability. However, the use of tyrosinase from this source is problematic as the enzyme exhibits relatively low solvent and temperature stability, as compared to some bacterial tyrosinases [78]. Almost all reported tyrosinases produced by bacterial and fungal strains, yield also other polyphenol oxidases such as peroxidase and laccase. Presence of polyphenol oxidases along with tyrosinase imposes serious problems for commercial usage. All these enzymes can use tyrosine as substrate but produce different products, resulting in reduced yield as well as increased cost for the downstream process. Effects of various key cultivation parameters for large-scale production of microbial tyrosinase are reported recently [79, 80, 81].

4.1. Sources of Tyrosinase

Tyrosinase activities are widely distributed in all domains of life from microorganisms to mammals. Tyrosinases have been purified, and their properties and functions have been extensively studied. They are found in whole cells, tissues from mushrooms, fruits, and vegetables and are mainly involved in the biosynthesis of melanin and other polyphenolic compounds [29].

4.1.1. Fungus as a Source of Tyrosinase

Different fungal strains of *Agaricus bisporus* have been investigated for isolation of tyrosinase [82, 83]. Other strains like *Neurospora crassa*, *Amanita muscaria*, *Lentinula edodes*, *Aspergillus oryzae*, Portabella mushrooms, *Pycnoporus sanguineus*, and *Lentinula boryana* are also studied for tyrosinase production [29].

4.1.2. Bacteria as a Source of Tyrosinase

Streptomyces tyrosinases are the most thoroughly characterized enzymes of bacterial origin [79, 84]. Bacterial tyrosinase is involved in the melanin production and is normally extracellular in origin. The enzyme has been reported in other species such as *Rhizobium, Symbiobacter iumthermophilum, Pseudomonas maltophilia, Sinorhizobium meliloti, Marinomonas mediterranea, Thermomicrobium roseum, Bacillus thuringiensis, Pseudomonas putida* [85], *Streptomyces castaneoglobisporus, Ralstonia solanacearum*, and *Verrucomicrobium spinosum* [79]. Tyrosinases are produced by marine bacteria as well [30, 33].

4.2. Pharmaceutical and Industrial Applications

Microbial tyrosinases are extensively utilized as promising enzymes for pharmaceutical, food bioprocessing, and environmental industry [29, 86]. Tyrosinases are used as a potential prodrug for treating melanoma where patients were successfully treated via tyrosinase activity [87, 88]. Methods for the recombinant production of bacterial tyrosinase, which utilized for the formation of biomaterials such as cross-linked proteins and melanin, are reported [70, 89]. Application of tyrosinase in cell culture has implicated the wide acceptance of its catalytic property, as it helps nerve cells to grow. The process includes stamping of the tyrosinase onto plastic surfaces, which causes *in situ* formation of thin films of melanin. It may be useful in prevention of bacterial contamination, as melanin has a bacteriostatic effect [29].

Tyrosinases are fundamental enzymes involved in several biological functions and defense mechanism (especially in melanogenesis). Control of tyrosinase activity is very important during melanoma. Un-controlling of tyrosinase results in increased melanin synthesis, causing hyper-pigmentation. Several polyphenols, including flavonoids, substrate analogues, free radical scavengers, and copper chelators have been known to inhibit tyrosinase [90]. Henceforth, the medical and cosmetic industries are focusing research on tyrosinase inhibitors to treat skin disorders [91]. Abnormal melanin can result from the changes anywhere in the pathway of synthesis either due to abnormal tyrosinase or due to the defects caused by the transfer of melanosomes to keratinocytes [92, 93]. Gelatin/chitosan blend was used as protein-polysaccharide matrix. Microbial transglutaminase and tyrosinase were employed as biological modifiers, where thermal stability, the stability in aqueous medium, and tensile strength of the modified material have been improved due to the combined effects of two enzymes that microbial transglutaminase can cross-link gelatin and tyrosinase can catalyze gelatin-chitosan conjugation [86, 94].

5. TREATMENT STRATEGIES FOR PIGMENTATION DISORDERS

Learning more about skin diseases and pigmentary disorders may increase the chance of an effective treatment. Although treatment of these disorders remains difficult, there is an increasing number of therapeutic options that could lead to an improvement in the cosmetic appearance of skin for patients. Most of the existing literature on safety and efficacy of these treatments consists of case reports and cohort studies; there is a lack of larger randomized controlled trials [95]. However, safe and predictable treatment options for skin discoloration are ranging from the simplest option sun avoidance to advanced techniques such as laser treatment are reported [95]. Numerous types of laser treatments are available to help for treating a variety of skin conditions. Laser treatment, or phototherapy, is most commonly used in patients with skin disorders, irritation, or other conditions related to an immunologic deficiency that increases the growth rate of skin cells. As a result, patients experience symptoms such as skin redness and irritation due to thick scaly patches that cover parts of their body [96, 97].

In response to years of sun exposure, skin tries to protect itself by producing an overabundance of melanin, the pigmented cells in the skin [98]. This hyper-pigmentation is most noticeable on face and hands, but it is also common on any other skin that is exposed to the sun. It can be triggered by inflammation, such as acne, or hormonal fluctuation, but most often by unprotected sun exposure. Although pigmentation disorders can affect all skin types, individuals with darker skin, including Asians, blacks, Latinos, and American Indians, are more susceptible [95].

Properties of effective natural ingredients for lightening age spots include the ability to: *a*) inhibit or block tyrosinase caused by sun exposure, plus a great powerful anti-inflammatory properties; *b*) help diminish skin discolorations, brighten skin, and improve elasticity as well as deliver powerful antioxidant protection; *c*) protect the skin against damage caused by free radicals; and *d*) treat allergic inflammation of the skin [98]. Vitamin C has been tested extensively and proven to inhibit the production of melanin. Skin lightening ingredients act as direct inhibitors of tyrosinase to reduce melanin production [99]. In addition to reducing production of melanin in the skin, some of the natural products contain also titanium dioxide, a mineral whitening pigment that helps lighten skin by absorbing the UV rays, thus preventing the sun from darkening the skin. Vitamin C is required for collagen synthesis, which declines in aging skin. Therefore, topical application of vitamin C in a skin-penetrating cream or serum can improve collagen production. Skin lightening products often perform best when used in combination with exfoliation of the outer, pigmented layer of dead skin cells. The skin renews itself every 28 days, gradually giving a lighter, more even skin tone in four to eight weeks [98]. In case of discontinue use of skin lighteners, the skin may return to its original color after a few weeks. Standardization of skin-depigmenting agents in either their efficacy or their safety is essential for skin health [100]. While it is clear that great progress has been made in the study of skin lightening, it is even more apparent that there is a great deal of work still left to be done [100].

6. SAFETY OF COSMETIC AGENTS

Mass-market cosmetic companies were touting for skin bleaching products with hydroquinone as a quick fix for hyper-pigmentation [98]. With hydroquinone, fast and effective results to whiten skin were obtained. However, drawback of the fast effect was the high cytotoxicity of hydroquinone and its huge irritation potential. The Food and Drug Administration reported that hydroquinone poses too many health risks, including irritation, toxicity, and allergic reactions [98]. Therefore, natural product companies are formulating creams and serums that contain mild natural lighteners [9, 101]. New active ingredients with a much-improved toxicology profile have been developed instead. They often provide the whitening effect in a reversible way, demonstrating that the biology of melanin production will only be suppressed during treatment but not destroyed [102]. Various processes in the complex mechanism of skin pigmentation include direct inhibition of tyrosinase enzymes, regulation of melanocyte homeostasis, alteration of facultative pigmentation, and down-regulation of melanosomes [103]. These processes can be regulated to alter complexion coloration and thus relieve skin complexion diseases [98]. Great advances have been made in understanding the cellular and biochemical mechanisms in pigment biology and the processes underlying skin pigmentation [104, 105]. This has led to the development of various skin lightening agents to reduce skin hyper-pigmentation. Natural extracts from plants demonstrate that skin whitening nowadays can be achieved by safe ingredients instead of toxicological drawbacks with extra benefits [9]. Topical agents in conjunction with or followed by procedural approaches such as light therapy and lasers can be employed and have shown promising results in treating skin discoloration.

CONCLUSION

Skin pigmentation disorders or abnormalities, and skin whitening have been extensively examined during the last decades. A number of different pathways were evaluated to achieve a reduction in skin color formation. Skin pigmentary abnormalities, seen as aesthetically unfavorable, have led to the development of cosmetic and therapeutic treatment modalities of varying efficacy. Tyrosinase catalyzes melanin biosynthesis in human skin and the epidermal hyper-pigmentation results in various dermatological disorders, such as melasma, freckles and age spots. Therefore, safe and effective tyrosinase inhibitors have become important for their potential applications in preventing pigmentation disorders and other melanin-related health problems in human skin. Pigmentation disorders, direct inhibition of tyrosinase, and the regulation of processes that control skin complexion coloration are extensively reported. Internet consumers became much better informed and aware of potential risks to avoid side effects of skin pigmentation disorders.

REFERENCES

[1] Sariri R, Seifzadeh S. and Sajedi R. 2009. Anti-tyrosinase and antioxidant activity of *Lavandula* sp. Extracts. *Pharmacology online*, 3: 319-326.

[2] Choi S, Lee Y, Kim S, et al. 2014. Inhibitory effect of corn silk on skin pigmentation. *Molecules*, 19: 2808-2818.

[3] Ollagnier M, Moran B. and Boo M. 2011. *Formulating towards fairer skin.* www.lubrizol.com/personalcare.

[4] Patel A. 2014. Post inflammatory hyper-pigmentation: Review of pathogenesis, prevention, and treatment. *Pigment Int.*, 1: 59-69.

[5] Yadav M, Ghonasgi S, Shah R, et al. 2012. Familial Progressive Hyperpigmentation: A Case Report. *Case Rep Dentis.*, doi.org/10.1155/2012/840167.

[6] Chen W, Tseng T, Hsiao N, et al. 2015. Discovery of highly potent tyrosinase inhibitor, T1, with significant anti-melanogenesis ability by zebrafish *in vivo* assay and computational molecular modeling. *Sci Reports*, 5: 7995-7802.

[7] Vivek K, Park Y, MinKyun N. et al. 2015. α-Glucosidase and tyrosinase inhibitory effects of an abietane type diterpenoid taxoquinone from Metasequoia glyptostroboides. *BMC Complem Altern Med.*, 15: 84-89.

[8]   Prakot P, Ninnaj C. and Aphichart K. 2015. *In vitro* anti-tyrosinase activity of protein hydrolysate from spotted Babylon (Babylonia areolata). *Food Appl Biosci J.*, 3: 109–120.

[9]   Hollinger J, Angra K, and Halder R. 2018. Are natural ingredients effective in the management of hyperpigmentation? A systematic review. *J Clin Aesthet Dermatol.*, 11: 28–37.

[10]   Thanigaimalai P, Namasivayam V, Manickam M, et al. 2018. Inhibitors of melanogenesis: An updated review. *J Med Chem.*, 17: 7395-7418.

[11]   Thanigaimalai P, Manoj M. and Vigneshwaran N. 2017. Skin whitening agents: medicinal chemistry perspective of tyrosinase inhibitors. *J Enz Inhib Med Chem.*, 32: 403-425.

[12]   King R, Hearing V, Creel D, Oetting W. Albinism. 1995. In: *The Metabolic and Molecular Bases of Inherited Disease,* edited by Scriver CR, Beaudet AL, Sly WS, Valle D. New York: McGraw-Hill.

[13]   Gupta A, Gover M, Nouri K, et al. 2006. The treatment of melasma: A review of clinical trials. *J Amr Acad Dermatol.*, 55:1048-1065.

[14]   Sapkota K, Roh E, Lee E, et al. 2011. Synthesis and anti-melanogenic activity of hydroxyphenyl benzyl ether analogues. *Bioorg Med Chem.*, 19: 2168-2175.

[15]   Briganti S, Camera E. and Picardo M. 2003. Chemical and instrumental approaches to treat hyperpigmentation. *Pigment Cell Res.*, 16:101-110.

[16]   Lee D, Cha B, Lee Y, et al. 2015. The potential of minor ginsenosides isolated from the leaves of *Panax ginseng* as inhibitors of melanogenesis. *Int J Mol Sci.*, 16: 1677-1690.

[17]   Tunzi M, Gray G. 2007. Common skin conditions during pregnancy. *Amr Family Physician*, 75: 211–218.

[18]   Tang H. and Chen Y. 2015. Identification of tyrosinase inhibitors from traditional Chinese medicines for the management of hyperpigmentation. *Springer Plus*, 4: 184-201.

[19]   Kumar V, Abbas A, Fausto N, et al. 2005. *Robbins and Cotran pathologic basis of disease.* Philadelphia: Elsevier Saunders.

[20]   James W, Timothy B. and Dirk E. 2006. *Andrews' Diseases of the Skin: Clinical Dermatology.* Saunders Elsevier. ISBN 0-7216-2921-0.

[21]   American Academy of Dermatology. *Pigmentation disorders.* www.galderma.com.

[22]   van Geel N, Mollet I, Brochez L, et al. 2012. New insights in segmental vitiligo: case report and review of theories. *British J Dermatol.*, 166: 240–246.

[23]   Chao N, Haji A. 2018. Upregulation of melanogenesis and tyrosinase activity: Potential agents for vitiligo. *Molecules*, 22: 1303-1329.

[24]   Hong E, Zeeb H. and Repacholi M. 2006. Albinism in Africa as a public health issue. *BMC Public Health*, 6: 212-218.

[25] Witkop CJ. 1979. Albinism: hematologic-storage disease, susceptibility to skin cancer, and optic neuronal defects shared in all types of oculocutaneous and ocular albinism. *Ala J Med Sci.*, 16: 327–330.

[26] McAllister J, Dubis A, Tait D, et al. 2010. Arrested development: high-resolution imaging of foveal morphology in albinism. *Vision Res.*, 50: 810-817.

[27] Gheibi N, Taherkhani N, Ahmadi A, et al. 2015. Characterization of inhibitory effects of the potential therapeutic inhibitors, benzoic acid and pyridine derivatives, on the monophenolase and diphenolase activities of tyrosinase. *Iran J Basic Med Sci.*, 18: 122-129.

[28] Robb DA. 1984. Tyrosinase. In: *Copper Proteins and Copper Enzymes*. R. Lontie, Ed., pp. 207–241, CRC Press, Boca Raton, Fla, USA.

[29] Zaidi K, Ayesha S, Sharique A, et al. 2014. Microbial tyrosinases: promising enzymes for pharmaceutical, food bioprocessing, and environmental industry. *Biochem. Res Intr.*, vol. 2014: 1-16.

[30] Neethu K, Kumar S. and Bhaskara K. 2015. Antioxidant and haemolytic activity of tyrosinase producing marine actinobacteria from salterns. *Der Pharmacia Lett.*, 7: 172-178.

[31] Kang P, Chung H, Cho C, Hong M, Shin K. and Bae H. 2006. Survey and mechanism of skin depigmenting and lightening agents. *Phytother Res.*, 20: 921–934.

[32] Sabudak, T., Khan, M, Choudhary M, et al. 2006. Potent tyrosinase inhibitors from Trifoliumbalansae. *Nat Prod Res.*, 20: 665-670.

[33] Bin W. 2014. Tyrosinase inhibitors from terrestrial and marine resources. *Cur Top Med Chem.*, 14: 1425-1449.

[34] Lueder M. 2013. *How to obtain skin whitening in a safe and effective way?* Qenax AG, Switzerland.

[35] Lee S, Baek N. and Nam T. 2015. Natural, semisynthetic and synthetic tyrosinase inhibitors. *J Enz Inhib Med Chem.*,16: 1-13.

[36] Dureja H, Kaushik D, Gupta M, et al. 2005. Cosmeceuticals: An emerging concept. *Ind J Pharmacol.*, 37: 155–159.

[37] Sinclair R. 1999. How credible is the science behind cosmetic skin creams? *West J Med.*, 171: 35–36.

[38] Grace R. 2002. Cosmeceuticals: Functional food for the skin. *Natural Foods Merch.*, 23: 92–9.

[39] Halder R, Nootheti P. 2003. Ethnic skin disorders overview. *J Amr Acad Dermatol.*, 48: 143–148.

[40] Sarkar R, Pooja A. and Garg K. 2013. Cosmeceuticals for Hyperpigmentation: What is Available? *J Cutan Aesth Sur.*, 6: 4-11.

[41] Findlay G. 1982. Ochronosis following skin bleaching with hydroquinone. *J Amr Acad Dermatol.*, 6: 1092–1093.

[42]   Jimbow K. 1991. N-acetyl-4-S-cysteaminylphenol as a new type of depigmenting agent for the melanoderma of patients with melasma. *Arch Dermatol.*, 127: 1528–1534.

[43]   Nordlund J, Grimes P. and Ortonne J. 2006. The safety of hydroquinone. *J Eur Acad Dermatol Venereol.*, 20: 781–287.

[44]   Tse T. 2010. Hydroquinone for skin lightening: Safety profile, duration of use and when should we stop? *J Dermatolog Treat.*, 21: 272–275.

[45]   Draelos ZD. 2007. Cosmetic therapy. In: Wolverton SE, editor. *Comprehensive Dermatologic Drug Therapy.* 2nd ed. Philadelphia: Saunders, pp. 761–74.

[46]   Jarratt M. 2004. Mequinol 2% tretinoin 0.01% solution: An effective and safe alternative to hydroquinone 3% in the treatment of solar lentigines. *Cutis.*, 74: 319–322.

[47]   Colby S, Schwartzel E, Huber F, et al. 2003. A promising new treatment for solar lentigines. *J Drugs Dermatol.*, 2: 147–52.

[48.]   Scientific Committee on Consumer Products SCCP/1182/08, *Opinion on Kojic acid.* http://ec.europa.eu/health/ph_risk/risk_en.htm.

[49]   Suttie J. 2012. *Morus alba. Plant Production and Protection.* Food and Agricultural Organization of the United Nations.

[50]   Pop C, Vlase L. and Tamas M. (2009). Natural Resources Containing Arbutin. Determination of Arbutin in the Leaves of *Bergenia crassifolia* (L.) Fritsch, acclimated in Romania. *Not Bot Hort Agrobot Cluj.*, 37: 129–132.

[51]   Lee H, Kim K. 2012. Anti-inflammatory effects of arbutin in lipopolysaccha-ride-stimulated BV2 microglial cells. *Inflamm Res.*, 61: 817-25.

[52]   Slominski A, Tobin D, Shibahara S, et al. 2004. Melanin pigmentation in mammalian skin and its hormonal regulation. *Physiol Rev.*, 84: 1155–1228.

[53]   *Merriam-Webster's Medical Dictionary,* 2007. Licorice. Merriam-Webster, Inc.

[54]   Natty N. 2013. *Green Papaya Salad Recipe.* ThaiTable.com.

[55]   Wolf G. (2001). The discovery of the visual function of vitamin A. *J Nutr.*, 131: 1647–1650.

[56]   Hakozaki T, Minwalla L, Zhuang J, et al. 2002. The effect of niacinamide on reducing cutaneous pigmentation and suppression of melanosome transfer. *Brit J Dermatol.* 147: 20–31.

[57]   Padayatty S, Katz A, Wang Y, Eck P, Kwon O, Lee J, Chen S, Corpe C, Dutta A, Dutta S, Levine M. 2003. Vitamin C as an antioxidant: evaluation of its role in disease prevention. *J Amr Coll Nutr.*, 22: 18–35.

[58]   Maeda K, Fukuda M. 1991. *In vitro* effectiveness of several whitening cosmetic components in human melanocytes. *J Soc Cosm Chem.*, 42: 361-368.

[59]   Ando H, Kondoh H, Ichihashi M, et al. 2007. Approaches to identify inhibitors of melanin biosynthesis via the quality control of tyrosinase. *J Invest Dermatol.*, 127: 751-761.

[60] Nurul I, Farediah A. and Taher M. 2015. *In vitro* antioxidant, cholinesterase and tyrosinase inhibitory activities of *Calophyllum symingtonianum* and *Calophyllum depressinervosum* (Guttiferae). *J Coastal Life Med.*, 3: 126-131.

[61] Tachibana M, Takeda K, Yoshitaka N, et al. 1996. Ectopic expression of *MITF*, a gene for Waardenburg syndrome type 2, converts fibroblasts to cells with melanocyte characteristics. *Nature Genetics*, 14: 50 –54.

[62] Kim K. 2015. Effect of ginseng and ginsenosides on melanogenesis and their mechanism of action. *J Ginseng Res.*, 39: 1-6.

[63] Muller G, Ruppert S, Schmid E, et al. 1988. Functional analysis of alternatively spliced tyrosinase gene transcripts. *EMBO J.*, 7: 2723–2730.

[64] Ujvari A, Aron R, Eisenhaure T, et al. 2001. Translation rate of human tyrosinase determines its N-linked glycosylation level. *J Biol Chem.*, 276: 5924–5931.

[65] Franco D, Senra G, Gonçalves C, et al. 2012. Inhibitory effects of resveratrol analogs on mushroom tyrosinase activity. *Molecules*, 17: 11816-11825.

[66] Samaneh Z, Asieh B, Tareq M, et al. 2019. A comprehensive review on tyrosinase inhibitors. *J Enz Inhibit Med Chem.*, 34: 279-309.

[67] Palumbo A, d'ischia M, Misuraca G. et al. 1991. Mechanism of inhibition of melanogenesis by hydroquinone. *Biochim Biophys Acta*, 1073: 85–90.

[68] Grimes PE. 1995. Melasma: Etiologic and therapeutic considerations. *Arch Dermatol.*, 131: 1453–1457.

[69] Petris M, Strausak D. and Mercer J. 2000. The Menkes copper transporter is required for the activation of tyrosinase. *Hum Mol Gen.*, 9: 2845–2851.

[70] Ren Q, Henes B, Michael F, et al. 2013. High level production of tyrosinase in recombinant *Escherichia coli*. *BMC Biotechnol.*, 13: 18- 28.

[71] Mayer AM. 2006. Polyphenol oxidases in plants and fungi: Going places? A review. *Phytochemistry*, 67: 2318–2331.

[72] Jus S, Kokol V. and Guebitz G. 2008. Tyrosinase-catalysed coupling of functional molecules onto protein fibres. *Enz Microb Technol.*, 42: 535–542.

[73] Fairhead M, Thöny-Meyer L. 2012. Bacterial tyrosinases: old enzymes with new relevance to biotechnology. *New Biotechnol.*, 29: 183–191.

[74] Agarwal P., Gupta R. and Agarwal N. 2016. A review on enzymatic treatment of phenols in wastewater. *J Biotechnol Biomater.*, 6: 249-254.

[75] Seetharam G, Saville B. 2002. L-DOPA production from tyrosinase immobilized on zeolite. *Enz Microb Technol.*, 31: 747–753.

[76] Shrirang I, Swati J, Vishwas B, et al. 2014. Purification and Characterization of RNA Allied Extracellular Tyrosinase from *Aspergillus* Species. *Appl Biochem Biotechnol.*, 172: 1183-1193.

[77] Selinheimo E, Autio K, Krijus K, et al. 2007. Elucidating the mechanism of laccase and tyrosinase in wheat bread making. *J Agr Food Chem.*, 55: 6357–6365.

[78] Suki R, Ishita D, Minki M, et al. 2014. Isolation and characterization of tyrosinase produced by marine actinobacteria and its application in the removal of phenol from aqueous environment. *Front Biol.*, 9: 306-316.

[79] Gare S, Kulkarni S. 2015. Isolation and characterization of tyrosinase producing *Streptomyces luteogriseus. World J Pharm Res.*, 4: 1385-1395.

[80] Jing G, ZhimingR, Yang T, et al. 2015. Cloning and identification of a novel tyrosinase and its overexpression in *Streptomyces kathirae* SC-1 for enhancing melanin production. *FEMS Microbiol. Lett.*, doi.org/10.1093/femsle/fnv041.

[81] Park K, Kwon K, and Lee S. 2015. Evaluation of the antioxidant activities and tyrosinase inhibitory property from mycelium culture extracts. In: *Evidence-Based Complementary Alternative Medicine.* pp., 1-7: doi.org/10.1155/2015/616298.

[82] Khan A, Akhtar S. and Husain Q. 2005. Simultaneous purification and immobilization of mushroom tyrosinase on an immune-affinity support. *Proc Biochem.*, 40: 2379–2386.

[83] Ismaya W, Rozeboom H, Weijn A, et al. 2011. Crystal structure of *Agaricus bisporus* mushroom tyrosinase: Identity of the tetramer subunits and interaction with tropolone. *Biochem.*, 50: 5477–5486.

[84] Popa C, Bahrim G. 2011. Streptomyces tyrosinase: production and applications. *Innov. Romanian Food Biotechnol.*, 8: 1-7.

[85] Claus H, Decker H. 2006. Bacterial tyrosinases. *Syst Appl Microbiol.*, 29, 3–14.

[86] Anghileri A, Lantto R, Kruus K, et al. 2007. Tyrosinase-catalyzed grafting of sericin peptides onto chitosan and production of protein-polysaccharide bioconjugates. *J Biotechnol.* 127: 508–519.

[87] Jordan A, Khan T, Osborn H, et al. 1999. Melanocyte-directed enzyme prodrug therapy (MDEPT): development of a targeted treatment for malignant melanoma. *Bioorg Med Chem.*, 7: 1775–1780.

[88] Zhang J, Kale V. and Chen M. 2015. Gene-Directed Enzyme Prodrug Therapy. *The AAPS J.*, 17: 102-110.

[89] Aksambayeva S, Zhaparova L, Shagyrova S, et al. 2018. Recombinant Tyrosinase from *Verrucomicrobium spinosum*: Isolation, Characteristics, and Use for the Production of a Protein with Adhesive Properties. *Appl Biochem Microbiol.*, 54: 780-792.

[90] Chang TS. 2009. An Updated Review of Tyrosinase Inhibitors. *Int J Mol Sci.,* 10: 2440-2475.

[91] Kumar C, Sathisha U, Dharmesh S, et al. 2011. Interaction of sesamol (3,4-methylenedioxyphenol) with tyrosinase and its effect on melanin synthesis. *Biochimie*, 93: 562–9.

[92] Solano F, Briganti S, Picardo M. et al. 2006. Hypopigmenting agents: an updated review on biological, chemical and clinical aspects. *Pigment Cell Res.*, 19: 550–571.

[93] Ray K, Chaki M. and Sengupta M. 2007. Tyrosinase and ocular diseases: some novel thoughts on the molecular basis of oculocutaneous albinism type 1. *Prog. Retinal Eye Res.*, 26: 323–358.

[94] Wang Y, Yanmei W, Guo X, et al. 2015. Microbial transglutaminase and tyrosinase modified gelatin-chitosan material. *Soft Mat.*, 13: 32-38.

[95] Moioli E, Bakus A, Yaghmai D, et al. 2011. Treatment strategies for pigmentation disorders in skin of color. *Cosm Dermatol.*, 24: 524-531.

[96] Brauer J, Kazlouskaya V, Alabdulrazzaq H, et al. 2014.Use of a picosecond pulse duration laser with specialized optic for treatment of facial acne scarring. *JAMA Dermatol.*, 3045, E1-E7.

[97] Emil T, Shamshik S. 2015. *PicoToning: A novel laser skin toning approach for the treatment of Asian skin types.* www.cynosure.com.

[98] Sherrie S. Spotless Skin. *Better Nutrition.* www.betternutrition.com

[99] Jennifer C, Stephie C, Abhishri S, et al. 2012. A review on skin whitening property of plant extracts. *Int J Pharm Bio Sci.*, 3: 332 – 347.

[100] Sun Y, Shin J, Na J, et al. 2010. The efficacy and safety of 4-n-butylresorcinol 0.1% cream for the treatment of melasma: A randomized controlled split-face trial. *Ann Dermatol.*, 25: 21-25.

[101] Chandrashekar B, Shenoy C, Narayana L. et al. 2018. Effectiveness and safety of a novel topical depigmenting agent in epidermal pigmentation: an open-label, non-comparative study. *Int J Res Dermatol.*, 4: 489-494.

[102] Smit N, Vicanova J. and Pavel S. 2009. The Hunt for natural skin whitening agents. *Int J Mol Sci.*, 10: 5326-5349.

[103] Chang T. 2012. Natural melanogenesis inhibitors acting through the down-regulation of tyrosinase activity. *Materials*, 5: 1661-1685.

[104] Jody P, Ebanks R, Randall W. et al. 2009. Mechanisms regulating skin pigmentation: The rise and fall of complexion coloration. *Int J Mol Sci.*, 10: 4066-4087.

[105] Serre C, Busuttil V. and Botto J. 2018. Intrinsic and extrinsic regulation of human skin melanogenesis and pigmentation. *Intr J Cosmetic Sci.*, 40: 328-347.

In: Microbial Catalysts. Vol. 2
Editors: Shadia M. Abdel-Aziz et al.

ISBN: 978-1-53616-088-8
© 2019 Nova Science Publishers, Inc.

Chapter 10

PECTINASES: PROPERTIES AND APPLICATIONS

Richa Soni[1],*, Kshipra Kapil (née Soni)[2] and N. K. Jain[1]
[1]Department of Life Science, University School of Sciences,
Gujarat University, Ahmedabad, India
[2]Department of Biotechnology, K.L. Mehta Dayanand College for Women,
M.D. University, Faridabad, India

ABSTRACT

Pectinase is a collective term that refers to a group of enzymes that degrade pectic substances in the cell wall of higher plants. The pectinase enzyme is broadly classified into two types on the basis of their mode of action: Esterase and Depolymerases. Pectin esterase catalyses the de-esterification of the methoxyl group of pectin, forming pectic acid. While, depolymerases act on pectic substances either by hydrolysis or by trans-elimination lysis. Pectinases hold a leading position among the commercially produced industrial enzymes. Pectinolytic enzymes are naturally produced by many organisms like bacteria, fungi, yeasts, insects, nematodes, protozoan and plants. This chapter discusses types of naturally occurring pectic substances, pectinolytic enzymes, their classification, properties, production and their applications in different industrial sectors.

Keywords: pectin, pectic substances, pectinases, pectinolytic enzymes, pectin esterase

1. INTRODUCTION

Pectinases constitute a heterogeneous group of related enzymes which hydrolyses the pectic substances, found in higher plants. They are significantly used in different

* Corresponding Author's E-mail: richasoniricha@gmail.com.

industries like fruits and juice processing, wine processing, tea and coffee processing, textile processing, paper industry and processing of animal feed [1]. Pectinases enzymes are present in higher plants as well as in mocroorganisms [2]. In plants, they assist cell wall extension [3] and softening of tissues during maturation [4]. They are also responsible for decomposition and recycling of waste plant materials to maintain ecological balance [1]. Plant pathogenicity and spoilage of fruits and vegetables by rotting are some other major manifestations of pectinolytic enzymes [2].

Microorganisms are the primary source of industrial enzymes as 50 percent originate from fungi and yeast, 35 percent from bacteria, while the remaining 15 percent are of plant and animal origin [5]. Pectinases are being produced by different microorganisms [6] and microbial pectinases account for 25 percent of the global food enzymes sales [1].

This chapter deals with the substrate i.e pectic substances, pectinolytic enzymes, their types and different method of production. It also concentrates on applications of these enzymes in industrial sector.

2. PECTIC SUBSTANCES

Pectic substances and celluloses are the most abundant carbohydrates present in plants. Pectic substances like pectin, protopectin and pectic acids, present in cell wall and middle lamella. These contribute firmness and structure to plant tissues (Table 1) and are largely responsible for the structural integrity and cohesion [7]. Pectic substances are un-branched polygalacturonides, that is, polymer of galactouronic acid composed of α-1, 4-linked D-galacturonic acid. D-galacturonic acid units are linked together by α-1, 4-glycosidic linkages and the carbonyl side groups are 60–90 percent esterified with methanol. Neutral monosaccharides, such as galactose, rhamnose, arabinose and xylose are also present in pectic substances. The rhamnose units can be inserted into the main uronide chain and often side chains of arabinan, galactan or arabinogalactan are linked to rhamnose. It indicates that various forms of pectic substances are present in plant cells and for this reason pectinases exist in various forms [8].

Pectic substances can be divided broadly into two groups: one is pectic acid, a polymer of galacturonic acid, and the other is pectin, a polymer of galacturonic acid whose carboxyl groups are methyl-esterified [9]. Pectin was discovered by Louis Nicolas Vauquelin, a French pharmacist, in 1790 and then characterized by Henri Braconnot, another French Chemist, in 1825 [10].

Table 1. Pectic substances in different plants [1]

Plant	Tissue	Pectic substance (%)
Apple	Fresh	0.5-1.6
Banana	Fresh	0.7-1.2
Peaches	Fresh	0.1-0.9
Strawberries	Fresh	0.6-0.7
Cherries	Fresh	0.2-0.5
Peas	Fresh	0.9-1.4
Carrort	Dry matter	6.9-18.6
Orange pulp	Dry matter	12.4-28.0
Potatoes	Dry matter	1.8-3.3
Tomatoes	Dry matter	2.4-4.6
Sugar beet pulp	Dry matter	10.0-30.0

3. STRUCTURE AND CLASSIFICATION OF PECTIC SUBSTANCES

Chemically, pectic substances are complex colloidal acid polysaccharides, with a backbone of galacturonic acid residues linked by α -1, 4 linkage. The side chains of the pectin molecule consist of L-rhamnose, arabinose, galactose and xylose. The carboxyl groups of galacturonic acid are partially esterified by methyl groups and partially or completely neutralized by sodium, potassium or ammonium ions. Based on the type of modifications of the backbone chain, pectic substances are classified by the American Chemical Society into pectin, protopectin, pectic acid, pectinic acid [11].

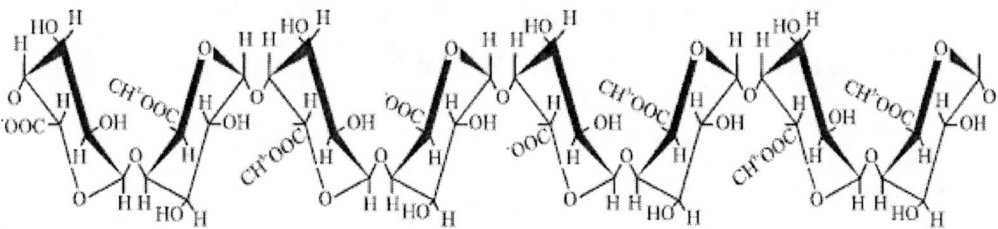

Figure 1: Primary structure of pectic substances [1].

Pectin is a polysaccharide found in the cell wall of plant cell. It is a jelly-like matrix which helps cement plant cells together and in which other cell wall components, such as cellulose fibrils, is embedded. Pectin is present in the plate (middle lamella) that is the first part of the wall to be formed during cytokinesis, following cell division [12]. As a part of the plant structure, pectin is a complex mixture of block of homogalacturonic acid called 'smooth regions' mixed with blocks of homogalacturonic acid containing many neutral sugars including rhamnose, galactose, arabinose and glucose called 'hairy regions' [13]. Pectin is known to contain neutral sugars which are present in side chains. The most common side chain sugars are xylose, galactose and arabinose [14]. In some species, pectins may be cross-linked to other pectins or non-cellulosic polysaccharides by ester linkages with dihydroxycinnamic acids such as diferulic acid. In plants, pectins are present in all stages of development. The composition depends not only on the species but also on tissue, stage of growth, maturity and growth conditions. Pectins are heterogeneous with respect to both chemical structure and molecular weight [11].

3.1. Proto-Pectin

Proto-pectin is water-insoluble pectic substances found in plant tissues [15]. Proto-pectin is the precursor and gives pectin or pectinic acid upon restricted hydrolysis. It is present in immature plant material.

3.2. Pectic Acids

Pectic acids are galacturonans with no methoxyl groups. Normal or acid salts of pectic acid are called pectates. These are found in overripe plant material.

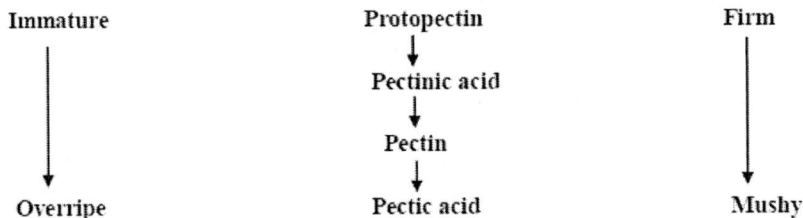

3.3. Pectinic Acids

Pectinic acid has the unique property of forming a gel with sugar and is present in mature plant material. Pectinic acids are the galacturonans with different amounts of methoxyl groups. Pectinates are normal or acid salts of pectinic acids [15].

4. PECTINOLYTIC ENZYMES

Pectinolytic enzymes or pectinases is a collective term for heterogeneous group of related enzymes that hydrolyses pectic substances.

Pectinases are a big group of enzymes that break down pectin a polysaccharide found in plant cell walls into simpler molecules like galacturonic acids. Pectic enzymes are widely distributed in nature and are produced by bacteria, yeast, fungi and plants. In plants, pectic enzymes are very important since they play a role in elongation and cellular growth as well as in fruit ripening. Pectolytic activity of microorganisms plays a significant role, firstly, in the pathogenesis of plants since these enzymes are the first to attack the tissue. In addition, they are also involved in the process of symbiosis and decay of vegetable residues [16]. Thus, by breaking down pectin polymer for nutritional purposes, microbial pectolytic enzymes play an important role in nature. These enzymes are inducible, produced only when needed and they contribute to the natural carbon cycle [17].

They can be extracted from fungi such as *Aspergillus niger*. The fungus produces these enzymes to break down the middle lamella in plants so that it can extract nutrients from the plant tissues and insert fungal hyphae. If pectinase is boiled it is denatured (unfolded) making it harder to connect with the pectin at the active site, and produce as much juice.

Pectinases are also used for retting. Addition of chelating agents or pre-treatment of the plant material with acid enhances the effect of the enzyme.

5. CLASSIFICATION

Different types of pectic enzymes vary in how they degrade pectin. Pectinase enzymes can be broadly classified into two groups as follows (Table 2, Figure 2).

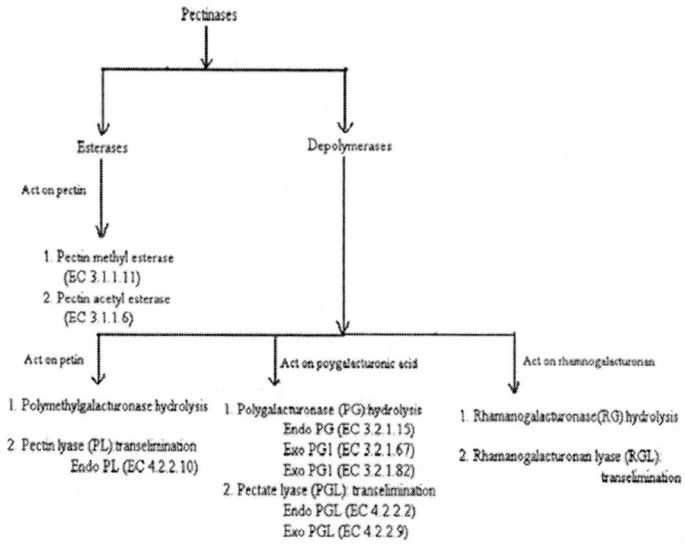

Figure 2. Classification of different pectinases based on their reaction with different pectin substances [35].

Table 2. Classification of pectinolytic enzymes [43]

Enzyme	Action mechanism	Primary Substrate	Product
Esterase 1.Pectin methyl estrase	Hydrolysis	Pectin	Pectic acid and methanol
Depolymerizing enzymes a. Hydrolases 1. Protopectinases	Hydrolysis	Protopectin	Pectin
2. Endopolygalacturonase	Hydrolysis	Pectic acid	Oligogalacturonates
3. Exopolygalacturonase	Hydrolysis	Pectic acid	Monogalacturonates
4. Exopolygalacturonan-digalacturonohydrolase	Hydrolysis	Pectic acid	Digalacturonates
5. Oligogalacturonate hydrolase	Hydrolysis	Trigalacturonate	Monogalacturonates
6. Δ 4:5 Unsaturated oligogalacturonatehydrolases	Hydrolysis	Δ4:5(Galacturonate)n	Unsaturated monogalacturonates & saturated (n-1)
7. Endopolymethyl-galacturonases	Hydrolysis	Highly esterified pectin	Oligomethylgalacturonates
8. Endopolymethyl-galacturonases	Hydrolysis	Highly esterified pectin	Oligogalacturonates
b. Lyases 1. Endopolygalacturonase lyase	Trans-elimination	Random Pectic acid	Unsaturated oligogalacturonates
2. Exopolygalacturonase lyase	Trans-elimination	Pectic acid	Unsaturated digalacturonates
3. Oligo-D-galactosiduronate lyase	Trans-elimination	Unsaturated digalacturonates	Unsaturated monogalacturonates
4. Endopolymethyl-D-galactosiduronate lyase	Trans-elimination	Unsaturated poly-(methyl-D-digalacturonates)	Unsaturated methyloligogalacturonates
5. Exopolymethyl-D-galactosiduronate lyase	Trans-elimination	Unsaturated poly-(methyl-D-digalacturonates)	Unsaturated methylmonogalacturonates

5.1. Esterases

These enzymes catalyse the de-esterification of pectin by the removal of methoxy esters. Pectin methyl esterase or pectinesterase (PE) (EC 3.1.1.11) catalyzes de-esterification of the methoxyl group of pectin forming pectic acid and methanol. The enzyme acts preferentially on a methyl ester group of galacturonate unit next to a non-esterified galacturonate unit. It acts before polygalacturonases and pectate lyases which need non-esterified substrates [11].

PE works differently according to its origin [18] like fungal PEs remove the methyl groups at random. Whereas plant PEs tends to act either at the non-reducing end or next to a free carboxyl group and proceed along the molecule. The reaction catalyzed by PE can be represented as follows [19]:

$$\text{Pectin} + nH_2O \xrightarrow{\text{PE}} \text{Pectate} + nC_2H_5OH$$

The presence of PE is reported in plants, plant pathogenic bacteria and fungi [20]. PE is produced by *Phytophthora infestans* [19], *Erwinia chrysanthemi* [21], *Saccharomyces cerevisiae* [22], *Pseudomonas solanacearum* [23], *Aspergillus niger* [24], *Penicillium frequentans* [25], *Penicillium occitanis* [26] and others. Among plants they are reported in, viz., *Carica papaya* [27], *Lycopersicon esculentum* [28], *Prunus malus* [29], *Vitis vinifera* [30], *Pouteria sapota* [31] and *Malpighia glabra* [32].

PE activity can be measured by gel diffusion assay and pH level [33, 34]. The molecular weights of most PEs are in the range of 35–50 kDa [1]. The active range of pH for PEs is from 4.0 to 8.0 where fungal PEs has a lower pH optimum than that of bacterial origin. Optimum temperature range for maximal activity for majority of PEs is 40–50°C [32].

5.2. Depolymerases

These enzymes act on pectic substances either by hydrolysis or by trans-elimination lysis.

5.3. Hydrolyases

These enzymes catalyze the hydrolytic cleavage with the introduction of water across the oxygen bridge. Polygalacturonase and polymethylgalacturonase breakdown pectate and pectin, respectively by the mechanism of hydrolysis. Depending upon the pattern of action, i.e., random or terminal, these are termed as Endo or Exo enzymes, respectively. Different hydrolases enzymes are as follows:

5.4. Protopectinases

Protopectinases (PPase) degrade the insoluble protopectin and give rise to highly polymerized soluble pectin. These were named by Briton et al. [36]. Protopectinase catalyzes the following reaction:

$$\text{Protopectin} + \text{H}_2\text{O} \xrightarrow{\text{PPase}} \text{Pectin}$$
$$\text{(Insoluble)} \qquad\qquad \text{(Soluble)}$$

The activity of PPase can be assayed by measuring pectin after the reaction of protopectin and the carbazole-sulphuric acid [37]. One unit of PPase activity is defined as the enzyme that liberates pectin corresponding to 1 mmol of D-galacturonic acid per millilitre of reaction mixture under assay conditions.

On the basis of reaction mechanism two types of PPases have been reported [38, 39].

5.4.1. A-Type PPases

These react with polygalacturonic acid region of protopectin inner site and found in the culture filtrates of yeast. They have been isolated from *Kluyveromyces fragilis* IFO0288, *Galactomyces reesei* L. and *Trichosporon penicillatum* SNO3 and are referred to as PPase-F, -L and -S, respectively [2]. These three A-type PPases have similar molecular weight (30 kDa) and are similar in biological properties but differ in chemical properties like PPase-F is an acidic protein whereas PPase-L and -S are basic in nature.

5.4.2. B-Type PPases

These react on the outer site, i.e., on the polysaccharide chains that may connect the polygalacturonic acid chain and cell wall constituents. They have been reported in *Bacillus subtilis* IFO 12113 [40], *B. subtilis* IFO3134 [41] and *Trametes* sp. and are referred to as PPase-B, -C and -T, respectively.

5.5. Polygalacturonases

Polygalacturonases (PGases) are the most comprehensively studied enzyme. They catalyse the hydrolytic cleavage of the polygalacturonic acid chain. Endo-PGases are present in fungi, yeast, bacteria [42] as well as in higher plants [41]. In plants, they are involved in plant-fungal interactions [43]. Among microorganisms, these are reported in *Aureobasidium pullulans, Rhizoctonia solani* Kuhn [44], *Fusarium moniliforme* [45], *Neurospora crassa* [46], *Rhizopus stolonifer* [47], *Aspergillus* sp. [48], *Thermomyces lanuginosus* [49] and *Peacilomyces clavisporus* [50].

Exo-PGases occur less frequently and reported in *Erwinia carotovora* [51], *Agrobacterium tumefaciens* [52], *Bacteroides thetaiotamicron* [53], *E. chrysanthemi* [54], *Alternaria mali* [55], *Fusarium oxysporum* [56], *Ralstonia solanacearum* [57] and *Bacillus* sp. [54].

The activity of PGase is determined by measuring the rate of increase in number of reducing groups; and the decrease in viscosity of the substrate solution [58]. One unit of enzyme activity is defined as the enzyme that releases 1 mmol/ min galacturonic acid under standard assay conditions. For viscosity, the unit of enzyme activity is mostly selected as the amount of enzyme required for attaining a certain decrease of viscosity per unit time.

Among the PGases obtained from different microbial sources, most have the optimal pH range of 3.5–5.5 and optimal temperature range of 30–50°C. But PGases reported from *Bacillus licheniformis* and *Fusarium oxysporum* worked on basic pH (11.0) [1].

5.6. Lyases

Pectate lyase cleaves glycosidic linkages preferentially on polygalacturonic acid forming unsaturated product (4, 5-D-galacturonate) through trans-elimination reaction. Pectate lyase has an absolute requirement of Ca^{2+} ions. Hence it is strongly inhibited by chelating agents as EDTA. Pectate lyases are classified as endo- (EC 4.2.2.2) that acts towards substrate in a random way, and exo- (EC 4.2.2.9) that catalyze the substrate cleavage from non-reducing end [7].

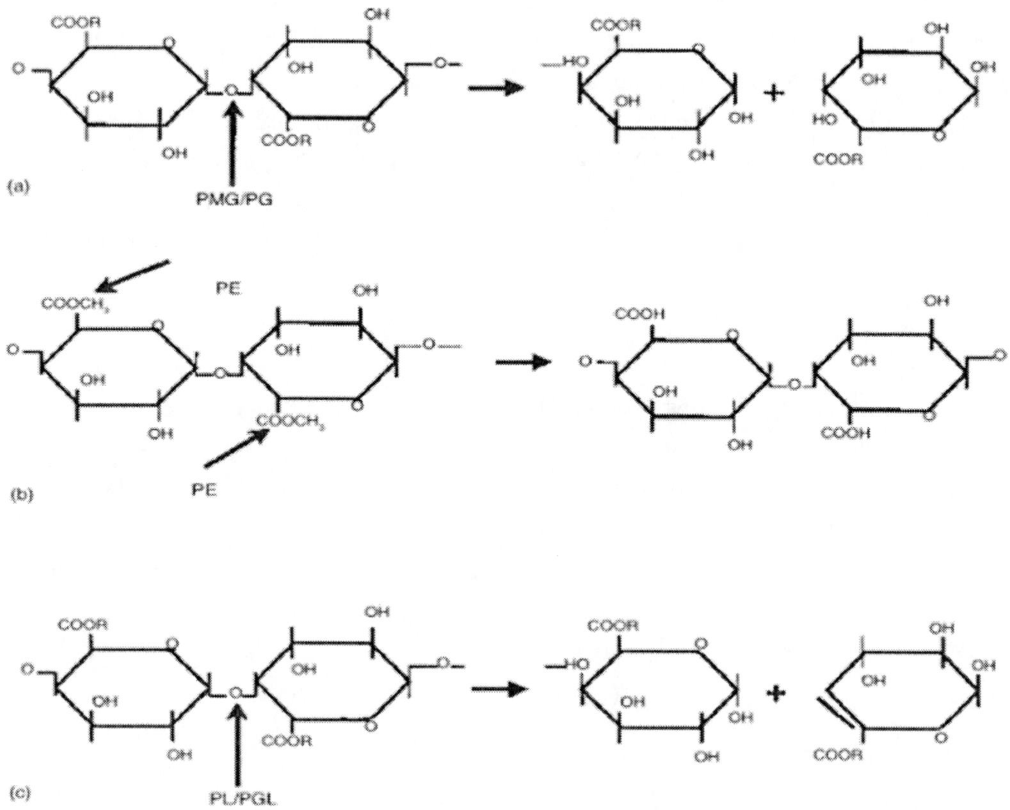

Figure 3. Mode of action of pectinases: (a) R = H for PG and CH3 for PMG; (b) PE; and (c) R = H for PGL and CH3 for PL. The arrow indicates the place where the pectinase reacts with the pectic substances. PMG, polymethylgalacturonases; PG, polygalacturonases (EC 3.2.1.15); PE, pectinesterase (EC 3.1.1.11); PL, pectin lyase (EC-4.2.2.10) [72].

On the basis of the pattern of action and the substrate, lyases can be classified into following type:

- Endopolygalacturonate lyase (EndoPGL)
- Exopolygalacturonate lyase (ExoPGL)
- Endopolymethylgalacturonate lyase (EndoPMGL)
- Exopolymethylgalacturonate lyase (ExoPMGL)

The reactions catalyzed by lyases can be illustrated as follows:

$$\text{Polygalactouronase} \xrightarrow{\text{PGL}} \Delta4:5 \text{ unsaturated galactouronates}$$

$$\text{Polymethylgalactouronate} \xrightarrow{\text{PMGL}} \text{unsaturated methyloligogalactouronates}$$

Many bacteria and fungi produce PGLs like different species of Colletotrichum such as *C. lindemuthionum* [59], *C. magna* [60] and *C. gloeosporioides* [61], *Bacteroides thetaiotaomicron* [62], *Erwinia carotovora* [63] *Pseudomonas syringae* [64].

As compared to PGLs, there are limited reports on the production of PMGLs and these are produced by *Aspergillus japonicus* [65], *Penicillium paxilli* [66], *Pythium splendens* [67], *Pichia pinus* [68] and *Thermoascus auratniacus* [69].

The activity of lyases can be measured by measuring the increase in absorbance at 235 nm due to formation of the Δ4,5 double bonds produced at the non-reducing ends of the unsaturated products [70]. One unit of enzyme activity is defined as the amount of enzyme that releases 1 mmol of unsaturated product per minute under assay conditions [71].

6. PRODUCTION OF PECTINOLYTIC ENZYMES

Pectinase production occupies about 10 percent of the overall manufacturing of enzyme preparations. The main sources for the pectinolytic complex enzymes are yeast, bacteria and a large variety of filamentous fungi, for which the most relevant one are *Aspergillus*.

Novozymes (Denmark), Novartis (Switzerland), Roche (Germany) and Biocon (India) are some important commercial producers of pectinases [8].

7. MICROBIAL PECTINASE

Most of the industrial enzymes are obtained from microorganisms. 50 percent of enzyme are obtained from fungi and yeast, 35 percent from bacteria and 15 percent are obtained from plant origin [17]. Microbes are variously abled to syntehsize different types of pectolytic enzymes wich have different mechanisms of action and biochemical properties [8].

The microbial production of pectinases became prominent for many decades. Many microorganisms, viz., bacteria, yeast and fungi could produce pectinases. Evidence showed that pectinases are inducible and they can be produced from different carbon sources [8]. Several studies on microbial enzymes have shown the production of multiple pectinase forms which differ on molecular mass and kinetic properties. There are many studies that have been conducted related to the characterization of different microbial pectic enzymes concerning their mechanisms of action and biochemical properties. The optimal pH that these enzymes may act range 3.5-11 while the optimal temperatures vary between 40-75°C.

Aspergillus niger pectinases are most widely used in industries because this strain posses GRAS (Generally Regarded as Safe) status so that metabolites produced by this strain can be safely used. This fungal strain produces various pectinases including polymethylgalacturonase (PMG), polygalacturonase (PG) and pectinesterase (PE) [8].

Members of actinomycetes [73] and yeast [74] have been used in pectinase production. Presence of inducer in the culture medium elevated the level of extra-cellularly induced enzymes [75].

Microbiologically derived pectinases find more uses due to their advantage over plant and animal derived pectinases. The reasons being cheap production, easier gene manipulations, faster product recovery, and further microbial enzymes are usually free of harmful substances [76].

There are two fermentation techniques for pectinases production.

1. Solid State Fermentation (SSF)
2. Submerged fermentation (SmF).

In solid state fermentation microbes are grown on moist solid substrate (innert carrier or insoluble susbstrate) that can be used as carbon source. Various susbstrates with different capacity for liquid retention can be used according to the requirement [17]. Whereas, in submerged fermentation microbes with the nutrients are submerged in water. SmF is used to produce 90 percent of all industrial enzymes. Enzymes produced by submerged fermentation are too expensive for agro-biotechnological applications [25]. Therefore, use of high yielding strains, optimal fermentation conditions and cheap raw materials as a carbon source can reduce the cost of enzyme production for subsequent applications in industrial processes [43]. SSF is beneficial than SmF on various parameters such as growth rate, productivity or volume activity [77]. Different advantages of SSF are:

- Low water demand
- High concentration of the end product
- Catabolite repression
- Utilisation of solid substrate
- Lower sterility demands
- Solid support for microorganism
- Simulation of the natural environment
- Fermentation of water-insoluble solid substrates
- Mixed culture of microorganisms
- High-volume productivity
- Low energy demand for heating
- Easy aeration
- Utilisation of otherwise unusable carbon sources
- No anti-foam chemicals

8. CONDITIONS OF OPTIMUM PRODUCTION

As with all enzymes, pectinases have an optimum temperature and pH at which they are most active. Several substances have been used to induce pectolytic enzymes sometimes it is pectin itself. Different complex media like wheat bran, beet sugar, ground nut meal and citrus fruit peels have also been used by different researchers [78]. Table 3 shows the optimal fermentation conditions for pectinase production by various micro organisms. Various studies related to the characterization of microbial pectic enzymes concerning their mechanisms of action and biochemical properties have been concluded that the optimal pHs for action range between 3.5-11 and the optimum temperatures vary between 40-75°C [8, 11]. Most commercial pectinase might typically be activated at 45 to 55°C and work well at a pH of 3.0 to 6.5.

9. PECINASE AND HUMAN NUTRITION: HUMAN GUT FLORA, PECTINASES AND DIETARY FIBER

Bacteria that colonize the mammalian intestine collectively possess a far larger repertoire of degradative enzymes and metabolic capabilities than their hosts [80]. The gastrointestinal microbiota is influenced by a number of factors including genetics, host physiology (age of the host, disease, stress, etc.) and environmental factors such as living conditions and use of medications [81].

Table 3. Fermentation conditions for pectinase production by various microorganisms [79]

S. No.	Microorganism	Substrate	Fermentation Type	Fermentation Temp.	pH
1	Aspergillus niger A138	Sucrose	SmF	32	4.5
2	Aspergillus niger 3T5B8	Wheat bran	SSF	32	–
3	Bacillus sp. DT7	Pectin	SmF	37	7.2
4	Penicillium veridicatum RFC3 Orange bagasse,	Wheat bran	SSF	30	–
5	Bacillus sp. DT7	Wheat bran	SSF	37	–
6	Aspergillus fumigates	Wheat bran	SSF	50	–
7	Aspergillus niger	Sunflower head	SSF	30	5.0
8	Aspergillus fumigatus MTCC 870	Wheat flour	SmF	30	5.0
9	Penicillium chrysogenum	Sucrose	SmF	35	6.5
10	Aspergillus heteromorphus	Orange peel	SmF	30	4.5
11	Thermomucor indicae-seudaticae	Wheat bran, Orange bagasse	SSF	45	–
12	Penicillium sp.	Pectin	SSF	35	6.0
13	Fomes sclerodermeus	Soy and Wheat bran	SSF	28	–
14	Bacillus subtilis	Pectin	SmF	50	7.0
15	Bacillus sp. AD 1	Pectin	SmF	37	7.0
16	Aspergillus niger	Pectin	SmF	37	5.5
17	Aspergillus sojae M3	Orange peel	SSF	22	–
18	Aspergillus flavus	Orange peel	SSF	40	5.5
19	Penicillium atrovenetum	Orange peel	SSF	40	5.0
20	Aspergillus oryzae	Orange peel	SSF	35	5.5
21	Bacillus subtilis	Date syrup	SmF	45	8.0
22	Pseudozyma sp. SPJ	Citrus peel	SSF	32	7.0
23	Mixed culture of Aspergillus fumigattus, Aspergillus sydowii	Pineapple residue	SSF	35	5.0
24	Aspergillus niger	Sour oranges peel	SSF	30	5.0
25	Streptomyces sp.	Pectin	SmF	30	8.5
26	Penicillium citrinum	Sugar beet pulp	SSF	30	5.5
27	Erwinia carotovora	Pectin	SmF	35	5.2
28	Bacillus firmus	Pectin	SmF	50	7.0
29	Aspergillus niger	Date pomace	SmF	–	6.18
30	Rhizomucor pusillus	Pectin	SSF	45	5.0
31	Rhodotorula glutinis MP-10	Citrus pectin	SmF	30	5.5
32	Aspergillus sojae	Wheat bran	SSF	37	6.0
33	Aspergillus niger HFD5A-1	Citrus pectin	SmF	30	4.5
34	Trichoderma viridi	Orange peel	SSF	30	5.5

It is well known that pectin (which is consumed mostly in the form of fruits and vegetables but also as hydrocolloid in "functional foods," jellies, milk products, etc.) is fermented in the large intestine to short chain fatty acid (SCFA) and gases. Decomposition of the polysaccharide pectin occurs during the following main steps: (i) macromolecular pectin, (ii) (unsaturated) oligoGalA, (iii) monogalacturonic acid (or its rearrangement products), and (iv) SCFA (and gases). The rate of enzymatic reactions is influenced by synergistic effects of bacteria and molecular parameters of the substrate (Pectin), such as the degree of esterification.

Recently, there has been increasing evidence that fermentable dietary fibers modulate various properties of the immune system as prebiotics, including those of the gut-associated lymphoid tissues. These fermentable prebiotic fibers potentially mediate immune changes through the colon bacteria [82]. Preparation of functional food ingredients by debranching and depolymerization of guar galactomannan using pectinases has also been suggested by Shobha et al. [83]. Pectin, which is not degraded in the human gut has application as a dietary fiber, increases the viscosity in the intestinal tract, which leads to reduced cholesterol absorption. In recent years, pectin methyl esterase (PME) modified pectin to obtain superior gelling agents, which are gaining attention as functional foods. The process involves enzymatically demethylating pectin and cross linking with added Ca^{2+} atoms to obtain modified pectins with various gelling and emulsion stabilizing properties. High-viscosity pectin is thought to lower cholesterol levels by raising the excretion of fecal bile acids and neutral sterols. It may also interfere with the formation of micelles and/or lower the diffusion rate of bile acid and cholesterol-containing micelles through the bolus, consequently diminishing the uptake of cholesterol and bile acids. Demethylated pectin has the ability to associate ions because of a high content of negative charges. Thus, it can behave as a weak cation exchange resin and, depending on the pH conditions, chelate toxic ions or make available minerals in the gut. It has been shown that demethylated pectins are more rapidly fermented by intestinal bacteria, forming short-chain fatty acids, acetate, propionate, and butyrate. These, apart from protecting the bowel against inflammatory diseases, also modulate the release of gut hormones that control insulin release and appetite [84]. Thus, pectin modified with pectinases find application as functional ingredients in different food products for probiotic applications [85].

10. COMMERCIAL APPLICATIONS OF PECTINASES

Pectinases were some of the first enzymes to be used in homes. Commercial application of pectinases was first observed in 1930 for the preparation of wines and fruit juices [7].

In 1960s, when the knowledge of chemical nature of plant tissues became apparent, scientists began to use a greater range of enzymes more efficiently [86]. Over the years, pectinases have been used in several conventional industrial processes, such as textile, plant fiber, tea, coffee processing oil extraction, treatment of industrial wastewater containing pectinacious material. They have also been reported to work on purification of viruses and in making of paper but are yet to be commercialized [12]. As a result, pectinases became one of the important upcoming enzymes of the commercial sector especially for fruit juice industry as a prerequisite for obtaining well clarified and stable juices with higher yields [86].

Table 4. Commercial applications of pectinases [76]

Area	Application
Juice industry	Juice clarification
Textile industry	They are capable of depolymerising the pectin breaking it into low molecular water soluble oligomers improving absorbency and whiteness of textile material andavoiding fiber damage
Pulp and paper industry	Effective in biobleaching of mixed hard wood and bamboo kraft pulp,
Wine industry	Improve wine characeristics of colour and turbidity, improvement of chromaticity and Stability of red wines
Coffee and tea fermentation	Fermentation by breaking pectins present in tea Leaves
Oil extraction	By avoiding emulsification formation

Pectinolytic enzymes can be applied in various industrial sectors wherever the degradation of pectin is favourable for a particular process. Pectinases can be divided into two groups based on the optimum activity pH of these enzymes: acidic and alkaline [87]. Fruit juice clarification/extraction, wine processing, saccharification of biomass, isolation of protoplasts requires acid pectinase whereas alkaline pectinases are important for retting and degumming of plant fibers as an ecofriendly alternative to the traditional chemical processes. Textile industry and paper and pulp industry require alkaline pectinases. Table 4 shows various industrial applications of commercial pectinases.

10.1. Processing of Animal Feed

Pectinolytic enzymes have a significant role in maintaining ecological balance by causing decomposition and recycling of plant materials [87]. Pectinases are used as additives in animal feeds, mainly for poultry and ruminants, as it helps in improving digestibility and nutritional value [88]. Pectinases are also used for the saccharification of pectin-rich agricultural residues, with the resulting sugars being used for the production of ethanol or platform chemicals [89].

Enzymes have been used for decades to improve the utilization of swine and poultry diets [90]. Usage of pectinases for ruminant's feed production can reduce the feed viscosity, which increases absorption of nutrients and liberates nutrients. By hydrolysis of non-biodegradable fibers or by liberating nutrients blocked by these fibers and reduces the amount of faeces [43]. In addition Jacob et al. [87] reported that the reducing sugar level was increasing gradually for banana fiber treatment with polygalacturonase indicating the effectiveness of the treatment and it was evident that the cells were separated after treatment as a result of pectin hydrolysis. This may be reflecting the importance of use pectinases as a fiber degrading enzymes in animal feed production. The mechanism by which exogenous fibrolytic enzymes improves feed by rumen microorganisms is still unknown and several potential modes of action have been proposed. These include: a) Increase in microbial colonization of feed particles. b) Enhancing attachment and improving access to the cell wall matrix by ruminal microorganisms and by doing so, accelerate the rate of digestion. c) Enhancing the hydrolytic capacity of the rumen due to added enzyme activities and synergy with rumen microbial enzymes [43].

10.2. Uses in Bio Refineries

Biomass is a plentiful resource, as it includes wood and its residues, agricultural crops and their corresponding residues, food waste, municipal solid waste as well as algae and microalgae [91]. Pectinases are also being used in bio refineries for hydrolyzing pectin present in pectin-rich agro-industrial wastes [88]. These wastes are processed into simple sugars so that they could be converted into bioethanol or used as fermentable sugars [92].

10.3. Extraction of Juice from Soft Fruits

Fruits are mainly water (75–90%), most located in vacuole causing turgor to the fruit tissue, and fruit juice is prepared by mechanically squeezing or macerating fresh fruits without the application of heat or solvents. Juices are products for direct consumption and are obtained by the extraction of cellular juice from fruit; this operation can be done by pressing or by diffusion [93].

Juices can be easily expressed from fruits like lemons, oranges, tomatoes, pineapple etc. But tropical fruits are usually too pulpy and pectinaceous to yield juice by simple pressing or centrifugation [76]. In soft fruits like guava, banana, papaya, mangoes etc., mechanical crushing results into highly viscous fruit puree from which it is difficult to extract juice directly by pressing. Mechanical crushing of the tissues gives juice that remains bound to the pulp to form a jellified mass. The pectin divides itself between the liquid phase and the pulp particles, causing an increase in the viscosity of the juice and facilitating water retention [94]. Thus, in order to improve yield of juice with high aromatic quality in a short processing time, to increase its nutritional quality and to reduce the amount of waste, there is a need to degrade the pectin. For this reason, addition of pectinolytic enzyme preparations to the fruit pulp prior to pressing is a prerequisite for obtaining a satisfactory juice yield [95].

Pectinases play a crucial role to reduce the viscosity, increase the yield and clarification of juice by liquification of pulps, and removal of the peels and in maceration of vegetables to produce various products like pastes and purées [79]. Pectinases in combination with other enzymes, viz., cellulases, arabinases and xylanases, have been used to increase the pressing efficiency of the fruits for juice extraction [96].

The largest industrial application of pectinases is in fruit juice extraction and clarification. A mixture of pectinases and amylases is used to clarify fruit juices. It decreases filtration time up to 50 percent [97].

10.4. Wine Production

Enzymes play a pivotal role in the wine making process; many of these enzymes originate from the grape itself. Since the endogenous enzymes of grapes, yeasts and other microorganisms present are often neither efficient nor sufficient to effectively catalyse, commercial enzyme preparations are widely used as supplements [98]. The most widely used enzymes available for commercial use in winemaking are:

- Pectinases, glucanases, xylanases and proteases: pectinases, glucanases, xylanases and proteases are used for improving the clarification and processing of wine
- Glycosidase: to release of varietal aromas from precursor compounds
- Urease: For reduction in ethyl carbamate formation
- Glucose oxidase: For eduction in alcohol levels

Pectinases have also been used in wine production since 1960s. The pectic enzymes play an important role in braking down grape pulp and skin cells and are able to split those chains and saccharide bonds between the chains. Pectinases are used predominantly on red varieties. Pectinase functions by breaking down the cell walls of red grape skins, thereby extracting anthocyanins (red color in red grapes) and tannin. This then helps to improve the overall color intensity, as well as the color stability of a wine, by allowing the anthocyanins to bind with tannin, as well as its structure. An additional benefit of pectinase treatment is that particles settle more quickly [99].

10.5. Tea and Coffee Production

Pectinases play an important role in coffee and tea fermentation. Fermentation of coffee using pectinolytic microorganisms is done to remove the mucilage coat from the coffee beans. Pectic enzymes are sometimes added to remove the pulpy layer of the bean, three-fourths of which consists of pectic substances. Cellulases and hemicellulases present in the enzyme preparation aid the digestion of the mucilage. A diluted commercial enzyme preparation is sprayed on to the beans at a dose of 2–10 g per ton at 15–20°C. The fermentation stage of coffee processing is accelerated and reduced from 80 hours to about 20 hours by enzyme treatment. Since large-scale treatment of coffee with commercial pectinases is costly and uneconomical, inoculated waste mucilage is used as a source of microbial pectic enzymes. The fermentation liquid is washed, filtered and then sprayed on to the beans [11].

10.6. Extraction of Vegetable Oil

Enzymes involved in the breakdown of plant cell wall polyssaccharides can be used to extract vegetable oils, coconut germ, palm, sunflower seed, rape seed olives and kernel oils which are traditionally produced by extraction with organic solvents, such as the potentially carcinogen hexane. By degrading cell wall components like pectin enzymes promote the oil liberation [7]. It is known that the addition of cell-wall-degrading enzyme preparations during the mechanical extraction of olive oil can improve the release of phenolic compounds in oil. In a study conducted by Vierhus et al. [100] on effect of enzyme treatment during mechanical extraction of olive oil on phenolic compounds and polysaccharides, researchers revealed that the use of pectinase changed part of the cell wall structure. Addition of pectinase also reduces the complexion of the phenolic compounds with the polysaccharides, thus it further improve the concentration of free phenols in the pastes and their release in the oil and vegetation water during processing. Also, the addition of pectolytic enzymes resulted in weakening and disruption of the cell wall, thus facilitating the release of phenolic compounds from the fruit.

10.7. Textile Industry

Pectinolytic enzymes have been applied to the retting and degumming of jute, sunn hemp, flax, ramie and coconut fibers for textile application. Retting is a fermentation process in which certain bacteria (e.g., *Clostridium*, *Bacillus*) and certain fungi (e.g., *Aspergillus*, *Penicillium*) decompose the pectin of the bark and release fiber. Commercially, retting is done by one of the two basic forms. In dew retting (an aerobic process) plant straw is thinly spread on the ground and exposed to the action of the fungi and aerobic bacteria for 2–10 weeks. Species of *Cladosporium, Penicillium, Aspergillus* and *Rhodotorula* have been isolated from dew-retted plants. Dew retting may produce fiber of a lower quality than the alternative anaerobic process. Anaerobic retting is achieved by submerging straw sheaves in water pits, concrete tanks or in running fresh water. Tank and other stagnant retts rapidly become depleted of dissolved oxygen encouraging the development of an anaerobic flora. Species of *Clostridium* especially *Clostridium butyricum* and *C. felsineum* are considered as the major retting agents [11]. Kapoor and co- workers [101] had run three treatments on ramie and sunn hemp bast fibers: enzymic, chemical and chemical associated with enzymic treatment. Of the three treatments, the third one was the most promising for degumming. The scanning electron microscopic studies revealed a complete removal of non-cellulosic gummy material from the surface of ramie and sunn hemp fibres [7].

10.8. Recycling of Paper

Pulp and paper mills are beginning to use enzymes to solve problems in their manufacturing processes. Paper making is essentially a continuous filtration process in which a dilute suspension of fibers, fiber fragments (fines), and inorganic filler particles, such as clay or $CaCO_3$, is formed into sheets [11].

During papermaking, alkaline peroxide bleaching of mechanical pulps solubilizes acidic polysaccharides which are troublesome interfering substances. Some of these acidic polysaccharides are pectins, or polygalacturonic acids. The ability of polygalacturonic acids to complex cationic polymers (cationic demand) depends strongly on their degree of polymerization, so monomers, dimers, and trimers of galacturonic acid did not cause measurable cationic demand, but hexamers and longer chains had high cationic demand. Pectinases can depolymerize polymers of galacturonic acid, and consequently lower the cationic demand of pectin solutions and the filtrates from peroxide bleaching [7].

10.9. Waste Water Treatment

The wastewater from the citrus-processing industry contains pectinaceous materials that are barely decomposed by microbes during the activated-sludge treatment. This waste water can be treated by using an alkalophillic microorganism. Alkalophilic *Bacillus sp.* (GIR 621) produces an extracellular endopectate lyase in alkaline media at pH 10.0. Treatment with this strain has proved to be useful in removing pectic substances from the wastewater [102].

Treatment of wastewater from citrus processing industries by conventional methods such as physical dewatering, spray irrigation, chemical coagulation, direct activated sludge treatment and chemical hydrolysis followed by methane fermentation possess some drawbacks such as low efficiency due to chemical resistance of the pectic substances, high treatment cost, long treatment periods and complexity of the process [16]. A soft-rot pathogen, *Erwinia carotovora* (FERM P-7576), which secrets endo-pectate lyase, has been reported to be useful in the pretreatment of pertinacious wastewater [102]. Pre treatment of these wastewaters with pectinolytic enzymes facilitates removal of pectinaceous material and renders it suitable for decomposition by activated sludge treatment [16].

CONCLUSION AND FUTURE PERSPECTIVES

Enzymes are extremely effective biological catalysts which perform all synthetic and degradative reactions in living organisms. The enzymes are favoured to chemicals in commercial endeavour mostly because of their high catalytic power, specific mode of action, stereo specificity, environmentally friendly nature and reduced energy demand. Pectinolytic enzymes are classified on the basis of their catalytic activity to pectin or its derivatives. Microbial pectinases are the leading enzyme of the industrial sector. There are a lot of industrial processes to which pectinases can be applied to improve the quality and the yield of final products. These enzymes are eco-friendly tool of nature that are being used extensively in various industries. Over the years pectinases have been used in several conventional industrial processes. Some of the established applications of pectinolytic enzymes are in fruit and vegetable processing industry, wine industry, paper industry for bleaching of pulp and waste paper recycling; fermentation of tea–coffee, processing of animal feed; extraction of vegetable oil and scouring of plant fibres.

The industry currently uses pectinases from mesophilic or thermophilic microorganisms but recently, there has been is a new trend in the food industry to adopt low-temperature processing and therefore, psychrophilic pectinases derived from cold-adapted microorganisms have been isolated and characterized in many recent studies. Many species of pectinase producing bacterial have also been found in the huam gut flora. Recent perspectives on the use of pectin and its derivatives as dietary fibers suggest enzymatic synthesis of the right oligomers from pectin for use in human nutrition. Purification and characterization of these enzymes can lead to a better understanding of its contribution in these processes.

REFERENCES

[1] Jayani, R. S., Saxena, S. and Gupta, R. 2005. "Microbial pectinolytic enzymes: A review." *Process Biochemistry* 40: 2931-2944.

[2] Whitaker, J. R. 1990. *Microbial pectinolytic enzymes, microbial enzymes and biotechnology.* London: Elsevier Science Ltd.

[3] Ward, O. P. and Moo-Young, M. 1989. "Enzymatic degradation of cell wall and related plant polysaccharides." *CRC Critical Reviews in Biotechnology* 8: 237–74.

[4] Sakai, T. 1992. Degradation of Pectins. In: *Microbial degradation of natural products.* Germany: WIley VCH.

[5] Anisa, S. K. and Girish, K. 2014. "Pectinolytic activity of *Rhizopus* sp. And *Trichoderma viride.*" *International Journal of Pure and Applied Microbiology* 4: 28–31.

[6] Mohamadi, A. S., Shahbazi, S., Behgar, M. and Fard, S. M. 2014. "A study of pectinase enzyme activity changes in gamma- irradiated *Trichoderma reesei* mutants." *International Journal of Farming and Allied Sciences* 3:555–561.

[7] Pedrolli, D. B., Monteiro, A. C., Gomes, E. and Carmona, E. C. 2009. "Pectin and pectinases: production, characterization and industrial application of microbial pectinolytic enzymes." *Open Biotechnology Journal* 3: 9-18.

[8] Gummadi, S. N. and Panda, T. 2003. "Purification and biochemical properties of microbial pectinases-a review." *Process Biochemistry* 38 7: 987-996.

[9] Mojsov, K. D. 2016. *Aspergillus enzymes for food industries. New and future developments in microbial biotechnology and bioengineering.* London: Elsevier publication.

[10] Caffall, K. H. and Mohnen, D. 2009. "The structure, function, and biosynthesis of plant cell wall pectic polysaccharides." *Carbohydrate Research* 344: 1879–1900.

[11] Kashyap, D. R., Vohra, P. K. and Tiwari, R. 2001. "Application of pectinases in the commercial sector: a review." *Bioresource Technology* 77: 215-27.

[12] Pasha, K. M., Anuradha, P. and Subbarao, D. 2013. "Applications of pectinases in industrial sector." *International Journal of Pure and Applied Sciences* 16: 89-95.

[13] Girma, E. and Worku, T. 2016. "Extraction and characterization of pectin from selected fruit peel waste." *International Journal of Science and Research* 6: 447-454

[14] Shembekar, V. S. and Dhotre, A. 2009. "Studies of pectin degrading microorganisms from soil." *MicroWorld* 11: 216-222.

[15] Neill, M. A. O., Dravill, A. G. and Albersheim, P. 2001. "Pectic Substances." *Encyclopedia of Life Science.* USA: Nature Publishing Group.

[16] Hoondal, G. S., Tiwari, R. P., Tiwari, R., Dahiya, N. and Beg, Q. K. 2000. "Microbial alkaline pectinases and their applications: A review." *Applied Microbiology and Biotechnology* 59: 409-18.

[17]   Oumer, O. J. 2017. "Pectinase: substrate, production and their biotechnological applications." *International Journal of Environment, Agriculture and Biotechnology* 2: 1007-1014.

[18]   Micheli, F. 2001. "Pectin methylesterase: cell wall enzymes with important roles in plant physiology." *Trends in Plant Science* 6: 414-419.

[19]   Forster, H. 1988. "Pectinesterase from *Phytophthora infestans.*" *Methods in Enzymology* 161: 355-357.

[20]   Hasunuma, T., Fukusaki, E. I. and Kobayashi, A. 2003. "Methanol production is enhanced by expression of an *Aspergillus niger* pectin methylesterase in tobacco cells." *Journal of Biotechnology* 106: 45-52.

[21]   Laurent, F., Kotoujansky, A. and Bertheau, Y. 2000. "Overproduction in *Escherichia coli* of the pectin methylesterase A from *Erwinia chrysanthemi* 3937: one-step purification, biochemical characterization and production of polyclonal antibodies." *Canadian Journal of Microbiology* 46: 474-480.

[22]   Gainvors, A., Frezier, V., Lemaresquier, H., Lequart, C., Aigle M. and Belarbi A. 1994. "Detection of polygalacturonase, pectin lyase and pectinesterase activities in *Saccharomyces cerevisiae* strain." *Yeast* 10: 1311-1319.

[23]   Schell, M. A., Denny, T. P. and Huang, J. 1994. Extracellular Virulence Factors of Pseudomonas Solanacearum: Role in Disease and their Regulation. In: *Molecular Mechanisms of Bacterial Virulence.* The Netherlands. Kluwer Academic Press.

[24]   Maldonado, M. C. and de Saad, A. M. C. 1998. "Production of pectinesterase and polygalacturonase by *Aspergillus niger* in submerged and solid-state systems." *Journal of Industrial Microbiology and Biotechnology* 20: 34-38.

[25]   Kawano, C. Y., Chellegatti, M., Said, S. and Fonesca, M. J. 1999. "Comparative Study of Intracellular and extracellular pectinases produced by *Penicillium Frequentans.*" *Biotechnology and Applied Biochemistry* 29: 133–40.

[26]   Hadj-Taieb, N., Ayadi, M., Trigui, S., Bouabodollah, F. and Gargouri, A. 2002. "Hyper production of pectinase activities by fully constitutive mutant (CT 1) of *Penicillium occitanis.*" *Enzyme and Microbial Technology* 30: 662-666.

[27]   Innocenzo, M. D. and Lajalo, M. 2001. "Effect of gamma irradiation on softening changes and enzyme activities during ripening of papaya fruit." *Journal of Food Biochemistry* 25: 425-438.

[28]   Warrilow, A. G. S., Turner, R. J. and Jones, M. G. 1994. "A novel form of pectinesterase in tomato." *Phytochemistry* 35: 863-868.

[29]   Macdonald, H. C. and Evans, R. 1996. "Purification and properties of apple pectinesterase." *Journal of the Science of Food Agriculture* 70: 321-326.

[30]   Corredig, M., Kerr, W. and Wicker, L. 2000. "Separation of thermostable pectinmethylesterase from marsh grapefruit pulp." *Journal of Agriculture and Food Chemistry* 48: 4918-4923.

[31] Arenas-Ocampo, M. L. S., Evangelista-Lozano, R., Arana-Errasquin, A. 2003. "Softening and biochemical changes of *Sapote mamey* fruit (*Pouteria sapota*) at different development and ripening stages." *Journal of Food Chemistry*. 27: 91-107.

[32] Assis, S. A., Fernandes, P., Ferreira, B. S. 2004. "Screening of supports for the immobilization of pectin-methylesterase from acerola (*Malpighia glabra*)." *Journal of Chemical Technology and Biotechnology* 79: 277-280.

[33] Downie, B., Dirk, L. M. A., Hadfield, K. A., Wilkins, T. A., Bennett, A. B. and Bradford, K. J. 1998. "A gel diffusion assay for quantification of pectinmethylesterase activity." *Analytical Biochemistry* 264: 149-157.

[34] Whitaker, J. R. 1984. "Pectic substances, pectic enzymes and haze formation in fruit juices." *Enzyme and Microbial Technology* 6: 341-349.

[35] Sharma, N., Rathore, M. and Sharma, M. 2013. "Microbial pectinases: sources, characterization and applications." *Reviews in Environmental Science and Bio/Technology* 12: 45-60.

[36] Brinton, C. S., Dore, W. H., Wichmann, H. J., Willaman, J. J. and Wilson, C. P. 1927. "Definitions written by the committee on nomenclature of pectin of the agriculture-food division." *Journal of American Chemical Society* 49: 38-40.

[37] Siebert, F. B. and Anto, J. 1946. "Determination of polysaccharides in serum." *Journal of Biological Chemistry* 163: 511-522.

[38] Sakai, T. and Okushima, M. 1982. "Purification and crystallisation of a protopectin- solubilizing enzyme from *Trichosporon penicillatum*." *Agricultural and Biological Chemistry* 46: 667-776.

[39] Sakamoto, T., Hours, R. A. and Sakai, T. 1994. "Purification, characterization and production of two pectic-transeliminases with protopectinase activity from *Bacillus subtilis*." *Bioscience Biotechnology and Biochemistry* 58: 353-358.

[40] Sakai, T. and Ozaki, Y. 1988. "Protopectin solubilizing enzyme that does not catalyze the degradation of polygalacturonic acid." *Agricultural and Biological Chemistry* 52: 1091-1093.

[41] Sakai, T., Sakamoto, T., Hallaert, J. and Vandamme, E. 1993. Pectin, pectinase and protopectinase: production, properties and applications. *Advances in Applied Microbiology* 39: 231-294.

[42] Luh, B. S. and Phaff, H. J. 1951. "Studies on polygalacturonase of certain yeasts." *Archives of Biochemistry and Biophysics* 33: 212-227.

[43] Murad, H. A. and Azzaz, H. H. 2011. "Microbial pectinases and ruminant nutrition." *Research Journal of Microbiology* 6: 246.269.

[44] Marcus, L., Barash, I., Sneh, B., Koltin, Y. and Finker, A. 1986. "Purification and characterization of pectolytic enzymes produced by virulent and hypovirulent isolates of *Rhizoctonia solani* Kuhn." *Physiological and Molecular Plant Pathology* 29: 325-336.

[45] De Lorenzo, G., Salvi, G., Degra, L., Dovidio, R. and Crevone, F. 1987. "Introduction of extracellular polygalacturonases and its mRNA in the phytopathogenic fungus *Fusarium moniliforme*." *Journal of General Microbiology* 133: 3365-3373.

[46] Polizeli, M. L. T. M., Jorge, J. A. and Terenzi, H. F. 1991. "Pectinase production by *Neurospora crassa*: purification and biochemical characterization of extracellular polygalacturonase activity." *Journal of General Microbiology* 137: 1815-1823.

[47] Manachini, P. L., Fortina, M. G. and Parini, C. 1987. "Purification and properties of an endopolygalacturonase produced by *Rhizopus stolonifer*." *Biotechnology Letters* 9: 219-224.

[48] Nagai, M., Katsuragi, T., Terashita, T., Yoshikawa, K. and Sakai, T. 2000. "Purification and characterization of an endo-polygalacturonase from *Aspergillus awamori*." *Bioscience Biotechnology and Biochemistry* 64: 1729-1732.

[49] Kumar, S. S. and Palanivelu, P. 1999. "Purification and characterization of an exoplolygalacturonase from the thermophilic fungus, *Thermomyces lanuginosus*." *World Journal of Microbiology Biotechnology* 15: 643-646.

[50] Souza, J. V. B., Silva, T. M., Maia, M. L. S. and Teixeira, M. F. S. 2003. "Screening of fungal strains for pectinolytic activity: endopolygalacturonase production by *Peacilomyces clavisporus* 2A.UMIDA.1." *Process Biochemistry* 4: 455-458.

[51] Palomaki, T. and Saarilahti, H.T. 1997. "Isolation and characterization of new C-terminal substitution mutation affecting secretion of polygalacturonases in *Erwinia carotovora* ssp. carotovora." *FEBS Letters* 400: 122-126.

[52] Rodrigues-Palenzuela, P., Burr, T. J. and Collmer, A. 1991. "Polygalacturonase is a virulence factor in *Agrobacterium tumefaciens* biovar 3." *Journal of Bacteriology* 173: 6547-6552.

[53] Tierny, Y., Bechet, M., Joncquiert, J. C., Dubourguier, H. C. and Guillaume, J. B. 1994. "Molecular cloning and expression in *Escherichia coli* of genes encoding pectate lyase and pectin methylesterase activities from *Bacteroides thetaiotaomicron*." *Journal of Applied Bacteriology* 76: 592-602.

[54] Kobayashi, T., Higaki, N., Yajima, N., Suzymastu, A., Hagihara, H. and Kawai S. 2001. "Purification and properties of a galacturonic acid-releasing exopolygalacturonase from a strain of *Bacillus*." *Bioscience Biotechnology and Biochemistry* 65: 842-847.

[55] Nozaki, K., Miyairi, K., Hizumi, S., Fukui, Y. and Okuno, T. 1997. "Novel exopolygalacturonases produced by *Alternaria mali*." *Bioscience Biotechnology and Biochemistry* 61: 75-80.

[56] Maceira, F. I. G., Pietro, A. D. and Roncero, M. I. G. 1997. "Purification and characterization of a novel exopolygalacturonase from *Fusarium oxysporum* f. sp. lycopersici." *FEMS Microbiology Letters* 154: 37-43.

[57] Huang, Q. and Allen, C. 1997. "An exo-poly-a-D-galacturonosidase, Peh B, is required for wild type virulence of *Ralstonia solanacearum.*" *Journal of Bacteriology* 179: 7369-7378.

[58] Rexova´-Benkova´, L., Markovic´, O. 1976. "Pectic enzymes." *Advances in Carbohydrate Chemistry* 33:323–85.

[59] Wijesundera, R. L. C., Bailey, J. A. and Byrde, R. J. W. 1984. "Production of pectin lyase by *Colletotrichum lindemuthionum* in culture and in infected bean (*Phaseolus vulgaris*) tissue." *Journal of General Microbiology* 130: 285-290.

[60] Wattad, C., Freeman, S., Dinoor, A. and Prusky, D. A. 1995. "A nonpathogenic mutant of *Colletotrichum magna* is deficient in extracellular secretion of pectate lyase." *Molecular Plant-Microbe Interactions* 8: 621-626.

[61] Drori, N., Kramer-Haimovich, K., Rollins, J., Dinoor, A., Okon, Y., Pines, O. and Prusky, D. 2003. "External pH and nitrogen source affect secretion of pectate lyase by *Colletotrichum gloeosporioides.*" *Applied and Environmental Microbiology* 69: 3258-3262.

[62] McCarthy, R. E., Kotarski, S. F. and Salyers, A. A. 1985. "Location and characteristics of enzymes involved in the breakdown of polygalacturonic acid by *Bacteroides thetaiotaomicron.*" *Journal of Bacteriology* 161: 493-499.

[63] Kotoujansky, A. 1987. "Molecular genetics of pathogenesis by soft-rot *Erwinia.*" *Annual Review of Phytopathology* 25: 405-430.

[64] Margo, P., Varvaro, L., Chilosi, G., Avanzo, C. and Balestra, G. H. 1994. "Pectinolytic enzymes produced by *Pseudomonas syringae* pv. Glycinea." *FEMS Microbiology Letters* 117: 1-6.

[65] Ishii, S. and Yokotsuka, T. 1975. "Purification and properties of pectin lyase from *Aspergillus japonicus.*" *Agricultural and Biological Chemistry* 39: 313-321.

[66] Szajer, I. and Szajer, C. 1985. "Formation and release of pectin lyase during growth of *Penicillium paxilli.*" *Biotechnology Letters* 7: 105-108.

[67] Chen, W. C., Hsieh, H. J. and Tseng, T. C. 1998. "Purification and characterization of a pectin lyase from *Pythium splendens* infected cucumber fruits." *Botanical Bulletin of Academia Sinica* 39: 181-186.

[68] Moharib, S. A., El-Sayed, S. T. and Jwanny, E. W. 2000. "Evaluation of enzymes produced by yeast." *Nahrung* 44: 47-51.

[69] Martins, E. S., Silva, D, Da Silva, R. and Gomes, E. 2002. "Solid state production of thermostable pectinases from thermophilic *Thermoascus aurantiacus.*" *Process Biochemistry* 37: 949-954.

[70] Liao, C. H., Revear, L., Hotchkiss, A. and Savary, B. 1999. "Genetic and biochemical characterization of an exo polygalacturonase and a pectate lyase from *Yersinia enterolitica.*" *Canadian Journal of Microbiology* 45: 396-403.

[71] Collmer, A., Ried, J. L. and Mount, M. S. 1988. "Assay methods for pectic enzymes." *Methods in Enzymology* 161: 329-335.

[72] Sathyanarayana, N. G. and Panda, T. 2003. "Purification and biochemical properties of microbial pectinases-a review." *Process Biochemistry* 38:987–96.

[73] Beg, Q. K., Bhushan, B., Kapoor, M. and Hoondal, G. S. 2000. "Production and characterization of thermostable xylanase and pectinase from *Streptomyces Sp. QG-11-3.*" *Journal of Industrial Microbiology and Biotechnology* 24: 396–402.

[74] Reid, I. and Michelle, R. 2000. "Pectinase in papermaking: solving retention problems in mechanical pulps bleached with hydrogen peroxide." *Enzyme and Microbial Technology* 26: 115–23.

[75] Alkorta, I., Garbisu, C., Llama, M. J. and Serra, J. L. 1998. "Industrial applications of pectic enzymes: a review." *Process Biochemistry* 33:21-28.

[76] Chovatiya, P. and Shaikh, N. 2013. "Ruminant and microbial pectinase." *CIBTech Journal of Microbiology* 2:1-9.

[77] Holker, U., Hofer, M. and Lenz, J. 2004. "Biotechnological advantages of laboratory-scale solid state fermentation with fungi." *Applied Microbiology and Biotechnology* 64: 175-186.

[78] Hoondal, G., Tiwari, R., Tewari, R., Dahiya, N. and Beg, Q. 2002. "Microbial alkaline pectinases and their industrial applications: a review." *Applied Microbiology and Biotechnology* 59: 408-418.

[79] Garg, G., Singh, A., Kaur, A., Singh, R., Kaur, J. and Mahajan, R. 2016. "Microbial pectinases: an ecofriendly tool of nature for industries." *3 Biotech* 6: 47.

[80] Flint, H. J., Scott, K. P., Duncan, S. H. and Louis, P. 2012. "Microbial degradation of complex carbohydrates in the gut." *Gut Microbes* 3:289–306.

[81] Holscher, H. D. 2017. "Dietary fiber and prebiotics and the gastrointestinal microbiota." *Gut Microbes* 8:172-184.

[82] Schley, P. D. and Field, C. J. 2002. "The immune-enhancing effects of dietary fibres and prebiotics." *British Journal of Nutrition* 87:S221–30.

[83] Shobha, M. S., Kumar, A. B. V., Tharanathan, R. N., Koka, R. and Gaonkar, A. K. 2005. "Modification of guar galactomannan with the aid of *Aspergillus niger* pectinase." *Carbohydrate Polymers* 62:267–273.

[84] Tolhurst, G., Heffron, H. and Larn, Y. S. 2012. "Short chain fatty acids stimulate glucagon –like peptide -1 secretion via the G protein- coupled receptor FFA2." *Diabetes* 61: 2364-71.

[85] Khan, M., Nakkeeran, E. and Sukumaran, U. K. 2013. "Potential application of pectinase in developing functional foods." *Annual Review of Food Science and Technology* 4:21-34.

[86] Tapre, A. R. and Jain, R. K. 2014 "Pectinases: enzymes for fruit processing industry." *International Food Research Journal* 21: 447-453.

[87] Jacob, N. 2009. "Pectinolytic enzymes." *Biotechnology for agro-industrial residues utilisation.* Dordrecht: Springer.

[88] Biz, A., Farias, F. C., Motter, F. A., De Paula, D. H., Richard, P., Krieger, N. and Mitchell, D. A. 2014. "Pectinase activity determination: an early deceleration in the release of reducing sugars throws a spanner in the works!" *PLoS ONE* 9: e109529.

[89] Edwards, M. C. and Doran-Peterson, J. 2012. "Pectin-rich biomass as feedstock for fuel ethanol production." *Applied Microbiology and Biotechnology* 95: 565–575.

[90] Ghorai, S., Banik, S. P., Verma, D., Chowdhury, S., Mukherjee, S., Khowala, S. 2009. "Fungal biotechnology in food and feed processing." *Food Research International* 42: 577-586.

[91] Velasco, D., Senit, J. J., Isabel, T. D. L., Santos, T. M., Yustos, P., Santos, V. E. and Ledero, M. 2017. "Optimization of the enzymatic saccharification process of milled orange wastes." *Fermentation* 3: 37.

[92] Collares, R. M., Miklasevicius, L. V., Bassaco, M. M., Salau, N. P., Mazutti, M. A., Bisognin, D. A. and Terra, L. M. 2012. "Optimization of enzymatic hydrolysis of cassava to obtain fermentable sugars." *Journal of Zhejiang University. Science B* 13:579-86.

[93] Lozano, J.E. 2006. "Fruit Manufacturing: Scientific basis, engineering properties and deteriorative reactions of technological importance." *Food Engineering Series,* USA: Springer.

[94] Lanzarini, G. and Pifferi, P. G. 1989. "Enzymes in the fruit juice industry." *Biotechnology applications in beverage production.* Elsevier applied food science series. New York: Springer.

[95] Kumar, S. 2015. "Role of enzymes in fruit juice processing and its quality enhancement." *Advances in Applied Science Research* 6: 114-124.

[96] Baker, R. A. and Wicker, L. 1996. "Current and potential application of enzyme infusion in the food industry." *Trends Food Science and Technology* 7:279-84.

[97] Blanco, P., Sieiro, C. and Villa, T. G. 1999. "Production of pectic enzymes in yeasts." *FEMS Microbiology Letters* 175: 1-9.

[98] Mojsov, K. 2013. "Use of enzymes in wine making: A review." *International Journal of Market Technology* 3: 112-127.

[99] Mojsov, K. D., Andronikov, D., Janevski, A., Jordeva, S. and Zezova, S. 2015. "Enzymes and wine – the enhanced quality and yield." *Advance Technology* 4: 94-100.

[100] Vierhuis, G., Servili, M., Baldioli, M., Schols, H. A. and Voragen, A. G. J. 2001. "Effect of enzyme treatment during mechanical extraction of olive oil on phenolic compounds and polysaccharides." *Journal of Agricultural and Food Chemistry* 49: 1218-1223.

[101] Kapoor, M., Beg, Q. K., Bhushan, B., Singh, K., Dadich, K. and Hoondal, G. 2001. "Application of an alkaline and thermostable polygalacturonase from *Bacillus sp.* MG-cp-2 in degumming of ramie (*Boehmeria nivea*) and sunn hemp (*Crotalaria juncea*) bast fibers." *Process Biochemistry* 36:803–807.

[102] Tanabe, H., Yoshihara, K., Tamura, K., Kobayashi, Y., Akamastu, I., Niyomwan, N. and Footeakul, P. 1987. "Pretreatment of pectic wastewater from orange canning process by an *Alkalophilic bacillus sp.*" *Journal of Fermentation Technology* 65: 243-246.

In: Microbial Catalysts. Vol. 2

Editors: Shadia M. Abdel-Aziz et al.

ISBN: 978-1-53616-088-8

© 2019 Nova Science Publishers, Inc.

Chapter 11

NARINGINASE: APPLICATIONS AND BIOTECHNOLOGICAL ASPECTS

Shadia M. Abdel-Aziz[1], Khalid T. Biobaku[2],
Ismail A. Odetokun[3],* and Abeer N. Shehata[4]

[1]Microbial Chemistry Department, Genetic Engineering and Biotechnology Research Division, National Research Centre, Dokki, Giza, Egypt

[2]Department of Veterinary Pharmacology and Toxicology, University of Ilorin, Ilorin, Nigeria

[3]Department of Veterinary Public Health and Preventive Medicine, University of Ilorin, Ilorin, Nigeria

[4]Biochemistry Department, Genetic Engineering and Biotechnology Research Division, National Research Centre, Dokki, Giza, Egypt

ABSTRACT

Industrial use of microbial enzymes has increased greatly in 21st century and continuously increases due to their significant potential. These microbial enzymes have gained global recognition for their widespread use in various sectors of industries. Naringinase, a multienzyme complex, possesses α-L-rhamnosidase and β-glucosidase active centers. Naringinase has debittering properties and have applied commercially for its ability to breakdown the compound naringin in citrus juices which confers a bitter taste. Firstly, rhamnosidase breaks naringin into prunin and rhamnose, followed by breaking prunin into glucose and naringenin by glucosidase. Naringinase had gained a great attention in recent years due to its numerous applications and safety. Hydrolytic activities of naringinase include production of rhamnose, prunin, and debittering of citrus fruit juices. Fungal naringinases have been extensively studied during the past decade and used industrially in large amounts. These enzymes improved the trend of usage as organic based feed additives which create a new ground in organic agriculture. This decreases the

---

* Corresponding Author's E-mail: odetokun.ia@unilorin.edu.ng, ismail23us@gmail.com.

usage of synthetic chemicals in animal and could have an appreciable impact on the national gross domestic product if adopted in agro-allied industry. Further research should be directed towards biomolecular engineering to get strains with high specificity and action directed for high safety degree.

Keywords: naringinase, α-L-rhamnosidase, debittering, feed additives, microbial enzymes

1. INTRODUCTION

Enzymes are one of the most important biomolecules that have a wide range of industrial applications. Industrial uses of microbial enzymes have increased greatly in 21st century and continuously increase as enzymes have significant potential for many industries. Enzymes had potential applications in various industrial sectors include pharmaceuticals, food, feed, beverages as well as paper and pulp, detergents, and leather processing [1, 2]. Microbial enzymes have significant importance in waste management, and consequently have a green biotechnology and environmentally-friendly features. Enzymes are currently among the well-established products in biotechnology. The global enzymes market was valued at $7,082 million in 2017, and is projected to reach $10,519 million in 2024, registering a CAGR (compound annual growth rate) of 5.7% from 2018 to 2024 [3]. Enzymes are biological molecules, proteinaceous in nature with the exception of catalytic RNA molecules (ribozymes), and act as catalyst to support almost all of the chemical reactions required to sustain life [4]. Enzymes are highly specific and accelerate the rate of particular biochemical reaction by lowering the activation energy without undergoing any permanent change in them, and therefore, are vital biomolecules that support life [5].

Microbial enzymes have gained recognition globally for their widespread use in diverse sectors of industries, e.g., food, agriculture, chemicals, medicine, and energy. Enzymes of safe organic origin such as naringinase are rapidly gaining the world's interest because of reduced process time, intake of low energy input, cost effective, nontoxic and eco-friendly characteristics [6, 7, 8]. Naringinase is a debittering enzyme used for commercial production of citrus juices. It breaks down the compound naringin that gives citrus juices its bitter taste [8]. Naringinase is a hydrolytic enzyme containing both α-L-rhamnosidase and β-glucosidase activities. Firstly, α-L-rhamnosidase hydrolyzes naringin into rhamnose and prunin (4,5,7-trihydroxy flavanone-7-glucoside), the prunin is then simultaneously converted into glucose and naringenin (4,5,7-trihydroxy flavanone) by the β-glucosidase activity [9, 10]. The enzyme code number of the naringinase and rhamnosidase are the same (EC 3.2.1.40).

Naringin and limonin are two bitter components of some citrus products such as grapefruit juice. Naringin could be removed by hydrolysis with immobilized naringinase and limonin might be removed by adsorption with cellulose monoacetate gel beads or cellulose triacetate gel beads [11]. Naringinases are applied in production of glycopeptide antibiotic, deglycosylation of flavonoids, and gellan depolymerization [12, 13]. The products of hydrolysis (rhamnose, prunin, and naringenin) showed biological activities and are considered as precursors for synthesis of substances applied in pharmaceutics, cosmetics, and food technology [14]. Recent reviews on the status of naringinases are somewhat scarce [15, 16, 17, 18]. This chapter has focused on applications of microbial naringinase in juice industry and its potential applications in biotechnology.

2. NARINGIN AND NARINGENIN

Naringin and naringenin are antioxidant constituents of many citrus fruits which are very beneficial in the cells of humans and animals by mopping up reactive oxygen, nitrogen oxygen species, and other radicals to prevent cell death. Naringenin is a metabolite of naringin [19]. Naringin is metabolized to the aglycone naringenin (has no bitter taste) by naringinase present in the gut, in humans. Both compounds are naturally found in high concentrations in grapefruits, citrus fruits, tomatoes, and grapes. Naringenin is the predominant flavanone in grapefruit while naringin is the major flavonoid glycoside in grapefruit, conferring the fruit its bitter taste. Naringin has two more sugar molecules attached to it, while naringenin does not possess these sugar molecules. Plants store their energy as sugars so by attaching sugars to molecules such as naringin. Solubility of naringin and naringenin in different solvents increases with an increase of temperature. Comparative studies on the solubility phenomena of naringin and naringenin in different solvents are less studied and often very scarce.

Several biological activities have been ascribed to the phytochemical naringenin, among them antioxidant, antitumor, antiviral, antibacterial, anti-inflammatory, antiadipogenic and cardioprotective effects. Nonetheless, most of the data reported have been obtained from *in vitro* or *in vivo* studies. Although some clinical studies have also been performed, the main focus is on naringenin bioavailability and cardioprotective action. In addition, these studies were done in compromised patients (i.e., hypercholesterolemic and overweight), whereas the effect on healthy volunteers is still debatable. In fact, naringenin ability to improve endothelial function has been well-established. Indeed, the currently available data are very promising, but further research on pharmacokinetic and pharmacodynamic aspects is encouraged to improve both available production and delivery methods and to achieve feasible naringenin-based clinical formulations [20, 21].

Use of natural antioxidants can play a vital role to extend the shelf life. Both *in vivo* and *in vitro* studies have recognized the worth of naringenin for several preclinical models of neurodegenerative disorders, cardiovascular diseases, osteoporosis, atherosclerosis, rheumatological disorders, and diabetes mellitus [22]. Moreover, naringenin plays a chief role in lowering cholesterol, triglycerides and improvements in immune functions and anti-oxidant status, as reported in human and different animal model studies. This flavonoid has faced limited research and usage in the poultry production industry, although it has many promising biological effects [22]. Nevertheless, few some studies have been carried out on citrus bioflavonoids and/or various citrus fruits for their possible applications in poultry. Naringin could boost humoral and mucosal immunity in animal and poultry.

2.1. Debittering of Naringin

Naringin must be removed or reduced from the processed juice products. The active principals responsible for bitterness in orange and grapefruit are flavonoids (naringin) and limonoids (limonin). This bitter taste could be decreased by naringinase which made more wholesome and acceptable in taste, especially naringinases that produced by microorganisms. Nowadays, microbial naringinase has completely replaced the chemical methods due to cost effective production and economically viable process [10]. Studies on the filamentous fungi, like *Penicillium* sp., *Aspergillus* sp., and *Rhizopus* sp., have confirmed the ability of these fungi to produce naringinase when grown on a bitter compound like naringin as a substrate [21].

2.2. Flavonoid Naringin

Recently, it is reported that citrus flavonoid naringin is a promising nutritional supplement that showed beneficial health implications in humans, animals and poultry [22]. Flavonoids are ubiquitous plant metabolites, that constitute an important group of natural compounds with various biologic activities and have been the subject of great interest for scientific research [22]. Citrus flavonoids have been established as an important sub-class of flavonoids. Citrus flavanones like naringin play an important role as anti-inflammatory, ant-oxidative, anti-apoptotic, antidepressant, hypolipidemic, immunoregulatory, hepatoprotective, wound healing, anti-diabetic and antihyperglycemic agent [22, 23]. Naringin plays a critical role in lowering triglycerides, cholesterol, improve immune functions, and anti-oxidative status. Naringin could be considered as a natural anti-oxidant due to being a strong scavenger of free radicals and preventing lipid peroxidation. It constitutes a major category of nature-derived bioactive compounds, has potent anti-oxidant and anti-inflammatory effects that render it as a promising dietary supplement in animal and poultry feeds. Naringin citrus flavonoid could be used as a natural feed additive to improve health and has potential to lower medicinal cost in animal and poultry industry. Overall biological activities of naringin are represented in Table 1.

3. NARINGINASE

Naringinase has been reported in the literature since 1938, initially in isolates from celery seeds and later in grapefruit leaves [24, 25]. In 1955, one of the first reports focused on naringinase production from molds was published by Kishi [26]. In this study, an estimated number of 96 strains were explored, and *Aspergillus niger* was established as the best producer of naringinase [26]. Subsequently, the enzyme has been isolated from several strains including bacteria [27], yeast [28, 29], and fungi [16, 30, 31, 32]. Citrus fruit juice has potential consumer demand due to its nutritional value and taste. Naringin is one of the bitter components that found in citrus fruits. During extraction of juice from the citrus fruits, naringin will impart the bitterness taste to the juice. Therefore, alteration of naringin into bitter-free component is an important step to achieve consumer acceptance of citrus juice [33]. On other side, citrus peel is considered as waste in the food industry, which has naringin as a principle component. Technology of exploiting citrus pulp with peel have potential application in food industry, pharmaceutical industry, debittering of citrus juice as well as production of citrus peel concentrate, production of antibiotics, anti-inflammatory and antiviral component from citrus peel [33]. Among the several methods employed, the enzymatic degradation was found to be a cost-effective method. Naringinase expresses activity on α-L-rhamnosidase and β-D-glucosidase.

Many natural glycosides, including naringin, rutin, quercitrin, hesperidin, diosgene, and ter-phenyl glycosides, containing terminal α-rhamnose and β-glucose can act as substrates of naringinase [15]. Naringinase, entrapped in alginate beads, was used in the debittering of grapefruit (*Citrus aurantium*) juice. Bitterness was reduced and resulted in 84% naringin hydrolysis [43]. The fenugreek seeds (*Trigonella foenum graecum*) upon enzymatic hydrolysis by naringinase produced sapogenins and diosgenin, precursors of clinically useful steroid drugs [15]. α-L-Rhamnosidase expressed by naringinase can be used in preparation of many drugs and drug precursors. α-L-rhamnosidase hydrolyzes the diosgene (a saponin) to produce α-L-rhamnose and diosgenin which is used in synthesis of clinically useful steroid drugs analogues such as progesterone [44].

Microorganism are the main sources of naringinase, although these enzymes have also been found in liver tissues of the marine gastropod *Turbo cornutus*, pig, and plant sources such as celery seeds and grapefruit leaves [16]. Processes based on microbial naringinases are, however, feasible for industrially practicable [16, 29, 31, 33]. Naringinase is produced by many microorganisms, mainly filamentous fungi (*Aspergillus, Circinella, Eurotium, Fusarium, Penicillium, Rhizopus*, and *Trichoderma*) [31, 32, 33, 45]. Microorganisms as a source for naringinase are represented in Table 2.

Table 1. Biological activities of the flavonoid naringin

Naringin	Effect	Reference
Feed additive: For rats, animals, rabbits. For production of healthier chicken meat in poultry industry.	Supplementation of naringin had significantly decreased the detrimental effect of the higher levels of liver enzymes in the plasma. Induced hepatotoxicity, significantly decreased the lipid peroxidation and restored the activities of anti-oxidative enzymatic indices like glutathione peroxidase, glutathione S-transferase, and catalase in the liver. Naringin has the ability to prevent the upsurge in hepatic biomarkers of enzyme activities and decreased the hepatic fibrosis and lipid deposition in the high-fat-diet and high- carbohydrate-fed, in obese rats. Both naringin and naringenin are strong free radical scavengers and have the potential as lipid peroxidation inhibitors *in vivo*. Dietary uptake of naringin had significantly alleviated antioxidant enzymes including catalase and peroxidase in diabetic rats. A research in rats, combined naringin treatment and vitamin C, had significantly play a vital role in treating diabetes related disorders. Naringin improves the wound healing process as they have anti-inflammatory, antimicrobial, anti-oxidant and many astringent properties. The mode of action of naringin in wound healing process is probably due to its' radical scavenging and anti-oxidant effects.	[34, 35, 36, 37, 38, 39, 40]
Human Supplement: Treatment of hypertension, obesity, and inflammations	Naringin can play an essential role in plasma cholesterol lowering, regulation of the anti-oxidant capacity in hypercholesterolemic subjects. Reduces plasma cholesterol by about 14% and LDL concentration by about 17%. Could be beneficial for inflammatory changes alleviation in the adipose tissue. Effective in cases of metabolic syndrome, diabetes and obesity associated with the inflammatory responses. Mitigates hypertension and reduce inflammations in a high carbohydrate/high fat ration fed rats.	[23, 41, 42]

Table 2. Microorganisms as a source for naringinase

Source	Microorganism	References
Bacteria	*Pseudomonas paucimobilis, Clostridium stercorarium, Sphingomonas paucimobilis, Penicillum ulaiense, Lactobacillus plantarum, Lactobacillus acidophillus, Staphylococcus xylosus* Bacillus methylotrophicus, Bacillus amyloliquefaciens 11568	[47], [48], [49], [50], [51], [52], [53], [17], [27]
Yeast	*Cryptococcus laurentii, Clavispora lusitaniae,* Clavispora lusitaniae	[54], [55], [29]
Fungi	*Penicillium decumbens, Aspergillus kawachii, Penicillium ulaiense, Aspergillus sojae, Rhizophus stolonifer, Aspergillus flavus, Aspergillus oryzae* 11250, Cryptococcus albidus	[56], [57], [50], [58], [30], [45] [31], [32]

4. BIOTECHNOLOGICAL APPLICATIONS OF NARINGINASE

An advantage of enzyme technology is that it presents an alternative to chemical processes, reducing both energy and material consumption, and thus minimizing the generation of waste. In this context, naringinases have been demonstrated to have applications in industrial uses. These applications are mainly based on the hydrolytic activity of naringinase [16]. Biotechnological applications of naringinase, with respect to the food and pharmaceutical industry are reported [16]. In food additives manufacture, rhamnosidase could be used in preparation of sweeteners [59].

Large amounts of citrus peel (rich in glycosylated poly-phenolic compounds) are generated as a byproduct of the juice processing industry. Hydrolysis of naringin extracted from citrus peel waste was potentially performed using the recombinant α-L-rhamnosidase in manufacture of rhamnose. Recombinant α-L-rhamnosidase has proved to possess industrial applicability as being an interesting candidate for the production of rhamnose and prunin from citrus peel waste [29]. The rhamnosidases from *L. plantarum* have been shown to convert flavonoid rutinosides (such as rutin from tomato) into well-absorbed glucosides. Such activity implies that probiotic lactobacilli when present in gut microflora may enhance flavonoid bioavailability [52]. Recently, it is reported that citrus flavonoid naringin is a promising nutritional supplement that showed beneficial health applications in humans, animals and poultry [22].

The flavonoid prunin, prepared using naringinase, possesses anti-inflammatory and antiviral activity against DNA/RNA viruses [60]. Pure prunin in high yield was obtained from naringin when immobilized naringinase pretreated with alkaline buffer. The flavonoid prunin possesses anti-inflammatory activity and may be used as sweetening agent in diabetes therapy [60]. Plant flavonoids may be useful for the treatment of cardiovascular disease as well as associated conditions such as obesity, hepatic steatosis, and type 2 diabetes [61]. Flavonoid naringenin-7-O-glucoside is a potential therapeutic agent for treating or preventing cardiomyopathy associated with doxorubicin [62, 63].

5. FURTHER PROSPECTS AND CHALLENGES

There is a great intense interest towards the application of naringinase in the biotech industry. However, more studies on fermentation processes on an industrial scale should be carried out to secure cost-effective availability of naringinases [16]. There is a scientific need to crystallize naringinase to obtain structural variation and complete hydrolysis of its natural substrate, naringin. Directed evolution may result in mutants with remarkable/unexpected changes that eventually lead to increased enzyme activity [64]. Thus, optimization of fermentation conditions and enzyme engineering will allow development of improved naringinases. The enzymatic debittering technology is regarded as the most promising method with the advantages of high specificity and efficiency, and a convenient method for removing the bitterness. Bio-enzymatic debittering by limonin dehydrogenase and naringinase can become the main direction of citrus juices debittering and enhance the flavor and aroma in citrus juices which could be stored for 30 days [29]. This technology can redress the problem by exploiting the ability of yeast species to produce low alcoholic naturally carbonated beverage from nutritive fruits thus making the fruit available throughout the year in the form of beverage [29]. The beverage offers advantages include: *a*) devoid of any chemical preservatives; *b*) minimally processed with high nutritive value; *c*) long shelf life; *d*) availability of new formulations and blends; *e*) tangy taste, effervescent, and antimicrobial due to carbonation [29]. The search for enzymes that have potential application in acidic environment, especially the debittering of acidic juices of diverse citrus species, should be essential. So, it will be important and attractive to summarize what the current market of the industrial enzyme is, and what the difficulties are? Moreover, demonstration of bottlenecks in the enzyme production and novel strategies to overcome the barriers are challenges [65].

CONCLUSION

Naringinase is an important enzyme and has potential application in food and pharmaceutical industries. Naringinases that show better stability in acidic pH and low temperatures are highly preferred for industrial and biotechnological applications. Therefore, search for enzymes have potential application in acidic environment, especially the debittering of acidic juices of diverse citrus species, should be essential. In addition, when enzymes are applied at low temperature, there would be no loss of nutritional and sensory qualities of juice. The debittering enzymes can offer advantages such as cost effectiveness, single step hydrolysis, short incubation, preservation of flavour, retention of color, vitamins and organoleptic components of juice. On other hand, naringin, a major flavanone glycoside mainly found in grapefruits and citrus family, had been reported to possess different promising pharmacological characteristics. These include: anti-inflammatory, antimicrobial, antimutagenic, anti-oxidant, cholesterol lowering, hepatoprotective, cardioprotective, antiatherogenic, antiulcer and neuroprotective effects. It could boost humoral and mucosal immunity in animal and poultry.

REFERENCES

[1]   Singh R, Kumar M, Mittal A, et al. 2016. Microbial enzymes: industrial progress in 21st century. *3 Biotech.*, 6: 174-188.

[2]   Yimer D, Tilahun A. 2018. Microbial biotechnology review in microbial enzyme production methods, assay techniques and protein separation and purifications. *J Nutr Health Food Eng.*, 8, 00249-00256.

[3]   *Allied Market Research.* 2018. https://www.prnewswire.com.

[4] Cech T, Bass B. 1986. Biological catalysis by RNA. *Ann Rev Biochem.*, 55: 599–629.

[5] Fersht A. 1985. *Enzyme structure and mechanism.* 2nd Edition. Publisher: W H Freeman & Co (Sd).

[6] Choi J, Han S. and Kim H. 2015. Industrial applications of enzyme biocatalysis: current status and future aspect. *Biotechnol Adv.*, 33: 1443–1454.

[7] Zhu Y, Jia H, Xi M, et al. 2017. Characterization of a naringinase from *Aspergillus oryzae* 11250 and its application in the debitterization of orange juice. *Process Biochem.*, 62: 114-121.

[8] Sindhu R, Binod P, Beevi S, et al. 2018. Applications of microbial enzymes in food industry. *FTB Food Technol. Biotechnol.*, 56: 16-30.

[9] Yusof S, Ghazali H. and King G. 1990. Naringin content in local citrus fruits. *Food Chem.*, 37: 113–121.

[10] Puri M, Banerjee U. 2000. Production, purification, and characterization of the debittering enzyme naringinase. *Biotechnol Adv.*, 18: 207–217.

[11] Hau-Yang T, Gee-Kaite Y. 1991. Lemonin and naringin removal from grapefruit juice with naringinase entrapped in cellulose triacetate fibers. *Food Sci.*, 56: 31-34.

[12] Kamiya S, Esaki S. and Tanaka R. 1985. Synthesis of some disaccharides containing an L-rhamnopyranosyl or L-mannopyranosyl residue, and the substrate specificity of alpha-L-rhamnosidase from *Aspergillus niger. Agric Biol Chem.*, 49: 55–62.

[13] Puri M, Kaur H. and Kennedy J. 2005. Covalent immobilization of naringinase for the transformation of a flavonoid. *J Chem Technol Biotechnol.*, 80: 1160–1165.

[14] Ellenrieder G, Blanco S. and Daz M. 1998. Hydrolysis of supersaturated naringin solutions by free and immobilized naringinase. *Biotechnol Technol.*, 12: 63–65.

[15] Puri M. 2012. Updates on naringinase: structural and biotechnological aspects. *Appl Microbiol Biotechnol.*, 93: 49-60.

[16] Pooya M. 2015. *Industrial enzymes (naringinase).* University of Tehran.

[17] Jian L, Qian L, Sun X, et al. 2019. The study of the characteristics and hydrolysis properties of naringinase immobilized by porous silica material. *RSC Adv.*, 9: 4514-4520.

[18] Yadav M, Sharma A, Singh R, et al. 2019. Aspartase, asparaginase, and naringinase: Current status and perspectives for the food industry. In: *Microbial Enzymes and Additives.* [Editors: Amit Kumar et al.], Nova Science Publishers.

[19] Hsiu S, Huang T, Hou Y, et al. 2002. Comparison of metabolic pharmacokinetics of naringin and naringenin in rabbits. *Life Sci.*, 70: 1481-9.

[20] Rashmi R, Magesh S, Kunka M, et al. 2018. Antioxidant potential of naringenin helps to protect liver tissue from streptozotocin-induced damage. *Repor Biochem Molecular Biol.*, 7: 76-84.

[21] Salehi B, Fokou P, Mehdi Sharifi-Rad M, et al. 2019. The Therapeutic Potential of Naringenin: A Review of Clinical Trials. *Pharmaceuticals*, 12: 11-29.

[22] Changxing L, Saeed M, Kamboh A, et al. 2018. Reconsidering a citrus flavonoid naringin as a promising nutritional supplement and its beneficial health applications in humans, animals and poultry. *Intr J Pharm*. DOI: 10.3923/ijp.

[23] Alam M, Subhan N, Rahman M, et al. 2014. Effect of citrus flavonoids, naringin and naringenin, on metabolic syndrome and their mechanisms of action. *Adv Nutr.*, 5: 404-417.

[24] Thomas D, Smythe C. and Labbee M. 1958. Enzymatic hydrolysis of naringin, the bitter principle of grapefruit. *Food Res*, 23: 591-598.

[25] Ting SV. 1958. Enzymatic hydrolysis of naringin in grapefruit. *J Agric Food Chem.*, 6: 546-549.

[26] Kishi K. 1955. Production of naringinase from *Aspergillus niger*. *Chem Ind Jpn.*, 29: 140.

[27] Zhu Y, Jia H, Xi M, et al. 2017*a*. Purification and characterization of a naringinase from a newly isolated strain of *Bacillus amyloliquefaciens* 11568 suitable for the transformation of flavonoids. *J Food Chem.*, 214: 39-46.

[28] Singh P, Sahota P. and Singh R. 2015. Evaluation and characterization of new α-L-rhamnosidase-producing yeast strains. *J General Appl Microbiol.*, 61: 149-156.

[29] Kaur M, Sahuta P, Sharma N, et al. 2018. Enzymatic production of debittered Kinnow juice and beverage. *Int J Cur Microbiol App Sci.*, 7: 1180-1186.

[30] Karuppaija S, Kapilan R. and Seevaratnam V. 2016. Characterization of best naringinase producing fungus strain isolated from palmyrah (Borrasus flabellifer) fruit pulp. *Intr. J. Biol. Res.*, 4: 97-101.

[31] Zhu Y, Jia H, Xi M, et al. 2017*b*. Characterization of a naringinase from *Aspergillus oryzae* 11250 and its application in the debitterization of orange juice. *Process Biochemistry*. http://dx.doi.org/10.1016/j.procbio.2017. 07.012.

[32] Borzova N, Olena G. and Lyudmila V. 2018. Purification and characterization of a naringinase from *Cryptococcus albidus*. *App Biochem Biotechnol*, 184: 953–969.

[33] Srikantha K, Kapilan R. and Seevaratnam V. 2016. Characterization of best naringinase producing fungus isolated from the citrus fruits. *Intr J Biol Res.*, 4: 83-87.

[34] Zhang J, Gao Z, Liu Z, et al. 2014. Systematic analysis of main constituents in rat biological samples after oral administration of the methanol extract of fructus aurantii by HPLC-ESI-MS/MS. *Iran. J. Pharm. Res.*, 13: 493-503.

[35] Hung U, Lee M, Jeong K, et al. 2004. The hypoglycemic effects of hesperidin and naringin are partly mediated by hepatic glucose-regulating enzymes in C57BL/KsJ-db/db mice. *J. Nutr.*, 134: 2499-2503.

[36] Pari L, Amudha K. 2011. Hepatoprotective role of naringin on nickel-induced toxicity in male Wistar rats. *Eur J Pharmacol.*, 650: 364-370.

[37] Renugadevi J, Prabu S. 2010. Cadmium-induced hepatotoxicity in rats and the protective effect of naringenin. *Exp Toxicol Pathol.*, 62: 171-181.

[38] Jeon S, Bok S, Jang M, et al. 2002. Comparison of antioxidant effects of naringin and probucol in cholesterol-fed rabbits. *Clin. Chim. Acta*, 317: 181-190.

[39] Kandhare A, Patil A, Sinha A, et al. 2016. Wound healing potential of naringin ointment formulation via regulating the expression of inflammatory, apoptotic and growth mediators in experimental rats. *Pharm Biol.*, 54: 419-432.

[40] Kamboh A, Zhu W. 2013. Effect of increasing levels of bioflavonoids in broiler feed on plasma anti-oxidative potential, lipid metabolites and fatty acid composition of meat. *Poult Sci.*, 92: 454-461.

[41] Alam M, Kauter K. and Brown L. 2013. Naringin improves diet-induced cardiovascular dysfunction and obesity in high carbohydrate, high fat diet-fed rats. *Nutrients*, 5: 637-650.

[42] Hirai S, Takahashi N. Goto T, et al. 2010. Functional food targeting the regulation of obesity-induced inflammatory responses and pathologies. *Mediators Inflam.* DOI:10.1155/2010/367838.

[43] Mishra P, Kar R. 2003. Treatment of grapefruit juice for bitterness removal by amberlite IR 120 and amberlite IR 400 and alginate entrapped naringinase enzyme. *J Food Sci*, 68: 1229-1233.

[44] Elujoba A, Hardman R. 1987. Diosgenin production by acid and enzymatic hydrolysis of fenugreek. *Fitoterapia*, 58: 299-303.

[45] Srikantha K, Kapilan R. and Seevaratnam V. 2017. Kinetic properties and metal ion stability of the extracellular naringinase produced by *Aspergillus flavus* isolated from decaying citrus maxima fruits. *Intr J Sci Res Env Sci.*, 5: 0071-0081.

[46] Thammawat K, Pongtanya P, Vitchuporn J, et al. 2008. Isolation, preliminary enzyme characterization and optimization of culture parameters for production of naringinase isolated from *Aspergillus niger* BCC 25166. *Kasetsart J Nat Sci*, 42: 61-72.

[47] Miake F, Satho T, Takesue H, et al. 2000. Purification and characterization of intracellular alpha-L-rhamnosidase from *Pseudomonas paucimobilis* FP2001. *Arch Microbiol.*, 173: 65-70.

[48] Zverlov V, Hertel C, Bronnenmeier K, et al. 2000. The thermostable alpha-L-rhamnosidaseR-amA of *Clostridium stercorarium*: biochemical characterization and primary structure of bacterial alpha-L-rhamnose hydrolase, a new type of inverting glycoside hydrolase. *Mol Microbiol.*, 35: 173-179.

[49] Miyata T, Kashige N, Satho T., et al. 2005. Cloning, sequence analysis and expression of the gene encoding *Sphingomonas paucimobilis* FP 2001 rhamnosidase. *Cur Microbiol.*, 51: 105-109.

[50] Rajal V, Graciela A, Guillermo E, et al. 2009. Production, partial purification and characterization of rhamnosidase from *Penicillium ulaiense*. *W J Microbiol Biotechnol.*, 25: 1025-1033.

[51] Avila M, Jaquet M, Moine D, et al. 2009. Physiological and biochemical characterization of the two rhamnosidase of *Lactobacillus plantarum*. *Microbiology*, 155: 2739-2749.

[52] Beekwilder J, Marcozzi D, Samuele V, et al. 2009. Characterization of rhamnosidases from *Lactobacillus plantarum* and *Lactobacillus acidophilus*. *App Env Microbiol.*, 75: 3447-3454.

[53] Puri M, Kaur A, Barrow C, et al. 2011. Citrus peel influences the production of an extracellular naringinase by *Staphylococcus xylosus* MAK 2. *Appl Microbiol Biotechnol.*, 89: 715–722.

[54] Lei S, Xu Y, Fan G, et al. 2011. Immobilization of naringinase on mesoporous molecular sieve MCM-41 and its application to debittering of white grapefruit. *App Surface Sci.*, 257: 4096-4099.

[55] Pandove G, Sahuta P, Gupta N, et al. 2016. Production of low-alcoholic naturally carbonated fermented debittered beverage from W. Murcott mandarin (*Citrus reticulata*) by naringinase producing yeast. *J Food Agr Env.*, 14: 30-34.

[56] Mamma D, Kalogeris E, Kekos D, et al. 2004. Biochemical characterization of the multi-enzyme system produced by *P. decumbens* grown on ruitn. *Food Biotechnol.*, 18: 1-18.

[57] Koseki T, Mese Y, Nishibori N, et al. 2008. Characterization of an α-L-rhamnosidase from *Aspergillus kawachii*. *Appl Microbiol Biotechnol.*, 80: 1007-1013.

[58] Chang H, Lee Y, Bae Y, et al. 2011. Purification and characterization of *Aspergillus sojae* naringinase: the production of pruning exhibiting markedly enhanced solubility with *in vitro* inhibition of HMG-CoA reductase. *Food Chem.*, 124: 234-241.

[59] Giavasis I, Harvey L. and McNeil B. 2000. Gellan gums. *Crit Rev Biotech.*, 20: 177–211.

[60] Kaul T, Middleton E. and Ogra P. 1985. Antiviral effects of flavonoids on human viruses. *J Med Virol.*, 15: 71-79

[61] Hartogh D, Tsiani E. 2019. Antidiabetic Properties of Naringenin: A Citrus Fruit Polyphenol. *Biomolecules*, 9: 99-123.

[62] Wood T, Bhat K. 1988. Methods for measuring cellulase activities. *Methods Enzymol.*, 160:87-112.

[63] Han X, Ren D, Fan P, et al. 2008. Protective effects of narigenin-7-O-glucoside on doxorubicin induced apoptosis in H9C2 cells. *Eur J Pharmacol.*, 581: 47-53.

[64] Jun L. 2018. Development and characterization of an α-l-rhamnosidase mutant with improved thermostability and a higher efficiency for debittering orange juice. *Food Chem.*, 245: 1070-1078.

[65] Nandy SK. 2018. Enzyme Use and Production in Industrial Biotechnology. In: *Research Advancements in Pharmaceutical, Nutritional, and Industrial Enzymology*. [Shashi L. Bharati, Editor]. DOI: 10.4018/978-1-5225-5237-6.ch015.

In: Microbial Catalysts. Vol. 2
Editors: Shadia M. Abdel-Aziz et al.

ISBN: 978-1-53616-088-8
© 2019 Nova Science Publishers, Inc.

Chapter 12

ASPARAGINASE: APPLICATIONS AND STRATEGIES FOR ACRYLAMIDE REDUCTION

Shadia M. Abdel-Aziz[1], Ram Prasad[2,3], Moataza M. Saad[1,] and Eman A. Karam[1]*

[1]Microbial Chemistry Department, Genetic Engineering and Biotechnology Research Division, National Research Centre, Dokki, Giza, Egypt
[2]Amity Institute of Microbial Technology, Amity University Uttar Pradesh, Noida, UP, India
[3]School of Environmental Sciences and Engineering, Sun Yat-Sen University, Guangzhou, China

ABSTRACT

Heating of foods causes formation of a toxicant like acrylamide in processed foods. The main concern regarding possible health effects of acrylamide in foods is its carcinogenicity and genotoxicity (DNA-damaging effects). Acrylamide, a byproduct of the Maillard reaction, arises worldwide concern for the safety measures of various food products. Maillard reaction occurs at high temperatures above 120°C and low moisture conditions. This reaction affects the development of flavor and color. Presently research is mainly focused on minimizing or completely removal of acrylamide formation in processed foods by using asparaginase. Pretreatment with asparaginase is a promising technique for preventing acrylamide formation by direct conversion of the precursor asparagine to aspartic acid and ammonia, while maintaining sensory quality of foods. Main acrylamide-containing food groups are cereal-based products and potato-based products. The major limiting factors responsible for formation of acrylamide in potato and cereal products are reducing sugars (glucose and fructose) and the free amino acid asparagine. Asparaginase from microbial sources has been reported as a method to effectively reduce formation of acrylamide in processed foods. Considering its properties

* Corresponding Author's E-mail: moataza_Saad@hotmail.com.

for processed foods, asparaginase has important chemotherapeutic effects for treatment of lymphoma diseases, particularly, for acute lymphoblastic leukemia and other diseases.

Keywords: acrylamide, asparaginase, microbial asparaginases, toxicity, food safety

1. INTRODUCTION

Foods are naturally processed in heat for enhancing the quality, promoting safety, and improving the sensory characteristic properties of foodstuff. However, heating of foods may also lead to the formation of a toxicant like acrylamide which is naturally occurring as contaminant in processed foods [1]. These include potato chips, French fries, cereal, and bakery products where acrylamide is formed via the Maillard reaction process. This reaction is a non-enzymatic browning reaction responsible for the golden color, desirable aroma, and tasty flavor of foods [1]. The free amino acid asparagine and the reducing sugars, fructose and glucose, have been identified as the main precursors for acrylamide formation in fried potato products, during the Maillard reaction. Fried potato products are rich in such main precursors and are responsible for the dietary exposure to acrylamide [2]. Extensive studies on animals have provided evidence that exposure to acrylamide causes cellular damage in both the nervous and reproductive systems, produces tumors in certain hormonally responsive tissues that is responsible for cumulative neurotoxicity [3]. A carcinogen in rodents and a suspected carcinogen in human can cause gene mutation and DNA damage [2-6].

Acrylamide presents at quite high levels in many food products consumed daily, thus it is essential to evaluate the "margin of exposure" for acrylamide. This value represents the ratio between a particular point on the dose–response curve, leading to tumors in experimental animals and the human intake [7]. The value of margin of exposure gives an indication about the possible extent of the risk. The higher the margin of exposure, the lower the risk of exposure to the component concerned. Daily exposure and intake of acrylamide varies globally depending upon local eating and cooking habits. Fried and baked potato products, biscuits, crisp bread, and coffee are among the foods accounting for the most significant proportion of dietary exposure, either because of high acrylamide content or because of a high daily intake [7, 8, 9]. Addition of the enzyme asparaginase, from microbial sources, has been reported as a method to effectively reduce formation of acrylamide in processed foods. Acrylamide reduction was occurred without affecting the shelf life and sensorial properties of French fries [7]. Another important aspect to consider is that implementation of such treatment requires modifications on the industrial lines and additional investments besides the costs of enzyme [9]. Baking conditions may dramatically affect acrylamide levels of the products. The correlation between Maillard browning and acrylamide formation has been demonstrated; a more intensive frying operation (temperature and time) produces more acrylamide [9]. On other hand, frying at lower temperatures (below 140°C) results in increased frying time and may enhance fat uptake.

Enzymes are proteins that catalyze biochemical reactions and a cornerstone of metabolism. Enzymes have a key role and focus in many manufacturing processes since dates back millennia. Similar to chemo-catalysts, enzymes speed up the rate of reactions without altering the thermodynamics [10]. Unlike chemo-catalysts, enzymes display high chemo- and regioselectivity, therefore minimizing the risk of side-reactions, reduce formation of by-products, and do easily downstream processing. Additionally, enzymes can be produced in large-scale fermentations, operate under mild temperature, and atmospheric pressure [11]. Moreover, enzymes are biodegradable, a feature makes enzyme-based processes to be considered eco-friendly. L-asparaginase is among the enzymes that can be obtained from different microorganisms [10]. This amidohydrolase acts on L-asparagine and produce L-aspartate and ammonia, and has an acknowledged chemotherapeutic application in acute lymphoblastic leukemia and tumors [10, 12]. Moreover, L-asparaginase is of interest in food industry as it reduces acrylamide formation [10]. In this regard, the present chapter is focused on technological strategies for application of asparaginase to reduce acrylamide levels in foods. Potential effect of the enzyme in medicine and therapy is also referred to herein.

2. HISTORICAL BACKGROUND

Importance for the development of L-asparaginase as a potential antineoplastic agent was made by "Clementi" in 1922, revealing presence of high activity of L-asparaginase in the serum of guinea pig [13]. Discovery and development of asparaginase as an anti-cancer drug began in 1953, when scientists first observed that lymphomas in rat and mice regressed after treatment with guinea pig serum [14]. Until 1961, it was found out that it is not the serum itself which provoke the tumor regression, but rather the enzyme asparaginase [15, 16]. After comparing different kinds of asparaginases, it was reported that asparaginases derived from *Escherichia coli* and *Erwinia chrysanthemi* turned out to have the best anti-cancer ability. *E. coli* has thereby become the main source of asparaginase due to the ease of production in large amounts. Since extraction of this enzyme from the guinea pig serum in sufficient amounts was difficult, other sources like microorganisms were looked into. Since 1964, purification of L-asparaginase from *E. coli* was reported and its tumoricidal activity similar to that of guinea pig sera was demonstrated [17, 18]. These findings provided a practical base for large-scale production of the enzyme for pre-clinical and clinical studies. It was not until 1967, efficacy of L-asparaginase in humans with leukemia was reported for the first time [19].

In 1994, the International Agency for Research on Cancer classified acrylamide as a potential carcinogen to humans based on its carcinogenicity in rodents. This classification was endorsed by the WHO (World Health Organization) consultation. In 2002, The Swedish National Food Administration reported relevant amounts of acrylamide in several carbohydrate rich foods when baked at high temperatures (> 120°C) upon frying, baking, and roasting. Toxicological studies demonstrated the carcinogenicity of acrylamide in animals and thus indicated potential health risks for humans. Mechanism of acrylamide formation in foods, risks associated for consumer, and possible strategies to lower acrylamide levels in foodstuffs were greater understanded. The Confederation of the European Food and Drink Industries established a Technical Acrylamide Expert Group in 2003 and created the 'Acrylamide Toolbox' [20]. This toolbox represents a regularly updated and robust medium for the categorization and summarization of formation and mitigation of acrylamide in various foods [20]. In January 2011, the European Commission published a recommendation regarding the acrylamide limits in several food categories. As of 2012, several research results have been published in approximately 850 papers. Use of L-asparaginase in treatment of leukemia and other lymphoproliferative disorders has expanded immensely [7, 20]. Studies have recently explored its administration for 20–30 weeks as consolidation treatment for lymphoid malignancies [10]. For these reasons L-asparaginase has established itself to be an indispensable component in modern procedures of combination chemotherapy.

3. ACRYLAMIDE FORMATION IN FOOD

Heating of foods is an important process for maintaining the quality and safety of foodstuff, but formation of acrylamide takes start in low moisture conditions and high temperatures. Acrylamide (C_3H_5NO), also known as 2-propenamide, is found in cooked foods, fried. and particularly baked starchy products [9]. Formation of acrylamide increases drastically towards the end of frying process due to: *a*) temperatures higher than 170-180°C; *b*) the higher content range of carbohydrates in foods (50-4000 µg Kg^{-1}); and *c*) a range of 5-50 µg Kg^{-1} in protein-rich foods [21, 22, 23]. Maillard reaction is typically occurred at temperatures above 120-165°C (280 to 330 °F), and is responsible for significant color, flavor, and acrylamide developments. At higher temperatures, caramelization becomes more pronounced. Water content, temperature, and pH have been identified as critical parameters for acrylamide formation, i.e., low water content and high temperature favor formation, and low pH reduces acrylamide formation [9]. Seared steaks, pan-fried dumplings, cookies, biscuits, breads as well as toasted marshmallows and many other foods, undergo the *Maillard* reaction. This reaction was named after French chemist Louis-Camille Maillard, who first described it in 1912, while attempting to reproduce biological protein synthesis [21].

Acrylamide exists in two forms: monomer and polymer. Polyacrylamide polymer finds many uses as a coagulant in waste water treatment, clarifying drinking water, grouting agents for the construction of dam foundations and tunnels, and as electrophoresis gels [24]. However, food scientists worldwide have been concerned about presence of acrylamide in food, because of its toxicity. Acrylamide is also an industrial chemical used in manufacture of polyacrylamides.

4. Occurrence of Acrylamide

Acrylamide is present in a wide range of processed foods, and found at very low levels in animal-based food products like meat and fish. It is not found in foods that are not fried or baked such as boiling or microwaving [24]. Exposure to this process contaminant is a public health concern and a priority for the National Food Safety Authorities. Food and Drug Administration in 2016 reported that: *a*) highest concentrations of acrylamide (>100 μg/kg) were found in sweet and savory biscuits; potatoes-based products, and other snacks (not potato based); and *b*) lowest amounts of acrylamide (≤10 μg/kg) were measured in takeaway fish-based meals, coffee and cocoa, branded food drinks, canned or jarred tomatoes, white unsliced bread, canned or jarred beans, mushrooms, cereal products, spreads and dressings as well as canned or frozen fruit, pears and pineapples [1].

Dietary intake of acrylamide is different in countries within individual food types, and variations in raw materials or processing conditions can contribute also to differences in the levels observed [25]. Acrylamide is likely to cause a small, but measurable, tumor incidence (called "neoplastic" effect) or other potential adverse effects such as neurological, pre- and post-natal development, and male reproduction [26].

5. Production of L-Asparaginase

Asparaginase (L-asparagine amidohydrolase, EC 3.5.1.1) catalyzes the hydrolysis of amide group in the side chain of asparagine to produce aspartic acid and ammonium. Asparaginases present in mammals and broadly distributed among living organisms, including birds, animals, and plants [9]. Microorganisms are considered, however, the mainly source for L-asparagine synthesis. L-asparaginases can be produced from bacteria, fungi, and yeast either by solid-state fermentation (SSF) or by submerged fermentation (SmF). Solid-state fermentation is preferred over SmF as it is cost effective, eco-friendly and it delivers high yield of enzyme [10, 27, 28, 29]. In SSF process, agricultural and industrial wastes are utilized as solid substrates, and the contamination level is substantially reduced through low moisture content. Chemistry and applications of L-asparaginase and various methods available for its production are reported in details, especially advantages and limitations of SSF and SmF processes [10, 11, 27, 29]. Submerged fermentation is widely reported for production of L-asparaginase from bacteria.

The relevance of asparaginase is not only limited as an anti-cancer agent, it also possesses a wide range of medical applications include antimicrobial property, treatment of infectious diseases, autoimmune diseases, canine and feline cancer. There is a huge market demand for asparaginase due to its wide range of applications. Industry is, therefore, still search for better-producing microbial sources. For industrial production of L-asparaginase, many factors need to be taken into account aiming to the higher yield and economic viability [27, 28, 30]. Several microorganisms are presented as L-asparaginase producers, however, *E. coli* and *E. chrysanthemi* are the current main bacteria for industrial-scale production in pharmaceutical area, while the fungus *Aspergillus oryzae* is the most used in food industry [22, 28, 30]. Different methods for downstream processes are reported such as centrifugation, filtration, liquid–liquid extraction, chromatography and protein precipitation. Regarding industrial production, protein precipitation is an advantageous technique due to features such as ease scale up, with simple equipment requirements, low costs, and possibility to use large number of precipitants. Additionally, the precipitant agent can be recycled in the final process, reducing the environmental impact associated to its disposal. Actually, precipitation and chromatography are the most steps in downstream process and it is usually combined with traditional techniques to enhance biomolecules purification and process yield [27, 28].

6. APPLICATIONS OF L-ASPARAGINASE

L-Asparaginase is able to reduce the amount of acrylamide in carbohydrate-rich fried and baked foods by deamination of asparagine. Implementing an asparaginase treatment into a range of food products is not a simple undertaking, since for each product the food matrix or components may influence enzyme action and reactivity [27]. Furthermore, the range of food products in which acrylamide formation might be remediated by asparaginase treatment varies greatly from dough-based products to intact vegetable or cereal products. The rate of enzymatic hydrolysis of asparagine is dependent upon physical process parameters, such as temperature, pH, water activity and time, as well as interactions among these parameters [27]. In addition, content of the precursors (free asparagine and reducing sugars) varies greatly in the affected food products, which will influence acrylamide formation and means that the reaction-limiting factor differs from a product to another.

6.1. Pharmaceutical Industry

Considering its properties, L-asparaginase has important chemotherapeutic effects for treatment of lymphoproliferative and lymphoma diseases. Particularly, L-asparaginase presents large importance in chemotherapeutic protocols for acute lymphoblastic leukemia [27]. Cancer cells, mainly lymphatic cells, require high amount of asparagine for fast and malignant growth. In this way, cancer cells require the amino acid from diet (blood serum) as well as amino acids produced by themselves. However, leukemic lymphoblasts and some other tumor cells do not have or present low quantity of L-asparagine synthetase used for L-asparagine syntheses. Thus, these malignant cells are dependent of asparagine from blood serum for their proliferation and survival [27]. L-asparaginase hydrolyzes asparagine from blood serum, leading tumor cells to death by lacking of an essential factor for protein synthetases (p53-dependent apoptosis). However, healthy cells are not affected, because they are able to produce asparagine using L-asparagine synthetase present in enough quantities [27].

6.2. Food Processing

Acrylamide, also known as 2-propenamide, acrylic amide, ethylene carboxamide, propenamide, propanoic acid amide, monomer of acrylamide, is presenting 71.08 g/mol of molecular mass [31]. Several studies show that L-asparagine is the main amino acid responsible for acrylamide production in fried and baked foods when reducing sugars are condensed with a carbonyl source. This phenomenon does not occur in boiled foods [22]. For L-asparagine reduction, several options have been investigated such as: *a*) selection of vegetal species with lower level of L-asparagine in their composition; *b*) deletion of important enzymes for L-asparagine biosynthesis control by suppression of specific genes; *c*) acid hydrolysis of L-asparagine leading formation of aspartic acid and ammonia; and *d*) acetylation process of L-asparagine to form N-acetyl-L-asparagine, preventing the formation of acrylamide from intermediate N-glycosides [31]. Use of the enzyme L-asparaginase before frying or baking food process could reduce more than 99% acrylamide level in the processed final product because the enzyme reduces more than 88% of L-asparagine concentration from the initial feedstock [32].

7. TOXICITY OF L-ASPARAGINASE

L-asparaginase has a distinct toxicity profile, which ranges from acute hypersensitivity and hyperglycemia to hepatocellular dysfunction and pancreatitis. Toxicity of asparaginase fall under two main categories: *a*) those pertaining to immunological sensitization (hypersensitivity) to a foreign protein; and *b*) is related to inhibition of protein synthesis [33]. There is a similarity in the frequency of toxicity with all commercially available asparaginase preparations. Unlike other chemotherapeutic agents of multi-agent treatment protocols, L-asparaginase causes little bone marrow depression and usually does not affect the gastrointestinal or oral mucosa or hair follicles [33]. Side effects of L-asparaginase can be minimized or prevented. In this regard, attempts have been made to reduce the potential immunogenicity of the enzyme while preserving its activity and prolonging its half-life period, so as to avoid the need for frequent intra muscular injections [10, 33]. Chemical modification, to some extent, appears to meet these requirements. In the mid-1970s, attempts were achieved to chemically modify L-asparaginase by adopting various methods to identify a form that is less immunogenic and retain good antitumor activity [33].

Although L-asparaginase is found to be very prominent in medical fields, its use is limited by serious side effects. The most relevant are (occur in 30% or more of patients): *a*) loss of appetite, nausea, vomiting, and Abdominal cramping; *b*) diabetes, leucopenia, hemorrhage, and some hypersensitivity reactions; *c*) edema, skin rash, swelling of face; allergic reactions and anaphylactic shock usually occur due to its dual substrate specificity towards asparagine and glutamine; *d*) dysfunction of liver and pancreas, fever, and central nervous system toxicity; and *e*) excessive fatigue or sleepiness, depression, and seizure [10, 34]. Other side effects of L-asparaginase include development of neutralizing antibodies (referred to as silent antibodies) and premature inactivation. The latter are countered by frequent injections and can further potentiate the side effects. These limitations prompt the need for serologically different L-asparaginases with enhanced therapeutic potential and suitable immunological features [10]. A summary of toxicities and side effects of L-asparaginase therapy are given in Table 1.

Table 1. Toxicity profile of L-asparaginase therapy

System	Complications
Immune	*Hypersensitivity* Anaphylaxis, urticaria, hypotension, hypotension, bronchospasm, serum sickness (rashes, joint stiffness, and fever).
Liver Pancreas Central nervous system Coagulation	*Immunesuppression* Hypoalbuminemia, elevations in transaminase and alkaline phosphatase, lipoprotein abnormalities, decrease in serum cholesterol. Pancreatitis, acute hemorrhagic pancreatitis, decreased serum insulin, diabetic ketosis. Mild depression, confusion and hallucinations. Hypofibrinogenemia, decrease in plasminogen, increase in prothrombin time (a protein present in blood plasma that is converted into active thrombin during coagulation).
Others	Parotitis, comma, seizure, skin rash, swelling of face, lethargy (periods of weakness and a lack of energy).

* Source: modified from Reference [33] (Umesh et al. 2007).

8. MODE OF ACTION

Unlike conventional cancer therapy, L-asparaginase therapy is highly preferential. However, usage of L-asparaginase as a remedial agent is restricted by the main following reasons: *a*) its premature inactivation; *b*) the more rapid plasma clearance and shorter duration of drug effect, and thus frequent injections are required to maintain a therapeutic level; and *c*) the number of side effects like allergies, development of immune responses, and anaphylactic shock that might be serious and life-threatening [35].

8.1. L-Asparaginases as Drugs

The amino acid asparagine is normally found in the body and involved in biological processes that are essential for cells to maintain life. The rationale behind usage of asparaginase as a drug is because the acute lymphoblastic leukemia cells and some suspected tumor cells are unable to synthesize asparagine, whereas normal cells are able to make their own asparagine [35, 36]. Thus, leukemic cells require high amounts of asparagine and depend on circulating asparagine. Becaue asparaginase catalyzes the conversion of L-asparagine to aspartic acid and ammonia, this deprives the leukemic cell of circulating asparagine, leading to cell death [36]. Asparaginase produces its anti-cancer effects by "breaking down" asparagine. Therefore, depletion of asparagine by asparaginase kills cancer cells, while healthy cells are not as affected [35].

L-asparaginase is widely used for treatment of haemopoietic diseases such as acute lymphoblastic leukemia in children that results from the monoclonal proliferation and expansion of lymphoid blasts in the bone marrows, blood, and other organs [37]. Acute lymphoblastic leukemia, correspond to the most common childhood acute leukemia, contributing to approximately 80% of childhood leukemias and 20% of adult leukemias [37]. The antineoplastic activity results from depletion of the circulating pool of L-asparagine by L-asparaginase which, in turn, inhibits protein synthesis, followed by apoptosis in susceptible leukemic cells, and subsequent death of the tumor cells [35, 38, 39].

Several studies are performed concerning pharmacokinetics of L-asparaginase. Pharmacokinetic of a drug is the branch of pharmacology concerned with the movement of drugs within the body, including the processes of absorption, distribution, localization in tissues, biotransformation, metabolism, and excretion. Pharmacokinetics of any drug greatly depends on the route of administration (i.m. or i.v., *i.e.,* intramuscular or intravenous) and the type of preparations used [33].

8.2. L-Asparaginases in Food Processing

In food processing, asparaginase can aid by addion before baking or frying the food, where asparagine is converted into aspartic acid and ammonium. As a result, asparagine cannot take part in the Maillard reaction, thereby formation of acrylamide is significantly reduced. Complete acrylamide removal is probably not possible due to other minor asparagine-independent formation pathways [9]. As a food processing aid, asparaginase can effectively reduce the level of acrylamide up to 90-99% in a range of starchy foods without changing the taste and appearance of the end product [32, 40].

9. STRATEGIES FOR ACRYLAMIDE REDUCTION

Several strategies may be considered for removal or reducing the formation of acrylamide in food products. Due to the high number of affected product categories and differences in acrylamide levels in these products, a range of methods and strategies have been developed [8, 9]. For reduction of acrylamide concentration in foods, two different technological strategies can be performed: *a*) mitigation strategies: aimed to prevent acrylamide formation during the heating process; or *b*) removal intervention: aimed to move away the already formed acrylamide from the finished product [20]. Most published papers deal with acrylamide mitigation strategies. These can be regarded both as agronomical and technological approaches. Mitigation strategies include selection of crop varieties with low reducing sugars and/or asparagine, post-harvesting interventions, and control of storage conditions [41].

9.1. Mitigation Strategies

9.1.1. Technological Strategies Based on Physical Approach

The physical approaches to minimize the content of toxic molecule in foods include: *a*) reduction of the thermal impact; *b*) acrylamide removal by means of low pressure treatments; and *c*) decomposition of acrylamide by degradation. Acrylamide formation is dramatically influenced by the heating temperature, time, modality of heat transfer during processing [42,43]. Thus, a key factor controlling acrylamide formation is represented by process parameters leading to the generation of a moderate thermal input [42]. This can be achieved by applying prolonged heating at lower temperatures; or by higher temperatures at the early stages of heating [44]. In addition, the water content is a limit factor for acrylamide formation. In fact, as long as the water evaporates (temperature does not exceed 100°C), no acrylamide is detected in the food.

Instead of preventing its formation, acrylamide can be removed, transformed or degraded. Acrylamide can be removed from a finished food product as vapor by exploiting its physical and chemical properties [45, 46]. Depending on the food product, significant acrylamide removal was achieved only in foods previously hydrated at high water activity values.

9.1.2. Technological Strategies Based on a Chemical Approach

Chemical approach strategies are relevant to the use of chemicals or ingredients to mitigate acrylamide formation. Such strategies include both soaking or blanching pre-treatments in additive solutions [20]. Organic acids (citric, acetic, L-lactic acid), free amino acids, and protein-based ingredients, as well as calcium and sodium ions resulted to effectively reduce acrylamide formation in the final product. Worthy mention, however, is that although quite effective, these ingredients may produce undesired effects and negatively may influence the products acceptability. For instance: *a*) organic acids may be responsible for sour taste; *b*) sulfur-containing amino acids may produce undesired off-flavours during heating; and *c*) mono and divalent cations, such as Na+ and Ca2+, may be responsible for the failure of dough to rise and a bitter aftertaste. Substitution of ammonium carbonate or bicarbonate with the corresponding sodium salt may lead to finished products often unacceptable for consumption [8, 47, 48, 49].

To overcome these negative effects on food sensory quality, a combination of treatments can be used. For instance, the combined use of glycine or soy protein hydrolysate with citric acid not only has an effect in acrylamide reduction but also a better flavor profile was achieved than by applying individually treatments [5]. Furthermore, sensory analysis of potato crisps blanched in water enriched with a combination of several additives, such acetic acid and glycine or L-lysine, showed that a reduction in acrylamide formation was possible while maintaining the expected product quality for the consumer [50].

9.1.3. Technological Strategies Based on a Biotechnological Approach

Biotechnological strategies for minimizing acrylamide formation showed also limited impact on the sensory properties of a final product. The biotechnological approaches for acrylamide reduction are represented by fermentation, operated by yeast or lactic acid bacteria, and pre-treatments by means of the enzyme asparaginase. Fermentation allows extensive acrylamide reduction (up to 80%) to be achieved, depending on the food type, composition, and the microbial strain [8, 20, 49]. Mitigation of acrylamide formation using asparaginases from various microbial sources is represented in Table 2.

A greater acrylamide reduction (up to 94%) can be obtained by combining the fermentation process with physical (blanching) or chemical (amino acid addition) interventions [20]. Asparaginase pre-treatment of raw potatoes and dough has been claimed to effectively reduce acrylamide levels without altering the appearance or taste of the final product [20]. Commercial enzymes based on cloning of *Aspergillus oryzae* and *Aspergillus niger* have received the generally recognized as safe (GRAS) status from the US Food and Drug Administration (FDA). Studies have shown that asparaginase effectiveness is influenced by many variables such as: *a*) enzyme dose, reaction time, and enzyme-substrate contact; *b*) water content of the reaction environment; and *c*) temperature and pH at which the reaction occurs as well as the different processing conditions [20].

**Table 2. Mitigation of acrylamide formation using asparaginases
from various microbial sources**

Food	Enzyme source	Enzyme Quantity	Processing condition	Acrylamide reduction	Ref.
Gingerbread	E. coli	4U/kg	Various time/temperature combinations	55%	[51]
Potato	E. coli	0.2-1U/g	180°C, 20 min	50-90%	[52]
French fries	A. oryzae	10000 ASNU/L	175°C, 3 min	67%	[53]
Fried dough model system	A. oryzae	100,500,1000 U	180 or 200°C; 4,6/8min	90%	[54]
Potato chips	A. oryzae	10000 ASNU/L	170°C, 5min	90%	[55]
Potato	B. licheniformis	30 IU/mL	175 °C, 15 min	80%	[56]
Cookies	A. oryzae	500U/kg	205°C,11 or 15 min	23-75%	[57]
Wheat-oat bread	A. niger	500U	220,230 and 250°C;10,30 and 40 min	90%	[58]
Potato crisp	B. subtilis	0-40U	170°C;90s	80%	[59]
French fries	Thermococcus zilligii	0-20U	175°C:5min	80%	[60]
Potato chips	B. subtilis KDPS1	500µl	170°C:6min	90-95%	[61]
Potato chips & Sweet bread	Fusarium culmorum	300U	170-180°C,90s	94% & 86%	[62]

*Source: Adapted from reference [1] (Sharma and Shubhi 2016).

9.2. Removal Intervention

Preventive strategies of acrylamide in foodstuff are aimed to minimize acrylamide formation during the heating process. Whereas the removal intervention strategies are aimed to remove or decompose already formed molecules in the finished product [40]. A huge number of interventions for removal or minimizing acrylamide levels in food have been suggested. Their efficacy depends on different variables such as food type, ingredients, interactions among food components or ingredients and process conditions [1]. Moreover, to find practical application, these interventions should satisfy process prerequisites such as: *a*) low cost; *b*) feasibility and compatibility with the existing industrial processc); and *c*) low impact on sensory and nutritional properties of food [20].

Overall, when implementing an acrylamide mitigation strategy on industrial level, many factors should be considered. Such factors include: *a*) seasonal variability of the raw material; *b*) raw material characteristics (before and after blanching); and *c*) dip tank parameters (temperature, pH, and duration) while maintaining the expected product quality for consumer [7, 24]. Other considerations such as feasibility, legislation, cost, effluent treatment, safety and comfort of the employees, ability to control dosage, etc. are equally relevant when considering the implementation of any change to an industrial process [24].

Cultivar selection, fertilization, and climatological conditions may have an impact on reducing sugar contents of the raw material. Long-term potato storage may also influence the levels of reducing sugars due to senescent sweetening [7]. It is known that storage temperatures below 8°C may induce low temperature sweetening; ideal storage temperature is about 8°C [7]. On other hand, during the winter season, potato storage at optimal temperatures may be difficult. These factors may lead to great variability in the raw material between different seasons and even within the same storage season [7].

Asparaginase treatment can be successfully applied for acrylamide reduction in a range of cereal-based recipes without changing taste and appearance of the final product. Thus, possible strategies of acrylamide reduction may include selection of raw materials, changing product composition without affecting the taste or sensory quality, pre-treatment procedures, and optimized processing conditions. Among which, pretreatment with asparaginase enzyme is most effective. Strategies developed so far to mitigate acrylamide formation are performed on lab conditions, which may not be suitable for commercial process [1, 10, 11, 43]. Further work to introduce new potential asparaginases with benefits for commercially food processing still remains a major challenge to avoid acrylamide formation in food.

CONCLUSION

Asparaginase is an important enzyme applied in pharmaceutical and food industries. However, its use requires specific properties to be safely used for human. In foods, asparaginase helps to reduce acrylamide concentrations, maintaining their nutritional and sensory properties. As a therapeutic agent, efficient action of asparaginase is required to reduce adverse effects such as hypersensitivity and immune inactivation. Recent technological advances have enable detailed pharmacokinetic and pharmacodynamic studies of various asparaginase preparations which have a significant impact on efficacy and in designing the optimum dosing schedule for clinical applications. There is still a need for further studies to achieve sufficient enzyme–substrate contact and approval results enable introduction of new asparaginase with potential commercially benefits for food processing and therapy fields.

CONFLICT OF INTEREST

The authors have declared no conflicts of interest.

REFERENCES

[1] Sharma A, Shubhi M. 2017. Asparaginase: A promising aspirant for mitigation of acrylamide in foods. *Intr J Food Sci Nutr,* 2:208-214.

[2] Claeys W, Baert K, Mestdagh F, et al. 2010. Assessment of the acrylamide intake of the Belgian population and the effect of mitigation strategies: Food Additives and Contaminants Part A-Chemistry Analysis Control Exposure & Risk. *Assessment*, 27:1199-1207.

[3] Kita A, Brathen E, Knutsen S, et al. 2004. Effective ways of decreasing acrylamide content in potato crisps during processing. *J Agr Food Chem,* 52:7011-7016.

[4] Gokmen V, Senyuva H. 2007. Acrylamide formation is prevented by divalent cations during the Maillard reaction: *Food Chem*, 103:196-203.

[5] Mestdagh F, Wilde T, Carel D, et al. 2008. Impact of chemical pretreatments on the acrylamide formation and sensorial quality of potato crisps. *Food Chem*, 106:914-922.

[6] Pedreschi F, Carel K, Kit G, et al. 2007. Acrylamide reduction under different pre-treatments in French fries. *J Food Eng.* 79:1287-1294.

[7] Vinci R, Mestdagh F. and Bruno M. 2011. *NutriFOODchem Unit, Department of Food Safety and Food Quality, Faculty of Bioscience Engineering*, University of Ghent, Belgium.

[8] Claus A, Mongili M, Georg W, et al. 2008. Impact of formulation and technological factors on the acrylamide content of wheat bread and bread rolls. *J Cereal Sci.* 7:546-554.

[9] Beate A, Stringer M, Lange N, et al. 2010. Asparaginase – an enzyme for acrylamide reduction in food products. In: *Enzymes in Food Technology*. Second edition (Robert J. and Maarten van Oort, eds.). Blackwell Publishing Ltd.

[10] Izadpanah F, Homaei A, Fernandes P, et al. 2018. Marine microbial L-asparaginase: Biochemistry, molecular approaches and applications in tumor therapy and in food industry. *Microbiol Res.* 208:99-112.

[11] Doriya K, Kumar D. 2018. Optimization of solid substrate mixture and process parameters for the production of L-asparaginase and scale-up using tray bioreactor. *Biocat Agr Biotechnol.* 13:244–250.

[12] Rigoldi F, Donini S, Alberto R, et al. 2018. Review: Engineering of thermostable enzymes for industrial applications. *APL Bioeng,* 2:011501-0115017.

[13] Clementi A. 1922. La desemidation enzymatique de l-asparagine chez les differentes especes animales et la signification physiologique de sapresence dans l organisma. *Arch Int Physiol.* 19:369–76.

[14] *Kidd JG. 1953.* Regression of transplanted lymphomas induced in vivo by means of normal guinea pig serum. I. Course of transplanted cancers of various kinds in mice and rats given guinea pig serum, horse serum, or rabbit serum. *J Exp Med. 98:565–582.*

[15] Broome JD. 1961. Evidence that the l-asparaginase activity in guinea pig serum is responsible for its antilymphoma effects. *Nature,* 191:1114–1115.

[16] Broome JD. 1963. Evidence that the l-asparaginase of guinea pig serum is responsible for its antilymphoma effects I. *J Exp Med.* 118:99–120.

[17] Mashburn L, Wriston J. 1964. Tumor inhibitory effects of l-asparaginase from *Escherchia coli. Arch Biochem Biophys.* 105:450–452.

[18] Campbell H, Mashburn L. 1969. L-Asparaginase EC-2 from Escherchia coli some substrate specificity characterstics. *Biochem.* 9:3768–3775.

[19] Oettgen R, Old L, Boyse E, et al. 1967. Inhibition of leukemias in man by l-asparaginase. *Cancer Res.* 27:2619–2631.

[20] Anese M. 2012. Occurrence, toxicology and strategies for reducing acrylamide levels in foods. Food Science Department, University of Udine, Italy.

[21] Maillard LC. *1912.* Action des acides amines sur les sucres; formation de melanoidines par voie méthodique [Action of Amino Acids on Sugars. Formation of Melanoidins in a Methodical Way]. *Compt. Rend. 154:66-68.*

[22] Tareke E, Per R, Patrik K, et al. 2002. Analysis of acrylamide, a carcinogen formed in heated foodstuffs. *J Agr Food Chem.* 50:4998-5006.

[23] Williams J. 2005. Influence of variety and processing conditions on acrylamide levels in fried potato crisps. *Food Chem.* 90:875-881.

[24] Krishnakumar T, Visvanathan R. 2014. Acrylamide in Food Products: A Review. *J Food Process Technol.* 5:343-351.

[25] Anese M, Sovrano S, and Bortolomeazzi R. 2008. Effect of radiofrequency heating on acrylamide formation in bakery products. *Food Res Technol.* 226:1197-1203.

[26] Fennell T, Sumner S, Snyder R, et al. 2005. Metabolism and hemoglobin adduct formation of acrylamide in humans. *Toxicol Sci.* 85:447-459.

[27] Jorge J. Antunes F, Peres G, et al. 2016. Current applications and different approaches for microbial l-asparaginase production. *Braz J Microbiol.* 47S:77–85.

[28] Lopes A, Oliveira-Nascimento L, Ribeiro A, et al. 2015. Therapeutic l-asparaginase: upstream, downstream and beyond. *Crit Rev Biotechnol.* 37:1–18.

[29] Doriya K, Nose N, Gowda M, et al. 2016. Solid-state fermentation *vs* submerged fermentation for the Production of l-Asparaginase. In: *Advances in food and nutrition research.* DOI: 10.1016/bs.afnr.2016.05.003.

[30] Vimal A, Kumar A. 2017. Biotechnological production and practical application of L-asparaginase enzyme. *Biotechnol Gen Eng Rev*. 33:40-61.

[31] Friedman M. 2003. Chemistry, biochemistry, and safety of acrylamide. A review. *J Agric Food Chem*. 51:4504–4526.

[32] Zyzak D, Saders R, Stojanovic M, et al. 2003. Acrylamide formation mechanism in heated foods. *J Agr Food Chem*. 51:4782–4787.

[33] Umesh K, Shamsher S. and Azmi W. 2007. Pharmacological and clinical evaluation of l-asparaginase in the treatment of leukemia. *Crit Rev Oncology/Hematol*. 61:208–221.

[34] *L-asparaginase*. CancerConnect.com.

[35] Shrivastava A, Khan A, Khorshid M, et al. 2016. Recent developments in l-asparaginase discovery and its potential as anticancer agent. *Crit Rev Oncology/Hematol*. 100:1–10.

[36] Broome JD. 1981. L-Asparaginase: Discovery and development as a tumor-inhibitory agent. *Cancer Treatment Reports. 65:111–114.*

[37] Fullmer A, O'Brien S, Kantarjian H, et al. 2010. Emerging therapy for treatment of acute lymphoblastic leukemia. *Expert Opin Emerg Drugs*, 15:1–11

[38] Killander D, Dhlwitz A, Engstedt L, et al. 1976. Hypersensitive reactions and antibody formation during L-asparaginase treatment of children and adults with acute leukemia. *Cancer*, 37:220–228.

[39] Kotzia G, Lappa K. and Nikolaos E. 2007. Tailoring structure-function properties of L-asparaginase: engineering resistance to trypsin cleavage. *Biochem J*. 404:337–343.

[40] Hendriksen H, Kornbrust B, Ostergaard P, et al. 2009. Evaluating the potential for enzymatic acrylamide mitigation in a range of food products using an asparaginase from *Aspergillus oryzae*. *J Agri Food Chem*. 57:4168-4176.

[41] Muttucumaru N, Elmore J, Curtis T, et al. 2008. Reducing acrylamide precursors in raw materials derived from wheat and potato. *J Agric Food Chem*. 56: 6167-6172.

[42] Taeymans D, Wood J, Ashby B, et al. 2004. A review of acrylamide: an industry perspective on research, analysis, formation and control. *Crit Rev Food Sci Nutr*. 44:323-347.

[43] Mausumi R, Adhikari S. 2017. Use of Microbial Asparaginase to Mitigate Acrylamide Formation in Fried Food. *Food Sci Nutr Technol*. 2:000133-000136.

[44] Biedermann M, Grob K. 2003. Model studies on acrylamide formation in potato, wheat flour and corn starch; ways to reduce acrylamide content in bakery ware. *Mitt Lebensm Hyg*. 94:406-422.

[45] Zhaoyang L. 2003. *Process and apparatus for reducing residual level of acrylamide in heat processed food*. Patent No US2003/0219518 A1

[46] Anese M, Suman M. and Nicoli M. 2010. Acrylamide removal from heated foods. *Food Chem*. 119:791-794.

[47] Graf M, Thomas M, Stephan G, et al. 2006. Reducing the acrylamide content of a semi-finished biscuit on industrial scale. *Lebensm Wiss Technol*. 39:724-728.

[48] Gokmen V, Senyuva H. 2007. Acrylamide formation is prevented by divalent cations during the Maillard reaction. *Food Chem*. 103:196-203.

[49] Sadd P, Hamlet C. and Liang L. 2008. Effectiveness of methods for reducing acrylamide in bakery products. *J Agric Food Chem*. 56:6154-6161.

[50] Low M, Koutsidis G, Parker J, et al. 2006. Effect of citric acid and glycine addition on acrylamide and flavour in a potato model system. *J Agric Food Chem*. 54:5976-5983.

[51] Amrein T, Barbara S, Felix E, et al. 2004. Acrylamide in gingerbread: Critical factors for formation and possible ways for reduction. *J Agri Food Chem*. 52:4282-4288.

[52] Ciesarova Z, Kiss E. and Boegl P. 2006. Impact of L-asparaginase on acrylamide content in potato products. *J Food Nutri Res*. 45:141-146.

[53] Pedreschi F, Kaack K. and Granby K. 2008. The effect of asparaginase on acrylamide formation in French fries. *Food Chem*. 109:386-392.

[54] Kukurova K, Morales F, Bednarikova A, et al. 2009. Effect of L- asparaginase on acrylamide mitigation in a fried-dough pastry model. *Mol Nut Food Res*. 53:1532-1539.

[55] Pedreschi F, Salome M, Kit G, et al. 2011. Acrylamide reduction in potato chips by using commercial asparaginase in combination with conventional blanching. *LWT–Food Sci Technol*. 44:1473-1476.

[56] Mahajan R, Saran S, Kameswaran K, et al. 2012. Efficient production of L-asparaginase from *Bacillus licheniformis* with low-glutaminase activity: optimization, scale up and acrylamide degradation studies. *Bio Techno*. 125:11-16.

[57] Kukurova K, Zuzana C, Burce A, et al. 2013. Raising agents strongly influence acrylamide and HMF formation in cookies and conditions for asparaginase activity in dough. *Euro Food Res Technol*. 237:1-8.

[58] Ciesarova Z, KuKurova K, Mikusova L, et al. 2014. Nutritionally enhanced wheat-oat bread with reduced acrylamide level. *Qual Assurance Saf Crops Foods*, 6:327-334.

[59] Onishi Y, Asep A, Shigekazu Y, et al. 2015. Effective treatment for suppression of acrylamide formation in fried potato chips using L-asparaginase from *Bacillus subtilis*. *Biotech*. 5:783-789.

[60] Zuo S, Zhang T, Jiang B, et al. 2015. Reduction of acrylamide level through blanching with treatment by an extremely thermostable L-asparaginase during French fries processing. *Extremophiles*, 19: 841-851.

[61] Sanghvi G, Bhimani K, Vaishnav D, et al. 2015. Mitigation of acrylamide by L-asparaginase from *Bacillus subtillis* KDPS1 and analysis of degradation products by HPLC and HPTLC. *SpringerPlus*, 5:533-543.

[62] Meghavarnam A, Janakiraman S. 2018. Evaluation of acrylamide reduction potential of L-asparaginase from *Fusarium culmorum* (ASP-87) in starchy products. *Food Sci Technol.* 89:32-37.

In: Microbial Catalysts. Vol. 2
Editors: Shadia M. Abdel-Aziz et al.

ISBN: 978-1-53616-088-8
© 2019 Nova Science Publishers, Inc.

Chapter 13

CHITINASE AS PROGNOSTIC MARKER: A MAJOR THRUST AREA OF RESEARCH FROM MEDICAL ASPECTS

Sneha C. Jha and Hasmukh A. Modi

Department of Life Sciences, University School of Sciences,
Gujarat University, Ahmedabad, Gujarat, India

ABSTRACT

Mammalian chitinase belongs to Family 18 of Glycosyl Hydrolase which catalyzes breakdown of chitin into either its oligomers through transglycosylation or to monomer through series of reaction. In the present natural environment, chitin acts as the second most important polysaccharide found mostly in fungi, bacteria, crustaceans, insects as well as parasitic nematodes but is absent in mammals. Stimulation of strong T-helper type -1 response occurs in mammals when exposed to antigens containing chitin or chitin like proteins from surrounding environment. This leads to induction of mammalian chitinase at sites of inflammation. Mammalian chitinase is categorized as True enzyme and chitinase like protein. True enzyme such as acidic mammalian chitinase (AMCase) and chitotriosidase (CHIT1) is involved in hydrolysis of chitin depicting chitinase activity while that of chitinase like protein is involved only in binding process and not showing any activity. Major cells producing mammalian chitinase as well as chitinase like proteins are neutrophils, epithelial cells, synovial cells, macrophages, chondrocytes and tumour forming cells. Enzymatically active chitinase have specific pathway which links in association with various disorders. Chitotriosidase acts as an indicator for macrophage-driven inflammatory processes in various organs. Chitotriosidase remains to be the biomarkers for various diagnostic approach and evaluation several disorders while acidic mammalian chitinase is induced in T-helper type 2 reaction which deals with host defence against parasites and failure to control it may lead to chronic as well severe diseased condition. Here, in this review, potential role of chitotriosidase as well as acidic mammalian chitinase is discussed with its mechanisms which could improve further in diagnosis as well as in therapeutics in various diseased conditions.

Keywords: chitinase, chitotriosidase, acidic mammalian chitinase, applications

1. INTRODUCTION

In living organisms, polysaccharides are backbone for the formation of structural component. Chitin is present as a structural component in outer shells of crustaceans, cellwall of fungi, exoskeleton of arthropods and protozoan cysts but is absent in mammalian cells. The evolutionarily conserved Glycoside Hydrolase family 18 (GH 18) consists of various enzymes such as chitinases (EC3.2.1.14), lysozyme (EC3.2.1.17), endo-β-N-acetylglucosaminidase (EC3.2.1.96); peptidoglycan hydrolase with endo-β-N-acetylglucosaminidase specificity (EC3.2.1.-); Nod factor hydrolase (EC3.2.1.-) [1]. The important function of this group is to carry hydrolysis of glycoside bond present between two or more carbohydrate moiety or between a carbohydrate and non-carbohydrate moiety. Davies and Henrissat suggested the pattern of hydrolysis of glycosidic bond is generally carried out by amino acids which are present on enzyme [2]. The function of chitinase is to hydrolyse chitin which is present mostly in pathogens as their structural constituent. It has mainly three functions: remodeling process, digestion purpose for lower organisms and secretion of enzyme during pathogenesis occurring in humans. Chitinase also plays vital role in innate as well as adaptive immunity [3]. Nowadays, scientists are working in fields of medicinal aspect, so that new immune properties and functions of mammalian chitinase can be known. Chitinase can activate receptors which are specific for particular inflammation and get involved in different signaling pathways which are responsible for diseased conditions.

2. STRUCTURAL OUTLINE OF MAMMALIAN CHITINASE

Mammalian chitinase was categorized under two categoriesbased on the presence of enzymatic activities: acidic mammalian chitinase (AMCase) and chitotriosidase (CHIT1) as well as chitinase like protein (CLP) such as chitinase3-like 2 (YKL39) and chitinase 3-like 1(YKL40or HC-go39). Presence of chitin binding cleft subjected with six cysteine residues in case of mammalian chitinase, helped in binding of chitin and thus showing enzymatic activity [4] while CLP donot have such cleft but still can bind with chitin [5]. Although, chitin was not found to be present in mammals, two enzymatically active chitinases, acidic mammalian AMCase and CHIT1 were expressed in various human cells as well as in mice [6, 7]. The first scientist [8] who discoveredchitotriosidase (CHIT1) in serum samples of patients suffering from Gaucher's disease which was structurally homologous in structure with chitinase was found in lower organisms. Hence, this became first enzyme to be purified as well as cloned with respect to chitinase [9, 10].

Acidic mammalian chitinase (AMCase) was considered as a compensatory enzyme and was optimally active enzyme in acidic range. This enzyme was required for functioning of CHIT1 enzyme [11].

AMCase and CHIT1 share sequence similarity with bacterial chitinase belonging to GH 18 family. Gene that express CHIT1 in human cells is found on chromosome 1q 31 and 1q32 as described by [12] consisting of 13 exons and lengths about 30 - 1055 basepairs [95] whereas another most active AMCase gene is located on chromosome 1p13[11]. Recently, there are few reports which link between expression of mammalian chitinases with inflammation. CHIT1 is expressed mostly in phagocytic cells depicting its role in various disorders such as Gaucher's disease, chronic obstructive pulmonary disease (COPD), Alzheimer's disease and bronchoalveolar lavage fluid of smokers [8, 13, 14, 15]. AMCase is expressed mostly in gastrointestinal tract, in lung, and conjunctiva (Figure 1). It is induced in epithelial cells as well as macrophages during process of pathogenesis and plays a significant role in the allergen-specific T-helper type 2 (Th2) - mediated diseases and anti-parasite responses occurring in asthma.

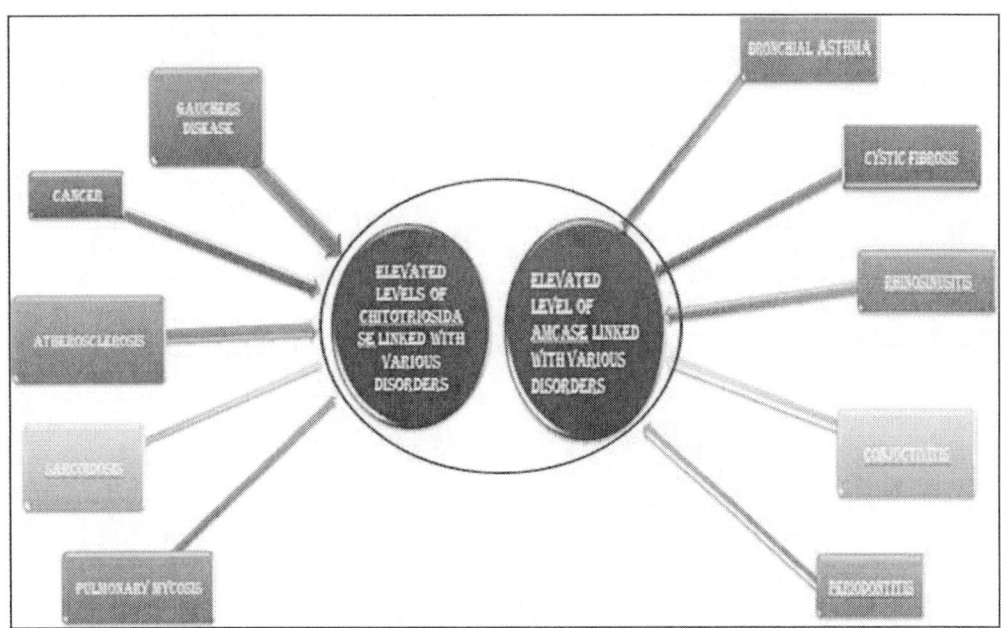

Figure 1. Types of mammalian chitinase: Chitotriosidase and Acidic mammalian chitinase and disorders occurring due to its enhanced level.

3. MECHANISM OF HYDROLYSIS PATTERN OF PATHOGEN AND SIGNALING PATHWAY

There are different pattern of hydrolysis of pathogen containing chitin as their structural moiety. Studies were reported by [16, 17] emphasizing direct interactions between chitin containing pathogen and specific receptors which caused inmmunological changes leading to activation of macrophages as well as natural killer cells (NK cells). Expression of different cytokines (IL-12, tumour necrosis factor TNF-α and IL-18) occurred due to activation of macrophages and production of gamma – interferon (INF-γ) mainly occurred due to NK cells. Another group of scientists Reese and his co-workers [18] studied direct administration of chitin beads into lungs of mice. Enhanced transcript of IL-4 was obtained which was expressed due to green fluroscence protein (GFP). Results of *in-vivo* experiment described that eosinophils and basophils were recruited in lungs due to exposure of chitin. These findings describe that chitin containing pathogen can give rise to allergic type-2 inflammation. As reported the size of chitin containing pathogens either in polymeric or fragmented form is important in causing effect on immune cell functioning [19]. They described that large polymers are generally inert and do not cause inflammation directly. In contrary to this, they described that smaller molecules or fragments have capacity to produce anti-inflammatory or pro-inflammatory cytokines (IL-10).

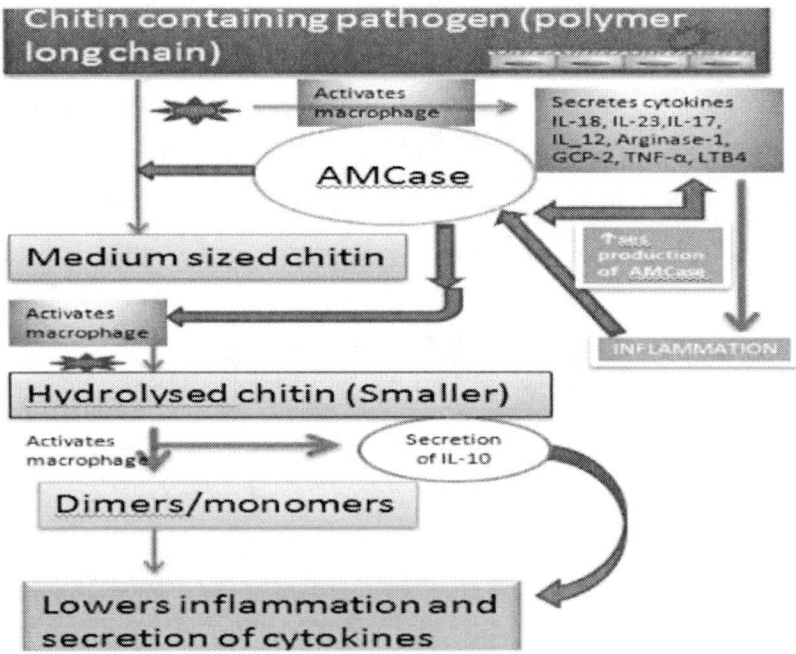

Figure 2. Size dependent pathway for chitin containing pathogen - Modified pathway adapted from [19].

Recognition of chitin containing pathogen (Figure 3) by epithelial cells stimulates production of IL-17, recruitment of eosinophils as well as secretion of TNF-α at site of inflammation which ultimately leads to expression of chitinase [17,18]. Stimulation of chitinase activity is caused when chitin containing pathogen is recognized by TLR-2 site or chitin-specific receptor site. As a result of it, prolactin stimulates macrophages directing it towards expression of chitinase through NF-κB pathway [20, 21, 22, 23]. NF-κB pathway is one of the most important pathway which is involved in recognization of fungal components [24, 25]. Activated macrophage leads into production of eosinophils, basophils, neutrophils and NK cells. In accordance, IL-4, IL-5, IL-13, TGF-β, Histamines, Alarmins are secreted by eosinophils and basophils causing enhanced chitinase activity against fungal cell component. NK cells is mainly involved in secretion of IFN-γ which leads to activation of macrophages and upregulateschitinase activity [16, 25, 26, 27, 28]. TNF-α, LPS and PMA secreted due to inflammation caused by chitin containing pathogen enhances chitinase expression. Secretion of IL-10 is mediated via by dectin-1 pathway and mainly responsible when chitin is hydrolysed into smaller fragments (di/mono saccharide) is helps in feedback control of local inflammatory response generated due to entry of pathogen [17] as described in Figure 2.

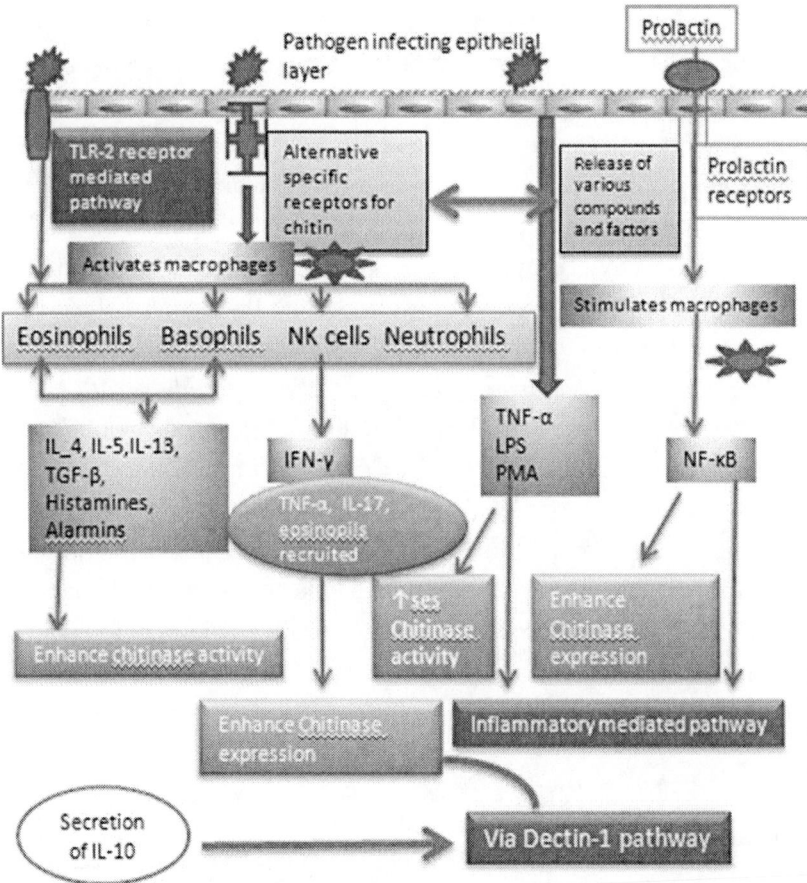

Figure 3. Signalling pathways which enhance chitinase expression when triggered by chitin containing pathogen – Modified pathway adapted from [25].
Abbrevations: TLR-2: Toll like receptor-2, NK cells: Natural killer cells, IL: Interleukin, TNF-α: Tumour necrosis factor – alpha, IFN-γ: Interferon – gamma, TGF-β:Transforming growth factor beta, LPS: lipopolysaccharide, PMA: phorbol 12-myristate 13-acetate, NF-κB, nuclear factor kappa-light-chain-enhancer of activated B cells.

4. CHITOTRIOSIDASE (CHIT1) AND ITS POTENTIAL RELEVANCE IN DISEASED CONDITIONS

CHIT1 is a vital enzyme to legalize the inclination to infection from those organisms which contains chitin as structural components [29]. Studies on recombinant technology by Wiesner and co-workers [30] provide an insight towards use of recombinant CHIT1 in inhibition towards hyphal growth of fungi which defines its role in the host defense mechanism against the attack of chitin-containing pathogens also known as alloalergens which acts as an adjuvant by invigorating the production of cytokines and chemokines[23]. An increased amount of CHIT1 was found in response to assorted pro-inflammatory signals in a complementary trend in case of neutrophils and macrophages [31]. The important inducer of CHIT1 in neutrophils is TLR signaling whereas in case of macrophage, NOD-2 signaling induces it [32]. Boot and his co-workers [12] reported that, if a defect occurs in CHIT1 gene, it may generate an abnormally spliced mRNA with an in-frame deletion of 87 nucleotides and encoding an enzymatically inactive protein which lacks an internal stretch of 29 amino acids. They reported mutation in CHIT1 gene relating it with high occurrence in different Caucasian populations while that in Africans, people residing in endemic areas where malarial parasites were dominant, CHIT1 mutations prevalence was comparatively lesser. Maintenance of wild-type CHIT1 gene in sub-Saharan regions proves its efficiency of having innate immunity against malarial infection [31]. Studies reported [34] suggested that mutant allele in individuals exhibiting higher susceptibility towards chitin-containing pathogens such as *Candida albicans, Plasmodium falciparum malaria, Wuchereriabancrofti filarial and Cryptococcus neoformans*whereas studies reported by [35] reveals that overexpression of CHIT1 in Kupffer cells is involved in tissue remodeling process. Another scientist[36]explainedthe role of CHIT1 produced by macrophages augments formation of atherosclerotic plaques and ultimately leading to thrombosis. Differentiation and maturation of of macrophages may cause damage to host tissues which may be associated with series of chronic inflammatory diseases [37]. CHIT1 is known to be active towards IL-13-driven alveolar fibrosis by enhancing transforming growth factor beta (TGF-β) and mitogen-activated protein kinase signaling in mice [38].

4.1. Gaucher's Disease

One of the major prognostic marker for patients suffering from Gaucher disease (GD) is CHIT1 which secreted by abnormal lipid-laden macrophages in their tissues [5]. Gaucher disease is genetic disorder which is caused by recessively inherited deficiency in activity of lysosomal hydrolase - glucocerebrosidase, and is marked by accumulation of glucosylceramide (glucocerebroside) in the lysosomes of macrophages [6]. This process is limited till tissue macrophages whichget converted into large swollen lipid-ladenGaucher cells. In-vitro studies revealed that production of CHIT1 by monocyte-derived macrophages occurs after approximately 7 days of culturing [39] as well as it is stalwartly stimulated by GM-CSF [34]. Some of the pathological implications of the accumulation of lipid-laden macrophages include bone lesions, hepatosplenomegaly, and less frequently neurological abnormalities. Hence, CHIT1 is considered as one of the easily identifiable biomarker for the GD. Hollak and his co-workers [8] reported average activity of chitotriosidase in plasma samples of patients suffering from GD which was more than 600 times higher in comparison to that of control. This was monitored by determining hydrolytic activity of CHIT1 towards fluorogenicchitooligosacharide substrates. Till date, there is no method available for assessment of individual contribution of each isoforms of CHIT1 depicting total activity but advances in proteomic study demonstrated possibility for quantitative imaging of isoforms of CHIT1.

Gaucher's disease is classified into three major types on the basis of symptoms of central nervous system such asType I, Type II and Type III showing non-neuropathic,neuronopathic/acute and neuropathic/chronic symptoms respectively. Amongst them, Type I is the most prevalent and is diagnosed by measurement of β-glucosidase activity in leukocytes and fibroblasts or inchorionic villi and cultured amniocytes (generally used in prenatal diagnosis incase of lamb cells) as reported by [37, 40, 41, 42, 43]. GD1 is differentiated by glucocerebroside laden macrophages, known as Gauchercells, which are surrounded by inflammatory phagocytes [44]. As reported by [45], Gaucher cells can be classified as a disparity of alternatively activated macrophages expressing CD68, C14, HLA class II, CD163, CCL18, and IL-1-receptor antagonist, but not expressing CD11b, CD40, and proinflamatory cytokines such as TNFα and MCP1. As a result of storage of glucocerebroside, proinflammatory activation of macrophages occurs with successiveballooning, decreased phagocytic activity, and destruction of viscera (e.g., spleen, liver, kidneys, lungs, brain, bone, and bone marrow). Thus, Gaucher cells secrete biomarkers into theblood which could be used in diagnosing Gaucher's disease, assessment of disease severity, andmonitoring the efficacy of enzyme replacement therapy (ERT) [46, 47]. ERT works best during first six months and after first year by decreasing CHIT1 activity. This may be due to the change taking place either during activation process or differentiation ofmacrophages and their precursors, rather than a decrease in Gaucher cell burden [48, 49].Stablity in level of CHIT1 may be achieved approximately two years afterstarting ERT.The continual high levels of CHIT1 could indicate that ERT isunable to treat and exclude some Gaucher cells in affected regions. In addition, after ERTcessation, macrophages became laden with glucocerebrosides again, indicating increasedsynthesis of CHIT1 and recurrence of Gaucher's disease.

4.2. Cancer

CHIT1 act as bio-marker for lung cancer as it could be differentiated from normal inflammation occurring in lungs. Recently, reports emphasize that the serum CHIT1 activity was useful parameter and can be useful in differentiation between the patients suffering from lungcancer and with inflammatory exudates [50]. They found that there was slightly higher CHIT1activities in patients with lung cancer as compared to patients with lung inflammation or with normal individuals. But, they also found that treatment with corticosteroid decreased these levelssignificantly. This finding indicates that for differentiation between inflammation and lung cancer, as well as progression of disease, CHIT1 activity can be utilized [50, 51]. Prostatecancer is the most commonly diagnosed non-skin cancer and also the second highest cause of death in men. Hence, monitoring for prostate cancer at an early phase is importantfor disease treatment. Comparision study for CHIT1 as a marker to be associated for screening between benign prostatichyperplasia and primary prostate cancer revealed that CHIT1 activity was significantly higher in benign prostatic cancer patientsas compared to that of primary prostate cancer. Another finding described that in primary prostatic cancerpatients, practically high CHIT1 activities were found only in patients having high Gleason scores,which has been most widely used diagnostic method for prostate cancer. Hence, there exists a correlation between Gleason scores and CHIT1 activity indicating the importance ofmacrophage involvement in cancer progression.

4.3. Atherosclerosis

Atherosclerosis is one of the inflammatory diseases which are characterized by excessive deposition oflipids and fibrous matrix in the arterial wall. The mechanism which occurs for cause of this disease during initial stage of atherosclerosis involvesactivation of endothelial cells, thus facilitating monocyte infiltration inside vessel.After differentiation of monocytes into macrophages, these cells collect lipids from the circulatory system, remain in the vessel wall, and become lipid-loaded foam cells. Secretion of certain growth factors as well as cytokines by macrophages and activated endothelial cells leads to activation of macrophages. This lipid-loaded cells can be characterized as a sign of atherosclerosis as there occurs overloading of lipids mainly cholesterol and cholesterol esters [52], resulting in thickening of arterial wall, narrowing space in vessel which leads in increased susceptibility to thrombosis [53, 54]. Elevated levels of CHIT1 in patient serum with atherosclerosis thus confirm relationship between CHIT1 expression and lipid-laden macrophages inside human atherosclerotic vessel walls. This elevation in serum CHIT1 activity occurs mainly due to presence of activated macrophages. High serum CHIT1 activity was also found to be dependent on age of patients [55]. This occurs due to accumulation of lipid-loaded macrophages specifically found in the supra-aortic and coronary vessels which signals in activation of macrophages [55, 39]. Studies were reported depicting higher CHIT1 activity in patients with atherothrombotic strokeand ischemic heart disease as compared with healthy individuals (control). Comparision of CHIT1 activity in between above two subjects revealed that subjects with atherothrombotic stroke have higher CHIT1 activity due to extended progression of disease while lower CHIT1 activity was noted for ischemic heart disease having limitation to only coronary vessels [56]. Ever since macrophages are found in all stage ofatherosclerosis and their number and also activity increased depends on the severity of that stage,so, CHIT1 activity in macrophages has been recognizeed as one of the biomarkers foratherosclerotic plaque formation [57].

4.4. Sarcoidosis

Sarcoidosis is a multisystem granulomatous disease of unknown origin however, it has been suggested that fungi may be the causative agent in sarcoidosis [58,59].The formation of these granulomatous indicates deficiency in cellular immune processing after exposure to certain chemical or infectious agent [60].The pathology is differentiated by non-necrotizing granulomas that results due to accumulation of proliferating mononuclear phagocytes and T lymphocytes mainly affecting the lungs [61].The most commonly affected organ is lungs, including hilar and mediastinal lymph nodes, but disease can be progressed further to various other body organs such as to eyes, skin, liver, peripheral lymph nodes, kidneys, joints, muscles, and thecentral nervous system. There are clinically different forms of Sarcoidosis such as acute, subacute, or chronic [62, 63] depending on disease progression.Higher serum as well as bronchoalveolar lavage CHIT1 activity was found in patients with active sarcoidosis and corresponding CHIT1 activity was found to raise in later stageof progression of sarcoidosis (stages 3 and 4) compared to early stages (stages 0 and 1) [64].Considering chronic sarcoidosis, which occurs due to slow progression of disease and almost constant involvement of organ such as lungs, takes place. Occurrence of severe hypoxemia and pulmonary hypertension may take place in case of advanced progression of lung fibrosis [65]. The important function of CHIT1 in innate immune system is probably due to defense mechanism against organisms such as fungi and insects whose outermost wall (layer) is made up of chitin. Lower level of CHIT1 was also found in patients who had deficiency in CHIT1 gene [29].

4.5. Pulmonary Mycoses

The pathogenic fungi which generally infect the respiratory system cause diseases such as histoplasmosis, blastomycosis, aspergillosis, and invasive candidiasis.Such fungal infection occurring in lungs is root cause of death in immunocompromised patients. Fungal spores are present in ubiquitous amount in our surrounding environment and hence, humans come in direct contact with pathogenic as well as opportunistic fungi in day to day life. Therefore, disease occurring in host (human) is dependent on various factors such as innoculum size, virulence factor and ultimately immunity of the host. For protection against persistent disease, immune system require CD4+ helper Th (T-helper) type 2 cells which acts as significant mediators against fungal spores. As chitin is present below layers of mannans and glucans in cellwall of fungi, hence chitinase secreted by host is significant component [19] and needs to penetrate inside the wall and degrade it into smaller fragments [66].Chitin as such is not present in human but instead, keratin is present for structural functions in human. Hence, when chitin containing pathogens come in contact with human immune system, Th - 2 cell activation occurs. Similar, studies has been suggested on mouse models [30] for pulmonary allergy which occurs due to macrophage activation, excessive Th-2 cell accumulation leading to enhancement in disease severity. Therefore, confirmation of pulmonary mycosis could be attained with the help of CHIT1 activity describing its importance in fungal infection in humans [30]. They suggested that*C. neoformans*chitin and the host-derived chitinase, chitotriosidase, enhance Th- 2 cell accumulation and thus cause disease.

5. ACIDIC MAMMALIAN CHITINASE(AMCASE) AND ITS POTENTIAL RELEVANCE IN DISEASED CONDITIONS

Acidic mammalian chitinase (AMCase) is second true chitinase enzyme and is expressed in gastrointestinal tract, in lungs epithelial cells, and conjunctiva. AMCase is a 50 kDa protein, a member of the glycosyl hydrolase 18 family (EC 3.2.1.14) and consists of a 30-kDa N-terminal catalytic domain that can hydrolyse chitin. The genefor AMCase is located on mouse chromosome3 and human chromosome 1 at p13.1–p21.3. The enzyme is known as acid stable, with a pH optimum of 2.0, and this resistance towards acidic pH differentiates AMCase from CHIT1, which has its optimum activity at pH 6. AMCase was found to be four times more active [67] at pH 2.0 compared to that of higher pH 5.2 and 7.5 respectively. Adaptation towards the extreme environment of the stomach may play an important role in defense and/or digestion of chitin-containing pathogens. In general, AMCase is expressed in different tissues (stomach, lung, and salivary glands) and hence is characterized to be associated with various inflammatory diseases [68].This enzyme is also for non-chitin-related mechanism and has various roles in fibroblast growth-promoting activity [69], Defensive action against pathogens and in food processing [70]. AMCase induction occurs when there is an inflammation at site of infection and remodeling process involved which leads to attraction of eosinophils and Th- 2 cells at site of infection. Recent studies reported that upregulation of AMCase was observed in response to Th- 2 inflammationin the lungs [5, 7, 71]. Genetic study reported implies that certain polymorphismsand haplotypes of AMCase are linked with bronchialasthma in humans [5]. Contradictory results were obtained by scientists [7] who studied on several United States ethnic populations revealed that a haplotype encoding an AMCase isoform displayed enhanced enzymatic activity was associated with protection from asthma. All these findings indicated that enhanced AMCase activity could be protective against asthma [7]. Thus, AMCase are thus known to modulate inflammation in tissues, remodeling process and immunity [72].

5.1. Asthma

Asthma is one of the most widespread chronic lung diseases which affect people of all genders, races and ages over the world [73]. Current estimation of people suffering from asthma over the world is about 300 million which increases by 50% globally every decade [74, 49]. There are different environmental factors [75] and genetic heredity in family responsible for asthma but show different susceptibility to asthma in same environment [71, 76].To exploit chitin containing parasites, worms or fungi (pathogen), host immune system needs chitinase which can control its growth and molting process.Basic mechanism involved in asthma infection was suggested[18] under mouse model in which they challenged 4 mice and administered them with chitin having knock-in IL-4 reporter cells which could help in easy detection of IL-4 by competetnt cells [77]. This challenge led to increase production of eosinophils as well as basophils in epithelial cells of lungs. Epithelial cell lining (either conjuctival or nasal) has a vital role as first line of defence in the immune system [45]. Production of chitinase takes place when an allergic response occurs due to non-existent parasites [45]. The COPD is linked with allergic asthma which produce various types of key cytokines such as Interleukin (IL)-4, IL-5 and IL-13 regulating allergen specific IgE and thus controlling eosinophilic airway inflammation.

5.2. Conjuctivitis

Allergic conjunctivitis is very common inflammation affecting mostly children that is caused due to seasonal variation. There are two types of allergic conjunctivitis: vernal conjunctivitis (VKC) and seasonal allergic conjunctivitis (SAC) and most often doctors consider both as a same entity. VKC is bilateral as well as frequently occurring inflammation and considered more severe allergic conjunctivitis (occurring generally at onset of spring and summer and decrease during winter) as they are resistant to anti-allergic as well as immunosuppressive treatment. AMCase plays an important part in diagnosis of human conjunctival allergic conditions. Significantly high AMCase activity was obtained in tears of allergic ocular conditions which was confirmed by checking its stability in acidic conditions (pH - 2) whereas source of AMCase was confirmed by extraction of RNA by RT-PCR technique. A Co-relation was obtained between expression of mRNA and AMCase activity. Various chemokines as well as Th- 2 dependent cytokines were found in tears of patients with VKC and SAC [78, 79].

Dry eye is another ocular disease which is characterized by a reduced stability of tear film stability which occurs due to decreased production of aqueous component or by an increased evaporation. Several symptoms such as burning of eye, foreign body sensation, impaired vision may be accompanied during dry eyes infection. Non-allergic ocular inflammation may sometimes be associated with secretion of AMCAse in tears as well as expression of conjunctival cells. A study in rabbit, where an experimental uveitis was induced injecting LPS in anterior chamber of eye, confirmed that the chitinolitic activity in tears collected at different time 0, 6, 24 h after injection, was due to AMCase [80].

5.3. Rhinosinusitis

Rhinosinusitis is an acute disease usually affecting more than 14% of adults and children [81, 82, 83, 84] and has high tendency to become chronic with time. The acute form of rhinosinusitis was initially thought to be unimicrobial, and chronic form was characterized by multiple microorganisms [85, 86, 87]. The chronic form of rhinosinusitis demonstrates antimicrobial resistance and hence is most challenging for the doctors[88]. Such chronically infected sinuses are often infected with fungi adding complications to the situation [89, 90] and itbecomes extremely difficult to eradicate [91]. Hence it needs to be detected and treatment for the same should be given at early diagnosed time period [92, 93]. Chronic rhinosinusitis is characterized with nasal polyps (CRSwNPs) that cause severe inflammation with eosinophilic infiltration, which is similar to that found in asthmatic inflammation. The role of AMCase in CRSwNPs has been studied [94] revealing that eotaxin-3 is a most effective eosinophil attractant and can provoke eosinophil recruitment and activation in the airways of asthmatics.They concluded that AMCase as well as eotaxin-3 may be chief mediators in the pathogenesis of nasal polyps.

CONCLUSION

True enzymes play vital role in pathogenesis of allergic or chronic conditions. Synthesis of chitinase enzyme takes place only during bacterial, fungal or non-viral pathogens is entering our body still its mechanism of production in response to pathogen is not fully understood. Under various pathological conditions, mammalian chtinase such as CHIT1 and AMCase are secreted as a result of inflammation and this could be bio-marker for several diseases in which false diagnosis method is used for confirmation. In this review, basic theory of involvement of mammalian chitinase in diseased conditions is discussed and this may help in improving research design and link consequences of elevated levels of chitinases with immune system. Overall, it can be concluded that Secretion of CHIT1 and AMCase depends on various other factors such as inflammation stage, organs involved and types of cell involved.

REFERENCES

[1] http://www.cazy.org/Glycoside-Hydrolases.html.

[2] Davies, G. and Henrissat, B. 1995. "Structures and mechanisms of glycosyl hydrolases." *Structure* 3: 853-859.

[3] Elias, J.A., Homer, R.J. and Hamid, Q. 2005. "Chitinases and chitinase-like proteins in TH2 inflammation and asthma." *J Allergy Clin Immunol* 116:497–500.

[4] Malaguarnera, L. 2006. "Chitotriosidase: the yin and yang." *Cell Mol Life Sci* 63: 3018-3029.

[5] Kzhyshkowska, J., Gratchev, A., and Goerdt, S. 2007. "Human chitinases and chitinase-like proteins as indicators for inflammation and cancer." *Biomark Insight,* 2: 128-146.

[6] Bussink, A.P., Speijer, D., Aerts, J.M. and Boot, R.G. 2007. "Evolution of mammalian chitinase(-like) members of family 18 glycosyl hydrolases." *Genetics* 177: 959–970.

[7] Lee, C.G., Da Silva, C.A., Dela Cruz, C.S., Ahangari, F. and Ma, B. 2011. "Role of chitin and chitinase/chitinase-like proteins in inflammation, tissue remodeling, and injury." *Annu Rev Physiol* 73: 479–501.

[8] Hollak, C.E., van Weely, S., van Oers, M.H. and Aerts, J.M. 1994. "Marked elevation of plasma chitotriosidase activity. A novel hallmark of Gaucher disease." *J Clin Invest* 93: 1288–1292.

[9] Renkema, G.H., Boot, R.G., Muijsers, A.O., Donker-Koopman, W.E. and Aerts, J.M. 1995. "Purification and characterization of human chitotriosidase, a novel member of the chitinase family of proteins." *J Biol Chem* 270: 2198–2202.

[10] Boot, R.G., Renkema, G.H., Strijland, A., van Zonneveld, A.J. and Aerts, J.M. 1995. "Cloning of a cDNA encoding chitotriosidase, a human chitinase produced by macrophages." *J Biol Chem* 270: 26252–26256.

[11] Boot, R.G., Blommaart, E.F., Swart, E., Ghauharali-vander, Vlugt K. and Bijl, N. 2001. "Identification of a novel acidic mammalian chitinase distinct from chitotriosidase." *J Biol Chem* 276: 6770–6778.

[12] Boot, R.G., Renkema, G.H., Verhoek, M., Strijland, A. and Bliek, J. 1998. "The human chitotriosidase gene- Nature of inherited enzyme deficiency." *J Biol Chem* 273: 25680–25685.

[13] Watabe-Rudolph, M., Song, Z., Lausser, L., Schnack, C. and Begus-Nahrmann, Y. 2012. "Chitinase enzyme activity in CSF is a powerful biomarker of Alzheimer disease." *Neurology* 78: 569–577.

[14] Letuve, S., Kozhich, A., Humbles, A., Brewah, Y. and Dombret, M.C. 2010. "Lung chitinolytic activity and chitotriosidase are elevated in chronic obstructive pulmonary disease and contribute to lung inflammation." *Am J Pathol* 176: 638–649.

[15] Seibold, M.A., Donnelly, S., Solon, M., Innes, A. and Woodruff, P.G. 2008. "Chitotriosidase is the primary active chitinase in the human lung and is modulated by genotype and smoking habit." *J Allergy Clin Immunol* 122: 944–950.

[16] Shibata, Y., Honda, I., Justice, J.P., Van Scott, M.R., Nakamura, R.M., and Myrvik, Q.N. 2001. "Th1 adjuvant N-acetyl-Dglucosamine polymer up-regulates Th1 immunity but downregulates Th2 immunity against a mycobacterial protein (MPB-59) in interleukin-10-knockout and wild-type mice." *Infection and Immunity* 69 (10):6123–6130.

[17] Da Silva, C.A., Hartl, D., Liu, W., Lee, C.G., and Elias, J.A. 2008. "TLR-2 and IL-17A in chitin-induced macrophage activation and acute inflammation." *Journal of Immunology* 181 (6): 4279–4286.

[18] Reese, T.A., Liang, H.E. and Tager, A.M. 2007. "Chitin induces accumulation in tissue of innate immune cells associated with allergy." *Nature* 447 (7140): 92–96.

[19] Lee, C.G., Da Silva, C.A., Lee, J.Y., Hartl, D. and Elias, J.A. 2008. "Chitin regulation of immune responses: an old molecule with new roles." *Curr Opin Immunol* 20: 684–689.

[20] Di Rosa, M., Zambito, A.M., Marsullo, A.R., Li Volti, G., and Malaguarnera, L. 2009. "Prolactin induces chitotriosidase expression in human macrophages through PTK, PI3-K, and MAPK pathways." *Journal of Cellular Biochemistry* 107 (5): 881–889.

[21] Malaguarnera, L., Musumeci, M., Di Rosa, M., Scuto, A., and Musumeci, S. 2005. "Interferon-gamma, tumor necrosis factor-alpha, and lipopolysaccharide promote chitotriosidase gene expression in human macrophages." *Journal of Clinical Laboratory Analysis* 19 (3): 128–132.

[22] Malaguarnera, L., Musumeci, M., Licata, F., Di Rosa, M., Messina, A., and Musumeci, S. 2004. "Prolactin induces chitotriosidase gene expression in human monocyte-derived macrophages." *Immunology Letters* 94 (1-2): 57–63.

[23] Di Rosa, M., Musumeci, M., Scuto, A., Musumeci, S. and Malaguarnera, L. 2005. "Effect of interferon-γ, interleukin-10, lipopolysaccharide and tumor necrosis

factor-α on chitotriosidase synthesis in human macrophages," *Clinical Chemistry and Laboratory Medicine* 43 (5): 499–502.

[24] Netea, M.G., Gow, N.A.R., and Munro,C.A. 2006. "Immune sensing of Candida albicans requires cooperative recognition of mannans and glucans by lectin and Toll-like receptors." *Journal of Clinical Investigation* 116 (6): 1642–1650.

[25] Vega, K. and Kalkum, M. 2012. Chitin, Chitinase Responses, and Invasive Fungal Infections. *International Journal of Microbiology* 2012: 1-10.

[26] Shibata, Y., Foster, L.A., Bradfield, J.F., and Myrvik, Q.N. 2000. "Oral administration of chitin down-regulates serum IgE levels and lung eosinophilia in the allergic mouse." *Journal of Immunology* 164 (3): 1314–1321.

[27] Shibata, Y., Foster, L.A., Metzger, W.J., and Myrvik, Q.N. 1997b. "Alveolar macrophage priming by intravenous administration of chitin particles, polymers of N-acetyl-D-glucosamine, in mice." *Infection and Immunity* 65 (5):1734–1741.

[28] Shibata, Y., Metzger, W.J., and Myrvik, Q.N. 1997a. "Chitin particle-induced cell-mediated immunity is inhibited by soluble mannan: mannose receptor-mediated phagocytosis initiates IL-12 production." *Journal of Immunology* 159 (5): 2462–2467

[29] Boot, R.G., Blommaart, E.F., Swart, E., Ghauharali-vander, Vlugt K., and Bijl, N. 2001. "Identification of a novel acidic mammalian chitinase distinct from chitotriosidase." *J Biol Chem* 276: 6770–6778.

[30] Wiesner, D.L., Specht, C.A., Lee, C.K., Smith, K.D., Mukaremera, L., Lee, S.T., Lee, C.G., Elias, J.A., Nielsen, J.N., Boulware, D.R., Bohjanen, P.R., Jenkins, M.K., Levitz, S.M. and Nielsen, K. 2015. "Chitin recognition via chitotriosidase promotes pathologic type-2 helper T cell responses to cryptococcal infection." *PLoSPathog* 11: e1004701.

[31] Malaguarnera, L., Simporè, J., Prodi, D.A., Angius, A., Sassu, A., Persico, I., Barone, R. and Musumeci, S. 2003. "A 24-bp duplication in exon 10 of human chitotriosidase gene from the sub-Saharan to the Mediterranean area: role of parasitic diseases and environmental conditions." *Genes Immun*4: 570-574.

[32] vanEijk, M., Scheij, S.S., van Roomen, C.P., Speijer, D., Boot, R.G. and Aerts, J.M. 2007. "TLR- and NOD2-dependent regulation of human phagocyte-specific chitotriosidase." *FEBS Lett* 581: 5389-5395.

[33] Boot, R.G., Renkema, G.H., Verhoek, M., Strijland, A. and Bliek, J. 1998. "The human chitotriosidase gene. Nature of inherited enzyme deficiency." *J Biol Chem* 273: 25680–25685.

[34] van Eijk, M., van Roomen, C.P., Renkema, G.H., Bussink, A.P., Andrews, L., Blommaart, E.F., Sugar, A., Verhoeven, A.J., Boot, R.G. and Aerts, J.M. 2005. "Characterization of human phagocyte-derived chitotriosidase, a component of innate immunity." *Int Immunol*17: 1505-1512.

[35] Di Rosa, M., Malaguarnera, G., De Gregorio, C., D'Amico, F., Mazzarino, M.C. and Malaguarnera, L. 2013. "Modulation of chitotriosidase during macrophage differentiation." *Cell Biochem Biophys* 66: 239-247.

[36] Artieda, M., Cenarro, A., Gañán, A., Lukic, A., Moreno, E., Puzo, J., Pocoví, M.and Civeira, F. 2007. "Serum chitotriosidase activity, a marker of activated macrophages, predicts new cardiovascular events independently of C-reactive protein." *Cardiology*108: 297-306.

[37] Kanneganti, M., Kamba, A. and Mizoguchi, E. 2012. "Role of chitotriosidase (chitinase 1) under normal and disease conditions." *J Epithel Biol Pharmacol*5: 1-9.

[38] Lee, C.G., Herzog, E.L., Ahangari, F., Zhou, Y., Gulati, M., Lee, C.M., Peng, X., Feghali-Bostwick, C., Jimenez, S.A., Varga, J. and Elias, J.A. 2012. "Chitinase 1 is a biomarker for and therapeutic target in sclerodermaassociated interstitial lung disease that augments TGF-ß1 signaling." *J Immunol* 189: 2635-2644.

[39] Boot, R.G., van Achterberg, T.A. and van Aken, B.E. 1999. "Strong induction of members of the chitinase family of proteins in atherosclerosis: chitotriosidase and human cartilage gp-39 expressed in lesion macrophages." *Arterioscler. Thromb. Vasc. Biol* 19 (3): 687–694.

[40] Daniels, L.B., Glew, R.H. and Diven, W.F. 1981. "An improved fluorimetric leukocyte beta-glucosidase assay for Gaucher disease." *Clin Chim Acta* 115: 369–375.

[41] Wenger, D.A., Clark, C. and Sattler, M. 1978. "Synthetic substrate β-glycosidase activity in leukocytes: a reproducible method for the identification of patients and carriers of Gaucher's disease." *Clin Genet* 13: 145–153.

[42] Staretz-Chacham, O., Lang, T.C. and LaMarca, M.E. 2009. "Lysosomal storage disorders in the newborn." *Pediatrics* 123: 1191–1207.

[43] Marsden, D. and Levy, H. 2010. "Newborn screening of lysosomal storage disorders." *Clin Chem* 56: 1071–1079.

[44] Beutler, E., Grabowsky, G.A., Scrivner, C.R., Beaudet, A.L., Sly, W.S. and Valle, D. 1995. *The metabolic and molecular basis of inherited disease*. McGraw Hill Inc; New York.

[45] Boven, L.A., van Meurs, M. and Boot, R.G. 2004. "Gaucher cells demonstrate a distinct macrophage phenotype and resemble alternatively activated macrophages." *Am. J. Clin. Pathol* 122(3):359–369.

[46] Vellodi, A., Foo, Y. and Cole, T.J. 2005. "Evaluation of three biochemical markers in the monitoring of Gaucher disease." *J Inherit Metab Dis* 28: 585–592.

[47] Deegan, P.B. and Cox, T.M. 2005. "Clinical evaluation of biomarkers in Gaucher disease." *Acta Paediatr* 94: 47–50.

[48] Harmanci, O. and Bayraktar, Y. 2008. "Gaucher disease: new developments in treatment and etiology." *World J Gastroenterol* 14: 3968–3973.

[49] Brady, R.O., Murray, G.J. and Barton, N.W. 1994. "Modifying exogenous glucocerebrosidase for effective replacement therapy in Gaucher disease." *J Inherit Metab Dis* 17: 510–519.

[50] Golab, K., Passowicz-Muszynska, E. and Jankowska, R. 2009. "Serum activity of chitotriosidase, lysozyme and cathepsin H in patients with lung cancer and patients with inflammatory exudate (preliminary report)." *Pol Merkur Lekarski* 26: 194–197.

[51] Tercelj, M., Salobir, B., Simcic, S., Wraber, B., Zupancic, M. and Rylander, R. 2009. "Chitotriosidase activity in sarcoidosis and some other pulmonary diseases." *Scand J Clin Lab Invest* 69 : 575–578.

[52] Febbraio, M., Podrez, E.A. and Smith, J.D. 2000. "Targeted disruption of the class B scavenger receptor CD36 protects against atherosclerotic lesion development in mice." *J Clin Invest* 105 : 1049 – 1056.

[53] Lusis, A.J., Mar, R. and Pajukanta, P. 2004. "Genetics of Atherosclerosis." *Annu Rev Genomics Hum Genet* 5: 189–218.

[54] Stary, H.C. 1989. "Evolution and progression of atherosclerotic lesions in coronary arteries of children and young adults." *Arteriosclerosis* 9:19–32.

[55] Sibal, L., Agarwal, S.C. and Home, P.D. 2010. "The role of asymmetric dimethylarginine (ADMA) in endothelial dysfunction and cardiovascular disease." *CurrCardiol Rev* 6 : 82–90.

[56] Artieda, M., Cenarro, A. and Ganan, A. 2003. "Serum chitotriosidase activity is increased in subjects with atherosclerosis disease." *Arterioscler Thromb Vasc Biol* 23: 1645–1652.

[57] Moreno, P.R., Falk, E. and Palacios, I.F. 1994. "Macrophage infiltration in acute coronary syndromes. Implications for plaque rupture." *Circulation* 90: 775–778.

[58] Teřcelj, M., Rott, T. and Rylander, R. 2007. "Antifungal treatment in sarcoidosis - a pilot intervention trial." *Resp Med* 101: 774–778.

[59] Teřcelj, M., Salobir, B. and Rylander, R. 2008. "Microbial antigen treatment in sarcoidosis – a new paradigm?" *Medical Hypotheses* 70: 831–834.

[60] Rossman, M.D. and Kreider, M.E. 2007. "Lesson learned from ACCESS (A case controlled etiologic study of sarcoidosis)." *Proc Am Thorac Soc* 4:453–456.

[61] Hunninghake, G.W., Costabel, U., Ando, M., Baughman, R., Cordier, J.F. and du Bois, R. 1999. American Thoracic Society; European Respiratory Society; World Association of Sarcoidosis and other Granulomatous Disorders. ATS/ERS/WASOG statement on sarcoidosis.*Sarcoidosis Vasc Diffuse Lung Dis.*16: 149–173.

[62] Lynch, J.P., Kazerooni, E.A. and Gay, S.E. 1997. "Pulmonary sarcoidosis."*Clin Chest Med* 18: 755–785.

[63] Judson, M.A., Baughman, R.P., Thompson, B.W., Teirstein, A.S., Terrin, M.L. and Rossman, M.D. 2003. "Two year prognosis of sarcoidosis: the ACCESS experience." *Sarcoidosis Vasc Diffuse Lung Dis* 20: 204–211.

[64] Bargagli, E., Bianchi, N., Margollicci, M. 2008. "Chitotriosidase and soluble IL-2 receptor: comparison of two markers of sarcoidosis severity." *Scand J Clin Invest* 68:479–483.

[65] Corte, T.J., Wells, A.U., Nicholson, A.G., Hansell, D.M. and Wort, S.J. 2011. "Pulmonary hypertension in sarcoidosis: a review." *Respirology* 16: 69–77.

[66] Gordon-Thomson, C., Kumari, A., Tomkins, L., Holford, P. and Djordjevic, J.T. 2009. "Chitotriosidase and gene therapy for fungal infections." *Cell Mol Life Sci* 66: 1116–1125.

[67] Goedken, E. R., O'Brien, R. F., Xiang, T., Banach, D. L., Marchie, S. C., Barlow, E. H., Hubbard, S., Mankovich, J. A., Jiang, J., Richardson, P. L., Cuff, C. A., and Cherniack, A. D. 2011. "Functional comparison of recombinant acidic mammalian chitinase with enzyme from murine bronchoalveolar lavage." *Protein Expr. Purif.* 75, 55–62.

[68] Chou, Y. T., Yao, S., and Czerwinski, R. 2006. "Kinetic characterization of recombinant human acidic mammalian chitinase." *Biochemistry* 45, 4444–4454.

[69] Guoping, C., Fan, P., Jingxi, S., Xiaoping, L., Shiqin, J., and Yuri, L. 1997. Purification and characterization of a silica-induced bronchoalveolar lavage protein with fibroblast growth-promoting activity. *J. Cell Biochem.* 67, 257–264.

[70] Muzzarelli, R. A. 2008. "New aspects of chitin chemistry and enzymology," in*Binomium Chitin-Chitinase Recent Issues*, eds. S. Musumeci and M. G. Paoletti (New York, NY: Nova Science Publishers), 1–25.

[71] Koppelman, G.H., Stine, O.C., Xu, J., Howard, T.D., Zheng, S.L., Kauffman, H.F., Bleecker, E.R., Meyers, D.A. and Postma, D.S. 2002. "Genome-wide search for atopy susceptibility genes in Dutch families with asthma." *J Allergy ClinImmunol* 109: 498-506.

[72] Zhu, Z., Zheng, T., Homer, R.J., Kim, Y.K., Chen, L. 2004. "Acidic mammalian chitinase in asthmatic Th2 inflammation and IL-13 pathway activation." *Science* 304: 1678-1682.

[73] Chen, Y., Wong, G.W. and Li, J. 2016. "Environmental exposure and genetic predisposition as risk factors for asthma in China." *Allergy Asthma Immunol Res* 8: 92-100.

[74] Masoli, M., Fabian, D., Holt, S. and Beasley, R. 2004. Global Initiative for Asthma (GINA) Program. The global burden of asthma: executive summary of the GINA Dissemination Committee report. *Allergy, 59*: 469-478.

[75] Skadhauge, L.R., Christensen, K., Kyvik, K. O. and Sigsgaard, T. 1999. "Genetic and environmental influence on asthma: a population-based study of 11,688 Danish twin pairs." *Eur Respir J* 13: 8-14.

[76] Thomsen, S.F., Ferreira, M.A., Kyvik, K.O., Fenger, M. and Backer, V. 2009. "A quantitative genetic analysis of intermediate asthma phenotypes." *Allergy* 64: 427-430.

[77] Mohrs, M., Shinkai, K., Mohrs, K. and Locksley, R.M. 2001. "Analysis of type 2 immunity in vivo with a bicistronic IL-4 reporter." *Immunity* 303–11.

[78] Ono, S.J. 2003. "Vernal keratocojunctivitis: evidence for immunoglobulin E-dependent and immunoglobulin E-independent eosinophilia." *Clin Exp Allergy* 33: 279-81.

[79] Metz, D.P., Hingorani, M., Calder, V.L., Buckley, R.J. and Lightman, S.L. 1997. "T-cell cytokines in chronic allergic eye disease." *J. Allergy Clin. Immunol* 100: 817-24.

[80] Bucolo, C., Musumeci, M., Maltese, A., Drago, F. and Musumeci, S. 2008. "Effect of chitinase inhibitors on endotoxin-induced uveitis in rabbits." *Pharmacology Research* 57 (3): 247-252.

[81] Schappert, S.M. 1998. "Ambulatory care visits to physician offices, hospital outpatient departments, and emergency departments: United States, 1996," *Vital and Health Statistics* 13 (134):1–37.

[82] Poole, M.D. 1999. "A focus on acute sinusitis in adults: changes in disease management," *American Journal of Medicine* 106 (5A).

[83] Masood, A., Moumoulidis, I., and Panesar, J. 2007. "Acute rhinosinusitis in adults: an update on current management," *Postgraduate Medical Journal* 83 (980): 402–408.

[84] Bruijnzeels, M.A., Foets, M., Van Der Wouden, J.C., Van Den Heuvel, W.J.A. and Prins, A. 1998. "Everyday symptoms in childhood: occurrence and general practitioner consultation rates," *British Journal of General Practice* 48 (426): 880–884.

[85] Biel, M.A., Brown, C.A. and Levinson, R.M. 1998. "Evaluation of the microbiology of chronic maxillary sinusitis." *Annals of Otology, Rhinology and Laryngology* 107 (11): 942– 945.

[86] Brook, I., 2004. "Microbiology and antimicrobial management of sinusitis," *Otolaryngologic Clinics of North America*, 37 (2):253–266.

[87] Brook, I., Frazier, E.H., and Foote, R.A. 1996. "Microbiology of the transition from acute to chronic maxillary sinusitis." *Journal of Medical Microbiology* 45 (5): 372–375.

[88] Wald, E.R., Bordley, W.C., Darrow, D.H. 2001. "Clinical practice guideline: management of sinusitis." *Pediatrics* 108 (3): 798–808.

[89] McCormick, A., Fleming, D., and Charlton, J. 1995. "Morbidity Statistics from General Practice." *Fourth National Study 1991– 1992*, HMSO, London, UK.

[90] Benninger, M.S. 1992. "Rhinitis, sinusitis and their relationship to allergy," *American Journal of Rhinology* 6: 37–43.

[91] DeShazo, R.D. 1998. "Fungal sinusitis," *American Journal of the Medical Sciences* 316 (1): 39–45.

[92] Pleis, J.R. and Coles, R. 2003. "Summary health statistics for U.S. adults: National Health Interview Survey, 1999." *Vital and Health Statistics* 10 (212): 1–137.

[93] Ray, N.F., Baraniuk, J.N., Thamer, M. 1999. "Healthcare expenditures for sinusitis in 1996: contributions of asthma, rhinitis, and other airway disorders." *Journal of Allergy and Clinical Immunology* 103 (3): 408–414.

[94] Gu, Z.,Cao, Z.,Jin, M. 2011. "Expression and role of acidic mammalian chitinase and eotaxin-3 in chronic rhinosinusitis with nasal polyps." *J Otolaryngol Head Neck Surg.*40 (1): 64-9.

[95] http://www.ncbi.nlm.nih.gov/gene/1118.

In: Microbial Catalysts. Vol. 2
Editors: Shadia M. Abdel-Aziz et al.

ISBN: 978-1-53616-088-8
© 2019 Nova Science Publishers, Inc.

Chapter 14

TRICHODERMA: A POTENT SOURCE OF FUNGAL CELL WALL DEGRADING ENZYMES

Kandaswamy Saravanakumar[1], Narayanasamy Rajendran[2], Aeron Abhinav[3], Kandasamy Kathiresan[4] and Myeong-Hyeon Wang[1,]*

[1]Department of Medical Biotechnology, College of Biomedical Sciences, Kangwon National University, Chuncheon, Gangwon, Republic of Korea
[2]Government Arts College, Tamil Nadu, India
[3]College of Environmental and Bioresource Sciences, Chonbuk National University, Jeonju, Republic of Korea
[4]CAS in Marine biology, Faculty of Marine Sciences, Annamalai University, India

ABSTRACT

The fungal strains belonging to *Trichoderma* have been extensively isolated from the diversified tropical and subtropical ecological habitats. These strains are accredited as a biotechnologically important cell factory for generation of industrially important enzymes. Particularly the cell wall degrading enzymes (CWDEs) such as chitinase, glucanases, cellulase and proteases enhance the mycoparasitism against fungal phytopathogens including *Pithium, Sclerotium, Fusarium, Botrytis,* and *Rhizoctnia*. This review has summarized the reported cell wall degrading enzymes from *Trichoderma* species, their molecular mechanisms, and CWDEs induced biocontrol activity byelicitor or receptor.

Keywords: biocontrol, cell wall, enzymes, phytopathogens, *Trichoderma*

---

* Corresponding Author's E-mail: mhwang@kangwon.ac.kr.

1. INTRODUCTION

The soil fungus *Trichoderma* is an important source of cell wall degrading enzymes (CWDEs) [1, 2]. This fungus has the vital role in triggering systemic resistance (ISR) against plant pathogens and promoting the plant growth. *T. harzianum* derived CWDEs (ThPG1) are also involved in the induced systemic resistance (ISR) through plant root cell wall degradation, and is called as the damage associated molecular pattern (DAMPs) [3]. *Trichoderma* derived elicitor triggers ISR through the generation of ET pathways and reactive oxygen species (ROS), and is called as the microbe associated molecular pattern (MAMPs). Approximately 10,000 different plant diseases are caused by microbial plant pathogens [4]. Ability of the phytopathogens in causing plant dieseases depends up on mode of pathogenesis and host plants adaptation. There are two types of CWDEs such as (i) plant CWDEs, and (ii) fungal CWDEs. Several phytopathogens can produce the plant CWDEs which are causing the plant disease through the degradation of plant cell wall components (pectin, hemicellulose and cellulose). The fungal CWDEs from *Trichoderma* are significantly involved in biological control of root-infecting and soft rot-inducing plant pathogens such as *Pithium, Sclerotium, Fusarium, Botrytis,* and *Rhizoctnia* in the agriculture farming [5]. However, *Trichoderma* spp. are known to synthesis the CWDEs which degrade the pathogenic fungal cell wall [6]. Moreover, the molecular mechanism of the lytic enzymes in the biological control has been reported by Haran et al. [7].

Biological control of plant pathogens using microbial derivatives is efficient over the chemical pesticides and it is an ecologically affable to reduce the plant pathogens. The biological control is accredited based on the antibiotics, nutrient competition, response of CWDEs and metabolites of mycoparasitism [8]. Similarly the CWDEs are involved in hydrolysis of pathogenic fungal cell wall through mycoparasitism in both *in vivo* and *in vitro* and causing the apoptosis [9, 10]. Several *Trichoderma* derived CWDEs such as Chitinase, β-1,3-glucanases, α-1,3-glucanases, β-1,6-glucanases, and protease are reported [11]. Recent advances in the research of proteomic, genomic and transcriptomic studies have begun to pinpoint knowledge about molecular mechanism of plant and *Trichoderma* elicitor or receptor involved in CWDEs related biocontrol activity. In this book chapter, we summarize the reports on CWDEs from beneficial fungi *Trichdoerma*, highlight the importance of CWDEs on biological control activity and their mechanism, and provide the information on transcriptional genes from *Trichoderma,* involved in CWDEs production.

1.1. *Trichoderma* and Its Enzymes

Trichoderma is an extensive source of industrial enzymes with enormous applications [11]. Its antagonistic potential is a biobase for the significant biocontrol activity of various phytopathogens. The antagonistic or biocontrol of pathogens through the mycoparasitic or competition process is highly based on the synthesis of a fungal CWDEs and their ability in hydrolysing the cell wall of host fungal pathogens [9]. A complex of CWDEs from *Trichoderma,* dangerous to the phytopahtogenic fungi has been reported [12].

1.2. Fungal Cell Wall Degrading Enzymes from *Trichoderma*

Several CWDEs, Chitinase [13-15], α-1,3-glucanase [16], β-1,3-glucanase [17], Cellulase, N-acetylglucosaminidases [18], Chitotriosidase[19], Endochitinases [20], Glucanases [21, 22], β-1, 4-glucanases[23], 1,3-d-glucanase [24], Endo-beta-1,3-glucanases [25], Exo-beta-1,3-glucanases [16], Glucan 1,3, β-glucosidase [19, 26], Proteases [12, 27], β-1,6-glucanases [22, 28] have been described as important components for the multi-enzymatic system of *Trichoderma* species. However, the biocontrol is highly related to the synthesis of chitinase, glucanase and protease in *Trichoderma* and it enhances the reduction of the phytopathogenic fungi [29].

Fungal CWDEs are involved in the mycoparasitism process in both saprophytic and antagonistic processes, which could provide the diversity of microbial isolates with the higher potential of antagonist capacity and degrading potential in fungal colonization in different ecological systems [30]. CWDEs are reported from several *Trichoderma* isolates such as *T. atroviride* [31], *T. harzianum* [32], *T. viride* [33], *T. asperellum* [34], *T. aureo viride.* The microbial production of the chitinase has captured the world wide attention.

1.3. Role of Cell Wall Degrading Enzymes in Induced Systemic Resistance

In an environment, both beneficial and harmful microbes are present, in which several microbes could positively interact with the plants and induce the systemic resistance towards the dangerous microbes and enhance the defence responses by activation of multiple signal transduction pathways for the survival of plants. The plants are resistant to most of dangerous microbes through the innate immunity [35]. Two type of plants defence system is activated based on the injury such as systemic acquired resistance (SAR), and induced systemic resistance (ISR). SAR is employed by plants to restrict pathogen expansion in tissue through necrosis at the local site upon primary infection [36]. This is typically characterized by studying the level of salicylic acid (SA) and activated SA responsible genes. ISR is employed to induce by root colonization of systemic mutualistic microbes [2, 37]. This is typically identified through ethylene (ET) and jasmonate (JA) dependent signalling pathways [2, 38].

Table 1. Mycoparasitism related genes from *Trichoderma* sp.

Species	Strain	Genes	References
Trichoderma spp.	-	MAPK	[51]
T. asperellum		Orthologs - TmkA	[52]
Trichoderma spp.		Heterotrimeric G-protein signaling	[53]
T. atroviride and *T. reesei*		G-alpha subunits TGA3 and GNA3	[54]
T. reesei		GNA3	[55]
Trichoderma spp.		VELVET	[56]
T. atroviride		Class I (adenylate cyclase inhibiting) G-alpha subunits TGA1	[53]
		TMK1	[57]
	P1	Th-En42	[58]
	IMI206040	ech42	[59]
	P1	nag1	[60]
T. harzianum	CECT 2413	chit43	[61]
	TM	chit36	[62]
	CECT2413	cbit33	[63]
T. virens	Gv29-8	cbt42	[64]
		TgaA	[65]
	Gv29-8	Tvbgn3	[66]
	Tv29-8	ech1, ech2, ech3, cht1, cht2	[67]
		MAP-kinase TVK1	[68]; [69]; [70]
		Lack of TVK1 considerably increases biocontrol effectivity	[68]
T. hamatum	Tam-61	th-cb	[71]
T.cf. harzianum	T24-1	ENC1	[72]
	T25-1	exc1	
	T25-1	exc2	

Innate immunity is employing the pattern-recognition receptors (PRRs) for sensitive and rapid detection of the potential dangers caused by pathogenic microbes [39]. PRR line encoded according to the pathogen-microbes associated molecular patterns (PAMPs/MAMPs) or damage associated molecular pattern (DAMPs) [40]. In general, when the plant is interacted with dangerous pathogens, the PAMPs trigger the plant immunity (PTI), and synthesis of protein or metabolites that control the pathogen [40]. Some virulent successful pathogens suppress the PTI through the production effectors and toxins. However, the existed knowledge of plants-microbes interactions is emphasized to study the plants receptor protein or microbial elicitors. In this context, previous work has identified the elicitor from *Trichoderma* responsible for the interaction with maize and induced systemic resistance [23, 41]. Plant cells are exposed to molecule or cell wall degrading enzymes of microbes as purified proteins or microbial metabolites. These influence on the plant innate defence responses have been characterized [42-44]. CWDEs of *Trichoderma* can interact with the plant roots and disturb the roots system through damaging of root cellobiose or chitin followed by the CWDEs attack. This can pass signal through plant internal system and activate the plant defence through the JA/ET/SA signal pathways [3, 45]. Xylanase Xyn/Eix derived from *T. viride* is identified as strong elicitor to activate the plant defence in tobacco and tomato [46]. Cellulose of *T. longibrachiatum* is known to induce the plant defence response in melon through activation of SA and ET signalling pathways [47]. Endopolygalacturonase ThPG1 of *T. harzianum* is proved to involve in the plant defence responses like ISR in tomato root [3].

2. CELL WALL DEGRADING ENZYMES

Enzymes are proteins that are involved in various biological and chemical reactions in internal and external systems of plants and animals. Several artificial and microbial enzymes are significantly used in industrial applications. However, their role in biocontrol of non beneficial plant pathogens is known as mycoparasitism. *Trichdoerma* mycoparasitism is a combined synergistic process of nutrient competition [19], CWDEs [48], secondary metabolites [49, 50] and formation of coiling around the host and development of appressorium like structure [29]. *Trichoderma* species can colonize or penetrate in host fungus cell wall through the utilization of cellular contents. CWDEs (chitinase, glucanases, cellulase and proteases) which encode genes from *Trichoderma* (Table 1) play a major role in mycoparasitism and biocontrol [9, 48].

2.1. Cellulase

Cellulase is one of the economically important enzymes and it is used in many industrial applications especially in bio-fuel production. *Trichoderma* species secretes the cellulase enzymes that cause lysis in cell wall of the plant pathogens and induce the plant defence responses [50, 73]. A study of the cellulase gene expression in *T. reesei*, *T. atroviride* and *T. virens* during mycoparasitism against *Rhizoctonia solani* indicates that the cellulase gene is up regulated during the interaction with pathogen [74]. This emphasizes the components of the pathogens and *Trichoderma* spp., are able to enhance the process and induce the role in cellulase regulator gene of *xyr*1 [75].

Table 2. Antifungal chitinases reported from of *Trichoderma* strains

Species	Strain	Chitinases types	References
Trichoderma harzianum	CECT 2413	42 kDa (CHIT42), 37 kDa (CHIT37) & CHIT33	[79]; [80]
T. cf hurziunum	Nottingham 39.1	40 kDa, Chitinases; 73 kDaN-acetyl-b-D-glucosaminidase; 46 endochitinase	[81]; [82] [63]
T. atroviride (T. harzianum P1)	P1	CHIT42 endochitinase) & e CHIT73 N-acetyl-b-D-glucosaminidase 73kDa; Ech30	[83]; [84]; [85]
T. harzianum Rifai	-	chitinase	[86]
T. harzianum	TUBF 781	chitinase	[87]; [88]
T. viride	-	28KDA chitinase	[89]
T.cf. harzianum	Nottingham T198	exochitinase	[90]
Trichoderma viride	N9	46 kDa chitinase	[91]

2.2. Chitinases

Chitin is the second most abundant polymer in nature, after cellulose. Chitinases can degrade the chitin, and possess several biological control and plant defense mechanisms [76]. Chitinases from *Trichoderma* spp. are extensively involved in fungal cell wall degradation in mycoparasitism [9, 77]. Chitinolytic enzymes can be classified in to three major types *viz.*, (a) 1,4-P-ILT-acetylglucosaminidases that splits the chitin in an exo-type fashion into GlcNAc monomers; (b) endochitinases that breakdowns the internal sites over the entire length of the chitin microfibril in an random manner; and (c) exochitinases that are involved in release of diacetylchitobiose in an sequential fashion. Quite a lot of chitinase proteins and genes has been isolated from *Trichoderma* spp. grown in media containing chitin as sole carbon source (Table 2). Gruber et al. [50] have analysed the gene expression of chitinases (subgroup C) in a mycoparasite *Trichoderma virens* and have revealed the nutritional stimulus that exhibits difficult expression patterns on various parts of fungal colony, while the cultivation manner results in the transcription levels of subgroup C chitinase. De Las Mercedes Dana [78] has worked on the up regulation of chit33 expression from *Trichoderma harzianum* CECT 2413 that encodes the genes, responsible for Chit33 endochitinase, and chief component of *Trichoderma harzianum* CECT 2413 chitinolytic enzyme system.

The production of chitinase by *Trichoderma* species is of increasing interest as it acts as a promising source of mycolytic enzymes [92, 93]. The *nag1* gene from *T. atroviride* encodes N-acetyl-beta-d-glucosaminidase with 73-kDa, that is secreted and partially bound to the cell wall, and the *nag1* gene products are responsible for N-acetyl-beta-d-glucosaminidase activity [8]. *T. harzianum* isolate 1051 has produced most of the hydrolytic enzymes especially N-acetylglucosaminidase, chitinase, cellulase, β-1,3-glucanase, endoglucanase, protease and amylase [18]. Cht2 an endochitinase is expressed from *T. virens* UKM1 in *Pichia pastoris* (methylotrophic yeast), and and it is also cloned and sequenced the cht2 gene, its cDNA and the endochitinase gene [94]. Naturally, *T. harzianum* produces only moderate amounts of endochitinase, and over-expression of the endochitinase that can generate more-effective biocontrol strains. An efficient expression system has been developed for the related species like *T. reesei*, which utilizes the promoter of highly expressed cellulase gene *cbh1*.

2.3. Glucanases

Glucanases are enzymes that are polysaccharides, made up of several glucose sub-units and breaks down a glucan, and it performs the hydrolysis of glucosidic bond. Glucanases from *Trichoderma* can degrade most of the fungal cell walls and play an important role in mycoparasitic action [23]. De La Cruz et al. [28] have characterized the enzyme and indicated that the enzyme releases soluble sugars and produces hydrolytic halli on yeast cell walls. The combined cell wall-degrading enzymes such as β-1,3-glucanases and chitinases are significantly hydrolyzing filamentous fungal cell walls. These enzyme act supportively with the later enzymes, preventing the growth of the fungi. Antibodies against the purified protein have also indicated that the two identified β-1,6-glucanases are not immunologically related and are probably encoded by two different genes. The enzyme is specific for β-1,6 linkages and it shows an endolytic mode of action on pustulan. *T. asperellum* produces two extracellular 1,3-β-d-glucanase upon induction with cell walls from *Rhizoctonia solani* [95]. Some of these proteins display strong antifungal activities when are applied *in vitro*, alone and/or combined, against plant pathogens [10]. Most of the CWDEs are involved in biodegrading, antagonistic and saprophytic processes. However, fungal proteases may also be significantly involved in cell wall degradation, since fungal cell walls contain chitin and glucan polymers, embedded in and covalently linked to a protein matrix [96].

2.4. Protease

Protease from fungi plays the crucial role in the morphogenesis and nutrient metabolism. The proteases are involved in the extracellular digestion of proteins and free amino acids and also promote the secretion of extracellular hydrolytic enzymes [97, 98].

Some researchers have reported the correlation between extracellular protease and CWDEs from *Trichoderma* such as chitinase [99], β-1,6- glucanase [100], β-glucosidase [101] and, cellulase [102, 103]. Protease encoding genes are reported from *T. ressei* [104] and *T. harzianum* [100]. Similarly, the gene *prb*1 from *T. harzianum* is reportedly involved in the mycoparasitism against fungal plant pathogen *Rhizoctonia solani* [105, 106]. Although, the molecular mechanism of protease encoding genes from *Trichoderma* spp. involved in mycoparasitism is not yet described in detail.

CONCLUSION

Trichoderma species are globally well-known fungi, significantly used in biological control of plant pathogens in both *in vitro* and *in vivo* agriculture farming systems and are commercially available as bio-control agents. Several scientists have reported the molecular mechanism of bio-control activity associated with combined production of CWDEs, secondary metabolites, peptaibols from *Trichoderma*. However, most importantly the expressions of CWDEs encoding genes from *Trichoderma* significantly trigger the mycoparasitism and interactions between *Trichoderma* and plant pathogens or host. The summarized information emphasizes that the CWDEs especially, chitinase followed by the glucanases from *Trichoderma* spp. are significantly involved in mycoparasitism and they can further increase the synergistic secretion of other enzymes and metabolites. Although the *Trichoderma* species are widely used as bio-control agent against plant pathogens, there is still a need to study the detailed mechanism and function in order to understand signalling pathways, involved during the interaction of plant-*Trichoderma*-pathogens. Therefore, further such studies can improve our better understating of molecular mechanism of biological control activity of *Trichoderma* from genetics to field.

ACKNOWLEDGMENTS

This work was supported by Korea Research Fellowship Program through the National Research Foundation of Korea (NRF) funded by the Ministry of Science, ICT and Future Planning (2017H1D3A1A01052610).

REFERENCES

[1]   Lorito, M. *Purification, Characterization, and Synergistic Activity of a Glucan 1,3-beta Glucosidase and an N Acetyl beta Glucosaminidase from Trichoderma harzianum,* 1994.

[2] van Loon, LC; Bakker, PA; Pieterse, CM. Systemic resistance induced by rhizosphere bacteria. *Annual review of phytopathology.*, 1998, 36, 453-83.

[3] Moran-Diez, E; Hermosa, R; Ambrosino, P; Cardoza, RE; Gutierrez, S; Lorito, M; et al. The ThPG1 endopolygalacturonase is required for the trichoderma harzianum-plant beneficial interaction. *Molecular plant-microbe interactions: MPMI.*, 2009, 22, 1021-31.

[4] Horbach, R; Navarro-Quesada, AR; Knogge, W; Deising, HB. When and how to kill a plant cell: infection strategies of plant pathogenic fungi. *Journal of plant physiology.*, 2011, 168, 51-62.

[5] Vos, CM; De Cremer, K; Cammue, BP; De Coninck, B. The toolbox of Trichoderma spp. in the biocontrol of Botrytis cinerea disease. *Molecular plant pathology.*, 2015, 16, 400-12.

[6] Naher, L; Yusuf, U; Ismail, A; Hossain, K. Trichoderma spp.: A biocontrol agent for sustainable management of plant diseases, 2014.

[7] Gajera, H; Domadiya, R; Patel, S; Kapopara, M; Golakiya, B. Molecular mechanism of Trichoderma as bio-control agents against phytopathogen system – a review, 2013.

[8] Brunner, K; Peterbauer, CK; Mach, RL; Lorito, M; Zeilinger, S; Kubicek, CP. The Nag1 N-acetylglucosaminidase of Trichoderma atroviride is essential for chitinase induction by chitin and of major relevance to biocontrol. *Current genetics.*, 2003, 43, 289-95.

[9] Kubicek, CP; Mach, RL; Peterbauer, CK; Lorito, M. Trichoderma: From Genes to Biocontrol. *Journal of Plant Pathology.*, 2001, 83, 11-23.

[10] Myths and Dogmas of Biocontrol Changes in Perceptions Derived from Research on Trichoderma harzinum T-22. *Plant Disease.*, 2000, 84, 377-93.

[11] Monte, E. Understanding Trichoderma: between biotechnology and microbial ecology. *International microbiology: the official journal of the Spanish Society for Microbiology.*, 2001, 4, 1-4.

[12] Geremia, R; Jacobs, D; Goldman, GH; Van Montagu, M; Herrera-Estrella, A. Induction and Secretion of Hydrolytic Enzymes by the Biocontrol Agent Trichoderma Harzianum. In: Beemster, ABR; Bollen, GJ; Gerlagh, M; Ruissen, MA; Schippers, B; Tempel, A, editors. *Developments in Agricultural and Managed Forest Ecology*, Elsevier, 1991. p. 181-6.

[13] de la Cruz, J; Hidalgo-Gallego, A; Lora, JM; Benitez, T; Pintor-Toro, JA; Llobell, A. Isolation and characterization of three chitinases from Trichoderma harzianum. *European Journal of Biochemistry.*, 1992, 206, 859-67.

[14] Seidl, V; Huemer, B; Seiboth, B; Kubicek, CP. A complete survey of Trichoderma chitinases reveals three distinct subgroups of family 18 chitinases. *The FEBS Journal.*, 2005, 272, 5923-39.

[15] Agrawal, T; Kotasthane, A. *Chitinolytic assay of indigenous Trichoderma isolates collected from different geographical locations of Chhattisgarh in Central India*, 2012.

[16] Ait-Lahsen, H; Soler, A; Rey, M; de La Cruz, J; Monte, E; Llobell, A. An antifungal exo-alpha-1,3-glucanase (AGN13.1) from the biocontrol fungus Trichoderma harzianum. *Applied and environmental microbiology.*, 2001, 67, 5833-9.

[17] El-Katatny, MH; Somitsch, W; Robra, KH; El-Katatny, MS; Gubitz, GM. *Production of chitinase and β-1,3-glucanase by Trichoderma harzianum for control of the phytopathogenic fungus, Sclerotium rolfsii*, 2000.

[18] Marco, JLD; Valadares-Inglis, MC; Felix, CR. Production of hydrolytic enzymes by Trichoderma isolates with antagonistic activity against Crinipellis perniciosa, the causal agent of witches' broom of cocoa. *Brazilian Journal of Microbiology.*, 2003, 34, 33-8.

[19] Lorito, M. *Chitinolytic Enzymes Produced by Trichoderma harzianum: Antifungal Activity of Purified Endochitinase and Chitobiosidase*, 1993.

[20] V S Saiprasad, G; B Mythili, J; Anand, L; Suneetha, C; J Rashmi, H; Naveena, C; et al. *Development of Trichoderma harzianum endochitinase gene construct conferring antifungal activity in transgenic tobacco*, 2009.

[21] H. El-Katatny, M; Somitsch, W; Robra, KH; S. El-Katatny, M; Guebitz, G. *Production of Chitinase and -1,3-glucanase by Trichoderma harzianum for Control of the Phytopathogenic Fungus Sclerotium rolfsii*, 2000.

[22] de la Cruz, J; Pintor-Toro, JA; Benitez, T; Llobell, A; Romero, LC. A novel endo-beta-1,3-glucanase, BGN13.1, involved in the mycoparasitism of Trichoderma harzianum. *Journal of Bacteriology.*, 1995, 177, 6937-45.

[23] Djonović, S; Pozo, MJ; Kenerley, CM. Tvbgn3, a β-1,6-Glucanase from the Biocontrol Fungus *Trichoderma virens*, Is Involved in Mycoparasitism and Control of *Pythium ultimum*. *Applied and Environmental Microbiology.*, 2006, 72, 7661-70.

[24] Liu, T; Wang, T; Li, X; Liu, X. Improved heterologous gene expression in Trichoderma reesei by cellobiohydrolase I gene (cbh1) promoter optimization. *Acta Biochimica et Biophysica Sinica.*, 2008, 40, 158-65.

[25] Harkki, A; Uusitalo, J; Bailey, M; Penttilä, M; Knowles, JKC. A Novel Fungal Expression System: Secretion of Active Calf Chymosin from the Filamentous Fungus Trichoderma Reesei. *Bio/Technology.*, 1989, 7, 596-603.

[26] Bruckner, H; Graf, H. Paracelsin, a peptide antibiotic containing alpha-aminoisobutyric acid, isolated from Trichoderma reesei Simmons. *Part A. Experientia.*, 1983, 39, 528-30.

[27] Geremia, RA; Goldman, GH; Jacobs, D; Ardiles, W; Vila, SB; Van Montagu, M; et al. Molecular characterization of the proteinase-encoding gene, prb1, related to

mycoparasitism by Trichoderma harzianum. *Molecular Microbiology.*, 1993, 8, 603-13.

[28] de la Cruz, J; Pintor-Toro, JA; Benitez, T; Llobell, A. Purification and characterization of an endo-beta-1,6-glucanase from Trichoderma harzianum that is related to its mycoparasitism. *Journal of Bacteriology.*, 1995, 177, 1864-71.

[29] Elad, Y. *Parasitism of Trichoderma spp. on Rhizoctonia solani and Sclerotium rolfsii -- Scanning Electron Microscopy and Fluorescence Microscopy*, 1983.

[30] Kullnig, C; Mach, RL; Lorito, M; Kubicek, CP. Enzyme diffusion from Trichoderma atroviride (= T. harzianum P1) to Rhizoctonia solani is a prerequisite for triggering of Trichoderma ech42 gene expression before mycoparasitic contact. *Applied and Environmental Microbiology.*, 2000, 66, 2232-4.

[31] Reiter, J; Herker, E; Madeo, F; Schmitt, MJ. Viral killer toxins induce caspase-mediated apoptosis in yeast. *The Journal of Cell Biology.*, 2005, 168, 353-8.

[32] Sreeramulu, K; Jayalakshmi, SK; Raju, S; Rani, SU; Benagi, V. *Trichoderma harzianum L1 as a potential source for lytic enzymes and elicitor of defense responses in chickpea (Cicer arietinum L.) against wilt disease caused by Fusarium oxysporum f. sp. Cicero*, 2009.

[33] Velmurugu, J; Wijeratnam, S; Wijesundera, R. Trichoderma as a Seed Treatment to Control Helminthosporium Leaf Spot Disease of Chrysalidocarpus lutescens, 2009.

[34] Wu, Q; Sun, R; Ni, M; Yu, J; Li, Y; Yu, C; et al. Identification of a novel fungus, Trichoderma asperellum GDFS1009, and comprehensive evaluation of its biocontrol efficacy. *PLOS ONE.*, 2017, 12, e0179957.

[35] Lorenzo, O; Chico, JM; Sánchez-Serrano, JJ; Solano, R. Jasmonate-Insensitive Encodes a MYC Transcription Factor Essential to Discriminate between Different Jasmonate-Regulated Defense Responses in Arabidopsis. *The Plant Cell.*, 2004, 16, 1938-50.

[36] Glazebrook, J. Contrasting mechanisms of defense against biotrophic and necrotrophic pathogens. *Annual Review of Phytopathology.*, 2005, 43, 205-27.

[37] Shoresh, M; Yedidia, I; Chet, I. Involvement of Jasmonic Acid/Ethylene Signaling Pathway in the Systemic Resistance Induced in Cucumber by Trichoderma asperellum T203. *Phytopathology.*, 2005, 95, 76-84.

[38] Djonovic, S; Pozo, MJ; Dangott, LJ; Howell, CR; Kenerley, CM., Sm1, a proteinaceous elicitor secreted by the biocontrol fungus Trichoderma virens induces plant defense responses and systemic resistance. *Molecular plant-microbe interactions: MPMI.*, 2006, 19, 838-53.

[39] Zipfel, C. Plant pattern-recognition receptors. *Trends in Immunology.*, 2014, 35, 345-51.

[40] Maekawa, T; Kufer, TA; Schulze-Lefert, P. NLR functions in plant and animal immune systems: so far and yet so close. *Nature Immunology.*, 2011, 12, 817-26.

[41] Viterbo, A; Harel, M; Horwitz, BA; Chet, I; Mukherjee, PK. Trichoderma mitogen-activated protein kinase signaling is involved in induction of plant systemic resistance. *Appl Environ Microbiol.*, 2005, 71, 6241-6.

[42] Nimchuk, Z; Eulgem, T; Holt, BF; 3rd; Dangl, JL. Recognition and response in the plant immune system. *Annual Review of Genetics.*, 2003, 37, 579-609.

[43] Hammond-Kosack, KE; Jones, JD. Resistance gene-dependent plant defense responses. *The Plant Cell.*, 1996, 8, 1773-91.

[44] Yang, Y; Shah, J; Klessig, DF. Signal perception and transduction in plant defense responses. *Genes & Development.*, 1997, 11, 1621-39.

[45] de Jonge, R; van Esse, HP; Kombrink, A; Shinya, T; Desaki, Y; Bours, R; et al. Conserved fungal LysM effector Ecp6 prevents chitin-triggered immunity in plants. *Science.*, 2010, 329, 953-5.

[46] Rotblat, B; Enshell-Seijffers, D; Gershoni, JM; Schuster, S; Avni, A. Identification of an essential component of the elicitation active site of the EIX protein elicitor. *The Plant Journal: For Cell and Molecular Biology.*, 2002, 32, 1049-55.

[47] Martinez, C; Blanc, F; Le Claire, E; Besnard, O; Nicole, M; Baccou, JC. Salicylic acid and ethylene pathways are differentially activated in melon cotyledons by active or heat-denatured cellulase from Trichoderma longibrachiatum. *Plant physiology.*, 2001, 127, 334-44.

[48] Lorito, M; Farkas, V; Rebuffat, S; Bodo, B; Kubicek, CP. Cell wall synthesis is a major target of mycoparasitic antagonism by Trichoderma harzianum. *Journal of bacteriology.*, 1996, 178, 6382-5.

[49] Saravanakumar, K; Yu, C; Dou, K; Wang, M; Li, Y; Chen, J. Synergistic effect of Trichoderma-derived antifungal metabolites and cell wall degrading enzymes on enhanced biocontrol of Fusarium oxysporum f. sp. cucumerinum. *Biological Control.*, 2016, 94, 37-46.

[50] Gruber, S; Seidl-Seiboth, V. Self versus non-self: fungal cell wall degradation in Trichoderma. *Microbiology (Reading, England).*, 2012, 158, 26-34.

[51] Zeilinger, S; Omann, M. Trichoderma biocontrol: signal transduction pathways involved in host sensing and mycoparasitism. *Gene regulation and systems biology.*, 2007, 1, 227-34.

[52] Viterbo, A; Chet, I. TasHyd1, a new hydrophobin gene from the biocontrol agent Trichoderma asperellum, is involved in plant root colonization. *Molecular plant pathology.*, 2006, 7, 249-58.

[53] Schuster, A; Schmoll, M. Biology and biotechnology of Trichoderma. *Applied Microbiology and Biotechnology.*, 2010, 87, 787-99.

[54] Mukherjee, PK; Latha, J; Hadar, R; Horwitz, BA. Role of two G-protein alpha subunits, TgaA and TgaB, in the antagonism of plant pathogens by Trichoderma virens. *Appl Environ Microbiol.*, 2004, 70, 542-9.

[55] C. A. da Silva, L; Honorato, T; Franco, T; Rodrigues, S. *Optimization of Chitosanase Production by Trichoderma koningii sp. under Solid-State Fermentation*, 2010.

[56] Mukherjee, PK; Kenerley, CM. Regulation of morphogenesis and biocontrol properties in Trichoderma virens by a VELVET protein, Vel1. *Appl Environ Microbiol.*, 2010, 76, 2345-52.

[57] Reithner, B; Schuhmacher, R; Stoppacher, N; Pucher, M; Brunner, K; Zeilinger, S. Signaling via the Trichoderma atroviride mitogen-activated protein kinase Tmk 1 differentially affects mycoparasitism and plant protection. *Fungal Genetics and Biology, FG & B.*, 2007, 44, 1123-33.

[58] Hayes, CK; Klemsdal, S; Lorito, M; Di Pietro, A; Peterbauer, C; Nakas, JP; et al. Isolation and sequence of an endochitinase-encoding gene from a cDNA library of Trichoderma harzianum. *Gene.*, 1994, 138, 143-8.

[59] Carsolio, C; Gutiérrez, A; Jiménez, B; Van, Montagu, M; Herrera-Estrella, A. Characterization of ech-42, a Trichoderma harzianum endochitinase gene expressed during mycoparasitism. *Proceedings of the National Academy of Sciences.*, 1994, 91, 10903-7.

[60] Peterbauer, C; Lorito, MK; Hayes, C; Harman, G; Kubicek, C. *Molecular cloning and expression of the nag1 gene (N-acetyl-?-D-glucosaminidase-encoding gene) from Trichoderma harzianum P1*, 1996.

[61] Garcia, I; Lora, JM; de la Cruz, J; Benitez, T; Llobell, A; Pintor-Toro, JA. Cloning and characterization of a chitinase (chit42) cDNA from the mycoparasitic fungus Trichoderma harzianum. *Current Genetics.*, 1994, 27, 83-9.

[62] Viterbo, A; Ramot, O; Chemin, L; Chet, I. *Significance of lytic enzymes from Trichoderma spp. in the biocontrol of fungal plant pathogens. Antonie van Leeuwenhoek.*, 2002, 81, 549-56.

[63] Limon, MC; Lora, JM; Garcia, I; de la Cruz, J; Llobell, A; Benitez, T; et al. Primary structure and expression pattern of the 33-kDa chitinase gene from the mycoparasitic fungus Trichoderma harzianum. *Current Genetics.*, 1995, 28, 478-83.

[64] Baek, JR; Howell, C; Kenerley, C. *The role of extracellular chitinase from Trichoderma virens Gv29-8 in the bio control of Rhizoctonia solani*, 1999.

[65] Rocha-Ramírez, A; Robles-Valderrama, E; Ramirez, E. *Invasive alien species water hyacinth Eichhornia crassipes as abode for macroinvertebrates in hypertrophic Ramsar Site, Lake Xochimilco, Mexico*, 2014.

[66] Djonovic, S; Vargas, WA; Kolomiets, MV; Horndeski, M; Wiest, A; Kenerley, CM. A proteinaceous elicitor Sm1 from the beneficial fungus Trichoderma virens is required for induced systemic resistance in maize. *Plant Physiology.*, 2007, 145, 875-89.

[67] Kim, WG; Weon, HY; Seok, SJ; Lee, KH. *In Vitro* Antagonistic Characteristics of Bacilli Isolates against Trichoderma spp. and Three Species of Mushrooms. *Mycobiology*, 2008, 36, 266-9.

[68] Mendoza-Mendoza, A; Pozo, MJ; Grzegorski, D; Martínez, P; García, JM; Olmedo-Monfil, V; et al. Enhanced biocontrol activity of *Trichoderma* through inactivation of a mitogen-activated protein kinase. *Proceedings of the National Academy of Sciences.*, 2003, 100, 15965-70.

[69] Mukherjee, AK; Sampath Kumar, A; Kranthi, S; Mukherjee, PK. Biocontrol potential of three novel Trichoderma strains: isolation, evaluation and formulation. *3 Biotech.*, 2014, 4, 275-81.

[70] Mendoza-Mendoza, A; Rosales-Saavedra, T; Cortes, C; Castellanos-Juarez, V; Martinez, P; Herrera-Estrella, A. The MAP kinase TVK1 regulates conidiation, hydrophobicity and the expression of genes encoding cell wall proteins in the fungus Trichoderma virens. *Microbiology (Reading, England)*, 2007, 153, 2137-47.

[71] Fekete, E; Karaffa, L; Karimi, Aghcheh, R; Németh, Z; Fekete, E; Orosz, A; et al. The transcriptome of lae1 mutants of Trichoderma reesei cultivated at constant growth rates reveals new targets of LAE1 function. *BMC genomics.*, 2014, 15, 447.

[72] Draborg, H; Christgau, S; Halkier, T; Rasmussen, G; Dalbøge, H; Kauppinen, S. *Secretion of an enzymatically active Trichoderma harzianum endochitinase by Saccharomyces cerevisiae*, 1996.

[73] Chen, S-C; Zhao, H-J; Wang, Z-H; Zheng, C-X; Zhao, P-Y; Guan, Z-H; et al. *Trichoderma harzianum-induced resistance against Fusarium oxysporum involves regulation of nuclear DNA content, cell viability and cell cycle-related genes expression in cucumber roots*, 2016.

[74] Atanasova, L; Druzhinina, I; Jaklitsch, W. *Two Hundred Trichoderma Species Recognized on the Basis of Molecular Phylogeny*, 2013. p. 10-42.

[75] Bischof, R; Fourtis, L; Limbeck, A; Gamauf, C; Seiboth, B; Kubicek, CP. Comparative analysis of the Trichoderma reesei transcriptome during growth on the cellulase inducing substrates wheat straw and lactose. *Biotechnology for Biofuels*, 2013, 6, 127-.

[76] Jach, G; Gornhardt, B; Mundy, J; Logemann, J; Pinsdorf, E; Leah, R; et al. Enhanced quantitative resistance against fungal disease by combinatorial expression of different barley antifungal proteins in transgenic tobacco. *The Plant Journal: For Cell and Molecular Biology.*, 1995, 8, 97-109.

[77] Jeffries, P. Biology and ecology of mycoparasitism. *Canadian Journal of Botany.*, 1995, 73, 1284-90.

[78] de las Mercedes, Dana, M; Limon, MC; Mejias, R; Mach, RL; Benitez, T; Pintor-Toro, JA; et al. Regulation of chitinase 33 (chit33) gene expression in Trichoderma harzianum. *Current Genetics.*, 2001, 38, 335-42.

[79] Galvez, J; De La Cruz, JP; Zarzuelo, A; De Medina, FS; Jimenez, J; De La Cuesta, FS. Oral administration of quercitrin modifies intestinal oxidative status in rats. *General Pharmacology: The Vascular System.*, 1994, 25, 1237-43.

[80] Lorito, M; Woo, SL; Garcia, I; Colucci, G; Harman, GE; Pintor-Toro, JA; et al. Genes from mycoparasitic fungi as a source for improving plant resistance to fungal pathogens. *Proceedings of the National Academy of Sciences of the United States of America.*, 1998, 95, 7860-5.

[81] Druzhinina, IS; Shelest, E; Kubicek, CP. Novel traits of Trichoderma predicted through the analysis of its secretome. *FEMS Microbiology Letters.*, 2012, 337, 1-9.

[82] Ulhoa, CJ; Peberdy, JF. Regulation of chitinase synthesis in Trichoderma harzianum. *Journal of General Microbiology.*, 1991, 137, 2163-9.

[83] Mach, RL; Peterbauer, CK; Payer, K; Jaksits, S; Woo, SL; Zeilinger, S; et al. Expression of two major chitinase genes of Trichoderma atroviride (T. harzianum P1) is triggered by different regulatory signals. *Appl Environ Microbiol.*, 1999, 65, 1858-63.

[84] Harman, G. *Chitinolytic Enzymes of Trichoderma harzianum: Purification of Chitobiosidase and Endochitinase*, 1993.

[85] Lorito, M; Peterbauer, C; Hayes, CK; Harman, GE. Synergistic interaction between fungal cell wall degrading enzymes and different antifungal compounds enhances inhibition of spore germination. *Microbiology (Reading, England).*, 1994, 140 (Pt 3), 623-9.

[86] El-Katatny, MH; Gudelj, M; Robra, KH; Elnaghy, MA; Gubitz, GM. Characterization of a chitinase and an endo-beta-1,3-glucanase from Trichoderma harzianum Rifai T24 involved in control of the phytopathogen Sclerotium rolfsii. *Applied Microbiology and Biotechnology.*, 2001, 56, 137-43.

[87] Nampoothiri, KM; Baiju, TV; Sandhya, C; Abdulhameed, S; Szakacs, G; Pandey, A. *Process optimization for antifungal chitinase production by Trichoderma harzianum*, 2004.

[88] Agrawal, T; Kotasthane, AS. Chitinolytic assay of indigenous Trichoderma isolates collected from different geographical locations of Chhattisgarh in Central India. *Springer Plus.*, 2012, 1, 73-.

[89] A Omumasaba, C; Yoshida, N; Ogawa, K. *Purification and characterization of a chitinase from Trichoderma viride*, 2001.

[90] E Deane, E; M Whipps, J; Lynch, J; Peberdy, J. *The purification and characterization of a Trichoderma harzianum exochitinase*, 1998.

[91] Jsr, SJ; Jeyasree, J; Kezia, Laveena, D; Manikandan, K. *Purification and Characterization of Chitinase From Trichoderma Viride N9 and Its Antifungal Activity against Phytopathogenic Fungi*, 2014.

[92] Berg, B; Pettersson, G. Location and formation of cellulases in Trichoderma viride. *The Journal of Applied Bacteriology*, 1977, 42, 65-75.

[93]   Binder, AK; Ghose, T. *Adsorption of cellulose by Trichoderma viride*, 1978.

[94]   Al-Rashed, SAA; Abu Bakar, F; Said, M; Hassan, O; Rabu, A; Illias, R; et al. *Expression and characterization of the recombinant trichoderma virens endochitinase cht*, 22010.

[95]   da Silva Aires, R; Steindorff, AS; Ramada, MHS; de Siqueira, SJL; Ulhoa, CJ. Biochemical characterization of a 27kDa 1,3-β-d-glucanase from Trichoderma asperellum induced by cell wall of Rhizoctonia solani. *Carbohydrate Polymers.*, 2012, 87, 1219-23.

[96]   Kapteyn, JC; Montijn, RC; Vink, E; de la Cruz, J; Llobell, A; Douwes, JE; et al. Retention of Saccharomyces cerevisiae cell wall proteins through a phosphodiester-linked beta-1,3-/beta-1,6-glucan heteropolymer. *Glycobiology.*, 1996, 6, 337-45.

[97]   Chen, H; Hayn, M; Esterbauer, H. Three forms of cellobiohydrolase I from Trichoderma reesei. *Biochemistry and Molecular Biology International.*, 1993, 30, 901-10.

[98]   Bussey, H. Proteases and the processing of precursors to secreted proteins in yeast. *Yeast (Chichester, England)*, 1988, 4, 17-26.

[99]   Margolles-Clark, E; Hayes, CK; Harman, GE; Penttila, M. Improved production of Trichoderma harzianum endochitinase by expression in Trichoderma reesei. *Appl Environ Microbiol.*, 1996, 62, 2145-51.

[100] Delgado-Jarana, J; Pintor-Toro, JA; Benítez, T. Overproduction of β-1,6-glucanase in Trichoderma harzianum is controlled by extracellular acidic proteases and pH. *Biochimica et Biophysica Acta (BBA) - Protein Structure and Molecular Enzymology.*, 2000, 1481, 289-96.

[101] Kubicek-Pranz, EM; Gruber, F; Kubicek, CP. Transformation of Trichoderma reesei with the cellobiohydrolase II gene as a means for obtaining strains with increased cellulase production and specific activity. *Journal of Biotechnology.*, 1991, 20, 83-94.

[102] Hagspiel, K; Haab, D; Kubicek, C. *Protease activity and proteolytic modification of cellulases from a Trichoderma reesei QM 9414 selectant*, 1989.

[103] Haab, D; Hagspiel, K; Szakmary, K; Kubicek, CP. Formation of the extracellular proteases from Trichoderma reesei QM 9414 involved in cellulase degradation. *Journal of Biotechnology.*, 1990, 16, 187-98.

[104] Bradner, JR; Gillings, M; Nevalainen, KMH. Qualitative assessment of hydrolytic activities in antarctic microfungi grown at different temperatures on solid media. *World Journal of Microbiology and Biotechnology.*, 1999, 15, 131-2.

[105] Harman, G; Björkman, T. *Potential and existing uses of Trichoderma and Gliocladium for plant disease control and plant growth enhancement.*, 1998. p. 229-65.

[106] Flores, A; Chet, I; Herrera-Estrella, A. Improved biocontrol activity of Trichoderma harzianum by over-expression of the proteinase-encoding gene prb1. *Current genetics.*, 1997, 31, 30-7.

ABOUT THE EDITORS

Prof. Shadia M. Abdel-Aziz, PhD
Microbial Chemistry Department
Genetic Engineering and Biotechnology Research Division
National Research Centre, Dokki, Giza, Egypt
Email: abdelaziz.sm@gmail.com

Prof. Neelam Garg, PhD
Department of Microbiology
Faculty of Life Sciences
Kurukshetra University, Kurukshetra, Haryana, India
Email: grgneelam@gmail.com; nlmgarg@rediffmail.com

Dr. Abhinav Aeron, PhD
Independent Researcher
Bhartiya Colony, Muzaffarnagar, India
Email: abhinavaeron@gmail.com

Dr. Chaitanya Kumar Jha, PhD
Assistant Professor (GES-II)
Microbiology Department
Gujarat Arts and Science College, Ahmedabad
Gujarat, India
Email: jha_chaitanya@rediffmail.com

S. Chandra Nayak, PhD
Professor & Officer Incharge -ICAR (AICRP- Mysore Centre)
Department of Studies in Biotechnology
University of Mysore
Mysore, Karnataka, India
Email: moonnayak@gmail.com

Vivek Kumar Bajpai, PhD
Department of Energy & Materials Engineering,
Dongguk University-Seoul, Seoul, Republic of Korea
Email: vbajpai04@yahoo.com

ABOUT THE CONTRIBUTORS

Shadia M. Abdel-Aziz

Microbial Chemistry Department, National Research Centre, Dokki - Giza, P.O-.12622, Egypt

abdelaziz.sm@gmail.com

Abhinav Aeron

Division of Biotechnology, College of Environmental and Bioresource Sciences, Chonbuk National University, Iksan, Republic of Korea; Department of Biosciences, DAV PG College, Muzaffarnagar-251001, Uttar Pradesh, India

abhinavaeron@gmail.com

Khalid T. Biobaku

Department of Veterinary Pharmacology and Toxicology, University of Ilorin, Nigeria

biobaku.kt@unilorin.edu.ng

Srinivas Chowdappa

Department of Microbiology and Biotechnology, Bangalore University, Jnanabharathi Campus Bangalore, Karnataka, India

srininvasbub@gmail.com

Sukmawati D

Department of Biology, Universitas, Negeri Jakarta, Indonesia

Mohan Chakrabhavi Dhananjaya

Department of Studies in Biotechnology, University of Mysore, Mysuru, Karnataka, India

Pinakin Dhandhukia

Ashok and Rita Patel Institute of Integrated Study and Research in Biotechnology and Allied Sciences, ADIT Campus, New Vidyangar-388121, Anand (Gujarat), India

pinakindhadhukia@aribas.edu.in, pinakin.dhandhukia@gmail.com

Greta Faccio

Empa, Swiss Federal Laboratories for Materials Science and Technology
Laboratory for Biomaterials, Lerchenfeldstrasse 5, 9014 St. Gallen, Switzerland

Greta.faccio@gmail.com

Neelam Garg

Department of Microbiology, Faculty of Life Sciences,
University of Kurukshetra 136119, Haryana, India

grgneelam@gmail.com

Dweipayan Goswami

Department of Biotechnology, P.D. Patel Institute of Applied Sciences,
Charotar University of Science and Technology, CHARUSAT Campus, Changa-388421, Anand (Gujarat), India

Muthusamy Govarthanan

Division of Biotechnology, College of Environmental and Bioresource Sciences,
Chonbuk National University, Iksan - 570752, South Korea
PG and Research Department of Biotechnology, Mahendra Arts and Science College,
Kalippatti, Namakkal - 637501, Tamil Nadu, India

gova.muthu@gmail.com

Vinod Gubbiveeranna

Department of Studies and Research in Biochemistry, Tumkur University,
Tumkur, Karnataka, India

Hoda A. Hamed

Microbial Chemistry Department, National Research Centre, Dokki - Giza
P.O-.12622, Egypt

Papia Haque

Department of Applied Chemistry and Chemical Engineering, Faculty of Engineering
and Technology, University of Dhaka, Dhaka 1000, Bangladesh

Md. Minhajul Islam

Department of Applied Chemistry and Chemical Engineering, Faculty of Engineering and Technology, University of Dhaka, Dhaka 1000, Bangladesh

Nayan K. Jain

Department of Biotechnology, K L Mehta Dayanand College for Women, M D University, Faridabad, Haryana, India

Sneha Chaitanya Jha

Department of Life Sciences, University School of Sciences, Gujarat University, Ahmedabad- 380009, Gujarat, India

jhasneha2690@gmail.com

Seralathan Kamala-Kannan

Division of Biotechnology, College of Environmental and Bioresource Sciences, Chonbuk National University, Iksan - 570752, Republic of Korea

Kshipra Kapil (née Soni)

Department of Life Science, University School of Sciences, Gujarat University, Ahmedabad, Gujarat, India

kshiprakapil@gmail.com

Eman A. Karam

Microbial Chemistry Department, Genetic Engineering and Biotechnology Research Division, National Research Centre, 33 El Bohouthst. (former El Tahrir st.) Dokki - Giza - P.O.12622, Egypt

Kandasamy Kathiresan

CAS in Marine Biology, Faculty of Marine Sciences, Annamalai University, India

M. Nuruzzaman Khan

Department of Applied Chemistry and Chemical Engineering, Faculty of Engineering and Technology, University of Dhaka, Dhaka 1000, Bangladesh

Ismat Z. Luna

Department of Applied Chemistry and Chemical Engineering, Faculty of Engineering and Technology, University of Dhaka, Dhaka 1000, Bangladesh

Hansmukh A. Modi

Department of Life Sciences, University School of Sciences, Gujarat University, Ahmedabad- 380009, Gujarat, India

El-Sayed E. Mostafa

Microbial Chemistry Department, National Research Centre, Dokki – Giza, P.O-.12622, Egypt

Venkataramana Mudili

Defence Research Developmemt Organization (DRDO) Centre, Coimbatore, Tamil Nadu, India

Shivaiah Nagaraju

Department of Studies and Research in Biochemistry, Tumkur University, Tumkur, Karnataka, India

Ismail A. Odetokun

Department of Veterinary Public Health and Preventive Medicine, University of Ilorin, Nigeria odetokun.ia@unilorin.edu.ng; ismail23us@gmail.com

Ram Prasad

Amity Institute of Microbial Technology, Amity University Uttar Pradesh, Sector 125, Noida-201303, UP, India
School of Environmental Sciences and Engineering, Sun Yat-Sen University, P. R. China

Mohammed Mizanur Rahman

Department of Applied Chemistry and Chemical Engineering, Faculty of Engineering and Technology, University of Dhaka, Dhaka 1000, Bangladesh
mizanur.rahman@du.ac.bd

Lakshmeesha Thimmappa Ramachandrappa

Department of Studies in Biotechnology, University of Mysore, Mysore, Karnataka, India

Michael Richter

Empa, Swiss Federal Laboratories for Materials Science and Technology - Laboratory for Biomaterials, Lerchenfeldstrasse 5, 9014 St. Gallen, Switzerland

Moataza M. Saad

Microbial Chemistry Department,Genetic Engineering and Biotechnology Research
Division, National Research Centre, 33 El Bohouthst. (former El Tahrir st.)
Dokki - Giza – P.O.12622, Egypt
Moataza_Saad@hotmail.com

Khandoker S. Salem

Department of Applied Chemistry and Chemical Engineering, Faculty of Engineering
and Technology, University of Dhaka, Dhaka 1000, Bangladesh

Kandaswamy Saravanakumar

Department of Medical Biotechnology, College of Biomedical Sciences,
Kangwon National University, Chuncheon, Gangwon, 24341, Republic of Korea
saravana732@gmail.com

Kandasamy Selvam

PG& Research Department of Biotechnology, Mahendra Arts and Science College,
Kalippatti, Namakkal - 637501, Tamil Nadu, India
Centre for Biotechnology, Muthayammal College of Arts and Science, Rasipuram,
Namakkal - 637408, Tamil Nadu, India

Thangasamy Selvankumar

PG and Research Department of Biotechnology, Mahendra Arts and Science College,
Kalippatti, Namakkal - 637501, Tamil Nadu, India
selvakumar75@gmail.com

Balakrishnan Senthilkumar

Centre for Biotechnology, Muthayammal College of Arts and Science, Rasipuram,
Namakkal - 637408, Tamil Nadu, India

Sadia Sharmeen

Department of Applied Chemistry and Chemical Engineering, Faculty of Engineering
and Technology, University of Dhaka, Dhaka 1000, Bangladesh

Abeer. N. Shehata

Biochemistry Department, Genetic Engineering and Biotechnology Research
Division, National Research Centre, Dokki, Giza, P.O.12622, Egypt

Chandra Nayaka Siddaiah
Department of Studies in Biotechnology, University of Mysore,
Mysore, Karnataka, India
moonnayak@gmail.com

Richa Soni
Department of Life Science, University School of Sciences, Gujarat University,
Ahmedabad, Gujarat, India
richasoniricha@gmail.com

Girish Kesturu Subbaiah
Department of Studies and Research in Biochemistry, Tumkur University,
Tumkur, Karnataka, India

Janki Thakker
Department of Biotechnology, P.D. Patel Institute of Applied Sciences,
Charotar University of Science and Technology, CHARUSAT Campus, Changa-388421,
Anand (Gujarat), India

Linda Thöny-Meyer
Empa, Swiss Federal Laboratories for Materials Science and Technology - Laboratory for
Biomaterials, Lerchenfeldstrasse 5, 9014 St. Gallen, Switzerland

Narayanasamy Rajendran
Government Arts College, Chidambaram-608 102, Tamil Nadu, India

Shobith Rangappa
Department of Studies in Microbiology, Maharani's College, Mysore, Karnataka, India

Taslim Ur Rashid
Department of Applied Chemistry and Chemical Engineering, Faculty of Engineering
and Technology, University of Dhaka, Dhaka 1000, Bangladesh

Fazilath Uzma
Department of Microbiology and Biotechnology, Bangalore University,
Jnanabharathi Campus Bangalore, Karnataka, India

Myeong-Hyeon Wang
Department of Medical Biotechnology, College of Biomedical Sciences,
Kangwon National University, Chuncheon, Gangwon, 24341, Republic of Korea
mhwang@kangwon.ac.kr

Asaduz Zaman
Department of Applied Chemistry and Chemical Engineering, Faculty of Engineering
and Technology, University of Dhaka, Dhaka 1000, Bangladesh

INDEX

α

α-L-rhamnosidase, 124, 287, 288, 291, 293, 296, 298

A

A. niger, 23, 118, 156, 157, 214, 215, 218, 219, 224, 312

A. oryzae, 20, 23, 118, 156, 159, 215, 312

acetylation, 25, 128, 307

acid, 19, 21, 24, 28, 37, 38, 40, 42, 59, 61, 62, 66, 68, 91, 94, 102, 113, 118, 120, 122, 123, 126, 127, 128, 130, 138, 140, 142, 148, 149, 155, 158, 160, 161, 162, 163, 164, 165, 167, 168, 172, 178, 179, 185, 188, 196, 197, 205, 210, 211, 212, 213, 215, 217,218, 223, 224, 225, 226, 227, 229, 230, 231, 243, 244, 245, 246, 247, 253, 254, 259, 260, 261, 262, 263, 264, 265, 266, 271, 273, 277, 281, 282, 297, 307, 311, 317, 328, 341, 348, 350

acidic, 7, 66, 125, 136, 138, 163, 172, 209, 211, 220, 222, 231, 266, 272, 276, 294, 319, 320, 328, 330, 331, 333, 336, 338, 354

acidic mammalian chitinase, 319, 320, 328, 331, 333, 336, 338

acne, 199, 207, 249, 257

acrylamide, vi, 19, 162, 163, 179, 180, 301, 302, 303, 304, 305, 306, 307, 310, 311, 312, 313, 314, 315, 316, 317, 318

acrylamide mitigation strategy, 313

acrylamide reduction, vi, 301, 302, 310, 312, 314, 317

activated carbon, 68, 80, 102, 109

activation losses, 77, 78, 95

active site, ix, 3, 4, 5, 6, 13, 38, 40, 241, 246, 263, 350

acute lymphoblastic leukemia, 27, 171, 302, 303, 307, 309, 316

additives, 20, 22, 156, 159, 160, 192, 194, 196, 215, 222, 247, 273, 287, 288, 311

adsorption, 68, 94, 289

adults, 316, 330, 337, 338

advancement, ix, 58, 174, 195

adverse effects, 243, 305, 313

aeromonascavi, 118

age, 219, 235, 236, 237, 238, 240, 249, 251, 271, 326

age spots, 235, 237, 238, 240, 249, 251

agriculture, 41, 246, 280, 287, 288, 340, 346

Agrobacterium tumefaciens, 127, 266, 282

albinism, 241, 252, 253, 257

alcohols, 21, 23, 158, 161, 163, 168

algae, 67, 102, 273

allergic reaction, 154, 250, 308

allergy, 328, 332, 337

Alternaria citri, 127

Alternaria mali, 127, 266, 282

amines, 8, 23, 24, 32, 33, 161, 315

amino, 3, 4, 5, 6, 8, 28, 29, 35, 38, 40, 147, 150, 155, 158, 160, 167, 198, 201, 202, 210, 221, 226, 227, 237, 301, 302, 307, 309, 311, 320, 324, 345

amino acid, 3, 4, 5, 6, 8, 28, 29, 35, 38, 40, 147, 150, 155, 158, 160, 167, 198, 201, 202, 210, 221, 226, 227, 237, 301, 302, 307, 309, 311, 320, 324, 345

ammonia, 23, 28, 35, 37, 40, 42, 120, 160, 261, 301, 303, 305, 307, 309, 310, 311

amylase, 28, 49, 154, 156, 158, 165, 170, 176, 177, 182, 187, 188, 189, 192, 193, 194, 197, 199, 200, 201, 202, 203, 204, 207, 212, 220, 345

anaerobic digestion, 83, 91, 93, 113, 114

C

D

G

H

I

J

K

L

R

U

V

W

Y

PLANT GROWTH PROMOTING MICROORGANISMS: MICROBIAL RESOURCES FOR ENHANCED AGRICULTURAL PRODUCTIVITY

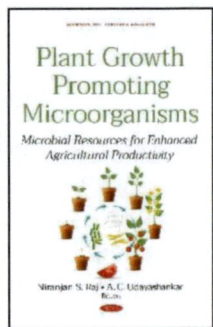

EDITORS: Niranjan S. Raj and A. C. Udayashankar

SERIES: Microbiology Research Advances

BOOK DESCRIPTION: Plant growth promoting microorganisms (PGPM) have gained acceptance and importance due to their dual benefits of promoting plant growth in addition to managing plant pests and diseases and are extensively used as microbial inoculants in improving agricultural productivity.

HARDCOVER ISBN: 978-1-53615-776-5
RETAIL PRICE: $230

ENTER THE WORLD OF MICROBIOLOGY: INTERVIEWS ABOUT THE WORLD'S MOST FAMOUS MICROBIOLOGISTS

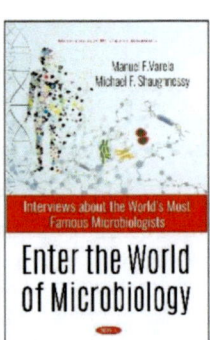

AUTHORS: Manuel Varela and Michael F. Shaughnessy

SERIES: Microbiology Research Advances

BOOK DESCRIPTION: Enter the world of microbiology: A dimension not only of viruses and bacteria, but also of contributions to medicine, health, and well-being. Enter the world of the most famous discoveries of scientists in the realm of biology from all over the world.

HARDCOVER ISBN: 978-1-53615-168-8
RETAIL PRICE: $195